P9-AER-370

# Windows of Opportunity

# Windows of Opportunity

## Mathematics for Students with Special Needs

*edited by*
Carol A. Thornton
Illinois State University
Normal, Illinois
*and*
Nancy S. Bley
Park Century School
Los Angeles, California



National Council of Teachers of Mathematics

Library of Congress Cataloging-in-Publication Data:

Windows of opportunity : mathematics for students with special needs / edited by Carol A. Thornton and Nancy S. Bley.
p. cm.
Includes bibliographical references and index.
ISBN 0-87353-374-7
1. Mathematics—Study and teaching—United States. 2. Gifted children—Education—Mathematics. 3. Handicapped children—Education—Mathematics. I. Thornton, Carol A. II. Bley, Nancy S.
QA13.W56 1994
510′.071′2—dc20 94-140
CIP

The publications of the National Council of Teachers of Mathematics present a variety of viewpoints. The views expressed or implied in this publication, unless otherwise noted, should not be interpreted as official positions of the Council.

Printed in the United States of America

# Contents

## Part 3: Promising Practices



## Part 4: The Challenge and the Promise

# Preface

*Windows of Opportunity: Mathematics for Students with Special Needs* is a professional resource for both regular classroom and special education teachers who work with—

- students with disabilities in mathematics;
- students who are gifted or talented in mathematics.

Clearly these individuals have "special needs" (defined in appendix B) that should be considered and met as part of a teacher's planning for mathematics and day-to-day decision making during instruction.

In response, the mathematics and special educators who have collaborated to write the chapters of this resource have voiced important concerns, faced critical issues, and shared practical, effective ideas for bringing the NCTM vision expressed in *Curriculum and Evaluation Standards for School Mathematics* and *Professional Standards for Teaching Mathematics* to reality for students with special needs. A particular focus that has emerged is nurturing the abilities of these students to *think mathematically* through appropriate, relevant, problem-centered instruction. This focus is threaded throughout the three major sections of this resource:

**Part 1** (chapters 1 through 7) raises and addresses current issues of concern relating to high-quality, broad-based, equitable school mathematics programs for students with special needs. The first chapter provides an overview to this section, viewing mathematics as a potentially "handicapping condition" that could be prevented if students are allowed to experience, firsthand, the power of mathematical thought. Other chapters in this section react to this challenge by focusing on the parameters of change affecting curriculum, teaching practice, assessment, and IEP planning that empower students with identified gifts or needs to use mathematics sensibly and productively.

**Part 2** (chapters 8 through 14) addresses major curriculum thrusts in mathematics. Different age levels of students from early childhood through secondary school are considered. Each chapter shares three classroom episodes and highlights the interaction and learning of "special needs" students. Throughout this section, the reader might note the effect of developing these students' abilities to explore mathematics beyond arithmetic, to conjecture, to reason logically, and to use a variety of approaches to solve relevant mathematical problems. Further, in these chapters, students spend less time working individually on computational skills and more time negotiating learning in interactive, problem-centered settings, justifying approaches that are proposed or taken, recording, and reporting results in different ways.

**Part 3** (chapters 15 through 20) presents *promising practices* of several existing mathematical programs that include or are designed for students with special needs. Mathematics as "sense making" is highly valued in these programs. Questions, problems, and mathematical tasks are presented that *stretch* students' thinking beyond their immediate grasp. The expectation is that with support from the teacher as facilitator and from collaborative work with peers, students will accept the challenge, reach out, and thereby "grow" mathematically. The story of each group of students is different but is indeed a promising model for those of us involved in the practicalities of creating appropriate mathematical programs for students with special needs.

**Part 4,** the concluding chapter, recaps major obstacles to needed change and invites us, one and all, to reshape barriers as bridges to quality mathematics instruction for students with special needs.

A recent incident in a special education classroom of eight primary-level children with severe problems characterizes the philosophy of what is recommended and suggests what can happen when teachers value and provide time and opportunity for mathematical problem solving. Jean Gomes, the special education teacher of the class at Margaretta Carey School in Waverly, Iowa, recognized that her children needed space and time to figure things out. She was patient, for she believed that her students would get better at thinking mathematically if she involved them in enough significant problem-solving situations. On this occasion, Jean took advantage of a "teachable moment" to create an unplanned problem that centered on the making of English muffin pizzas.

The children had been keeping a "Days in School" graph and had collected 100 soda-pop cans in celebration of 100 days of school. A field trip was then planned so the children could trade the cans for cash and use the money to buy the makings of English muffin pizzas. Given that each child wanted two pizzas and each of the three adults wanted one, the problem was to figure out how many English muffin pizzas were needed. (The notion that each pizza was only a half of a muffin was disregarded.)

Jean suggested that students work with a partner on the problem and paired the two autistic children. Although cooperative learning had been part of other mathematical tasks, this was the first time the two autistic children had worked together. One child had virtually no sense of number; the other, although verbal, lacked logical reasoning skills.

This mathematics problem occasioned a monumental "breakthrough" for the two autistic children. In what was atypical of all prior performance, they became involved in the problem, communicated with each other, and made drawings to correctly solve the problem (fig. 1). Though solution approaches differed, the final results obtained by all the pairs, including the autistic pair, were the same: 16 pizzas for the children and 3 for the adults—19 pizzas in all were needed.

This professional resource, dedicated to the interests of teachers like Jean Gomes and all students with special needs, reflects the experiences, exper-

Fig. 1

tise, and commitment of many individuals. The Editorial Panel created the content profile of the book, developed guidelines for authors, reviewed and selected manuscripts, made valuable suggestions for improving manuscripts, and shaped the book's overall direction. Our sincere thanks and deep appreciation go to these individuals:

| | |
|---|---|
| Paul Trafton (senior panel member) | University of Northern Iowa, Cedar Falls, Iowa |
| James Bruni | Lehman College, Bronx, New York |
| Portia Elliott | University of Massachusetts, Amherst, Massachusetts |
| Francis "Skip" Fennell | Western Maryland College, Westminster, Maryland |

Although Marty Hopkins was unable to continue throughout the project, we appreciate her input in helping us conceptualize the original structure of the book.

We also gratefully acknowledge the NCTM Educational Materials Committee, who suggested that this reference be budgeted, and the assistance of the NCTM staff during the developmental, editing, and production phases of the book. We are particularly indebted to Cynthia Rosso, publications business manager, for her administrative guidance and support; Charles Clements, senior editor, for his outstanding and sensitive editorial preparation of the manuscript; Sheila Gorg, for her capable editorial assistance; and Jo Handerson, editorial production assistant, for the appealing design and layout. The efforts of these NCTM teams contributed greatly to the quality and usefulness of this professional reference.

We and the authors also gratefully acknowledge the contributions of the following photographers:

Jerry Liebenstein, Illinois State University Photo Service
(chapters 1 and 10; parts of other chapters)

Carl Tolino, teacher of media, and Joe Jencik, Pittsburgh School District
(chapter 4)

Kalman Zabarsky, Boston University Photo Service
(chapter 6)

C. Kurt Holter, professional photographer, Frederick, Maryland
(chapter 9)

Oscar Rivera, Rivera Photos
(chapter 11)

Dan Beam, student at Anderson University, Anderson, Indiana
(chapter 17)

Brad Groves, recent Holt High School graduate, photography scholarship recipient to Albion College, Michigan
(chapter 20)

Without the thoughtful contributions of the authors themselves, of course, this professional reference would not be a reality. Finally, we are deeply appreciative of their willingness to share their thinking and experiences and for the time they have taken and the dedication they have shown in developing their chapters. The many and varied perspectives they bring furnish a broad foundation for reflection and action that we believe will open many windows of opportunity in mathematics for students with special needs.

# Part 1

## Current Issues

T O

1

# Mathematics Power for All

*Portia Elliott*
*Cynthia Garnett*

***I ain't never gonna learn no math.***
***—Walter P., age 14***

***No lo comprende matemáticas!***
***—Carmen R., age 11***

***I hate math.***
⠊ ⠓⠁⠞⠑ ⠍⠁⠞⠓⠲

***—Amy M., age 12***

***I HATE MATH, TOO.***

***Jesse S., age 9***

THOUGH the expressions above differ in form, they share the same sentiment. Our young learners are all expressing negative dispositions toward mathematics. What should distress us as much as the frequency of these negative expressions is that society seems to accept these sentiments without dissent. To their younger counterparts' comments, many adults proudly add, "I was never good at mathematics either." No such confessions are ever made about literacy limitations in reading or writing, and no such defeatist statement is considered acceptable if made in response to complaints about handicapping conditions.

## A SOCIETY'S SENTIMENTS

These sentiments, so commonly expressed, should be regarded as cries for help and should conjure up images of human suffering. These pleas for help are the presenting symptoms of a debilitating ailment—mathematics illiteracy—that if left untreated will advance and cripple present and future genera-

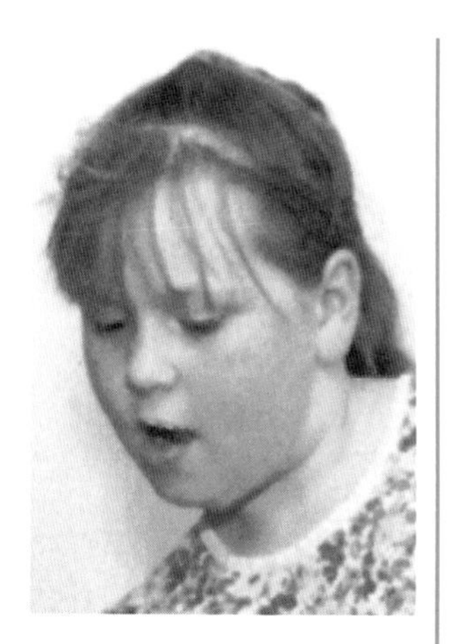

tions of learners. The highest-risk groups for this mathematics malady are the poor, ethnic minorities, and differently abled learners.

Demographers tell us that the fastest-growing segment of the American population consists of children like Walter, Amy, Carmen, and Jesse, who appear to experience difficulties in learning mathematics at an early age, attempt to memorize mathematics by rote, fall behind their classmates, get labeled as "intellectually impaired" (Baroody and Hume 1991, p. 55), and eventually "drop out of the 'mathematics pipeline'" (Steen 1990, p. 131). This situation should trouble us greatly, especially when this phenomenon is coupled with demographers' predictions of the increasing need for mathematical literacy. Steen (1990) predicts that the percentage of new jobs necessitating four years of secondary school mathematics will be 60 percent greater in the 1990s than in the 1970s. Secada (1990, pp. 137–38), echoes Steen's predictions:

> The level of mathematical literacy needed to participate fully in that world, its jobs, its economic and social orders, and its democratic institutions is steadily increasing.

For present classrooms of learners we ask, "Is this mathematics malady curable?" For future generations we ask, "Is it preventable?" Most mathematics educators would answer yes to both questions but would add this caveat: Only if we change the perceptions of dysfunctionality, only if we ratify the belief that "all can learn mathematics," and only if we change the ways in which we teach mathematics (Baroody and Hume 1991; Wilmot and Thornton 1989).

But altering perceptions and changing mathematics classroom-teaching practices so as to reduce the dropout rates will require energetic, well–trained teachers, especially from underrepresented minorities. Demographers tell us that the pool of well-trained teachers is diminishing and that there may not be enough interested and qualified replacement teachers in the "mathematics pipeline" if the present trend continues. This is painfully apparent when it comes to ethnic minorities and differently challenged individuals who either drop out of, are forced out of, or deselect mathematics-related disciplines and careers in education, leaving the profession with a paucity in the variety and number of interested and qualified role models. Steen (1990, p. 133) warns—

> unless we improve the attractiveness of mathematics at *all* levels of schooling and for *all* socioeconomic groups, we shall never be able to attract enough young people into science and engineering [and teaching] careers to retain our competitive position into the next century.

The exercise presented in figure 1.1 will help the reader get a clearer picture of the gravity of the problems inherent in trying to change a society's sentiments about who can and should be mathematics literate. These sentiments directly affect the mathematics pipeline and therefore our nation's economic, political, and educational lifeline. We ask that you read each proposition, test your number sense by trying to determine the correct percent for each statement, and compare your predictions with those suggested by demographic experts. See if your sentiments about the gravity of the problem correspond with the forecasts of these futurists.

## Test Your Percent Sense

Complete each statement below using one of the following options:

| | |
|---|---|
| Less than 5 percent | About 45 percent |
| About 10 percent | About 50 percent |
| About 25 percent | About 60 percent |
| About 30 percent | About 80 percent |
| About 35 percent | More than 95 percent |

Statements:

1. _________ of handicapped children are either classified as learning disabled, speech impaired, mentally retarded, or emotionally disturbed. (Reynolds and Birch 1988, p. 39)

2. _________ of students identified as "educable mentally retarded" (EMR) are African American. This percent is more than twice that of African Americans in the total school population. (Sailor 1991, p. 15)

3. _________ of the mathematics teachers in the next decade will be from ethnic minorities (National Research Council 1989, p. 21). Slightly fewer than this percent are classified as gifted in the total population of exceptional students. (Reynolds and Birch 1988, p. 39)

4. _________ of all high school students drop out (NRC 1989, p.12). This is the same percent who graduate marginally literate and virtually innumerate. (Steen 1990, p. 31)

5. _________ of college mathematics enrollees are in courses ordinarily taught in high school. (NRC 1989, p. 13)

6. _________ of all students identified nationally as needing special education services are described as learning disabled (LD). (Sailor 1991, p.15)

7. _________ of persons between the ages of 16 and 64 with handicapping conditions are employed. (Marozas and May 1988, p. 260)

8. _________ of the nation's 200 000 secondary school teachers of mathematics do not meet current professional standards for teaching mathematics. (NRC 1989, p. 28)

9. _________ of the Ph.Ds granted in the United States go to other than African American and Hispanic males and females. (Steen 1990, p.130)

10. _________ of the nation's elementary school teachers meet contemporary standards for their mathematics teaching responsibilities. (NRC 1989, p. 28)

*Answers: 1. 80%; 2. 35%; 3. 5%; 4. 25%; 5. 60%; 6. 45%; 7. 30%; 8. 50%; 9. 95%; 10. 10%*

Fig. 1.1

The nation can ill afford to consign any of its citizens to mathematics wastelands.

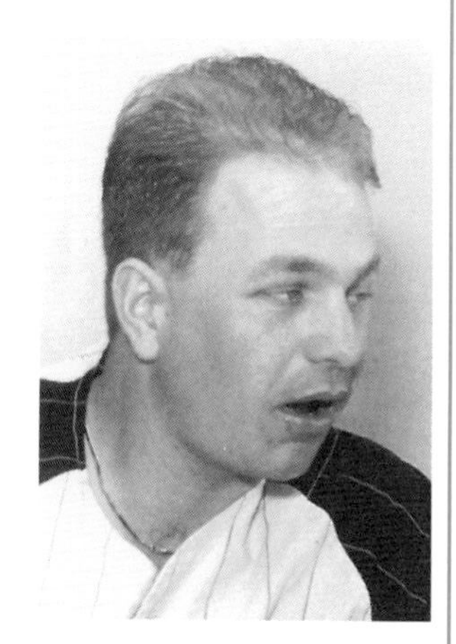

## A SOCIETY SUFFERS

Judging from the despondency about mathematics and from the demographic data reported in the previous section, it seems apparent that more and more individuals are succumbing to mathematics illiteracy and that our society is suffering from "static mathematics power syndrome," that is, mathematical power is held too tightly, by far too few. Signs of this syndrome are visible everywhere: on the faces of children (rich and poor) who muster Herculean strength to memorize isolated mathematics facts; on the brows of diners (black and white) who cannot carry out simple restaurant calculations; in the eyes of learners (disabled and gifted) who are reprimanded for using devices like computers and calculators to aid in laborious calculations; and in the frowns of workers (novice and veteran) when called on to pose and answer "what if" questions when their orientation has been to find bottom-line "right" answers.

Walter's, Amy's, Jesse's, and Carmen's only recourse to mathematics-power deprivation, they think, is to lash out against this perceived barrier to plenty, namely, mathematics itself—its hurdles, its humiliations, and its hurts. For immediate relief from their suffering, they choose to leave the take-it-or-leave-it mathematics they are being taught. But with this exodus, they find themselves open to becoming truly handicapped because none is able to "exercise the power inherent in his or her mind to reach conclusions with full assurance" (National Research Council 1989, p. 4). Our quartet of despondent learners will continue to fall behind the privileged and have their problems interpreted from the vantage point of the elite and addressed only if the powerful feel compelled to do so.

That such a schism exists, separating the privileged and powerful from the illiterate and impotent, is undeniable. That the nation can ill afford to consign any of its citizens to mathematics wastelands is likewise without dispute. Ensuring the survival of all the Walters, Amys, Jesses, and Carmens is an economic necessity and a political imperative if our experiment in democracy is to continue and our leadership in the free world is to endure.

## A SOCIETY SHIFTS

Baroody and Hume (1991, p. 55) remind us that most children who experience learning difficulties in mathematics—including those labeled as learning disabled—are not intellectually impaired but are "curriculum disabled." They suggest that many learners seem victimized by instruction that is ill suited to the ways children think and learn. To meet the intellectual needs of all children, instruction must be developmentally appropriate. Progressive educators, Baroody and Hume among others, suggest that instruction should have understanding as its focus, learning that is both active and purposeful should be fostered, informal knowledge should be supported, formal instruction should be linked to that informal knowledge, and reflection and discussion must be encouraged.

The National Council of Teachers of Mathematics in the *Professional Standards for Teaching Mathematics* has offered strategies to accomplish these instructional goals that call for dramatic and necessary shifts (NCTM 1991, p. 3 [emphasis added])—

- *toward* classrooms as mathematical communities—*away from* classrooms as simply a collection of individuals;
- *toward* logic and mathematical evidence as verification—*away from* the teacher as the sole authority for right answers;
- *toward* mathematical reasoning—*away from* merely memorizing procedures;
- *toward* conjecturing, inventing, and problem solving—*away from* an emphasis on mechanistic answer-finding;
- *toward* connecting mathematics, its ideas, and its applications—*away from* treating mathematics as a body of isolated concepts and procedures.

This new way of thinking advocated by the Council will require a paradigmatic shift away from transmittal-fashioned teaching to a more constructivist-oriented approach to the teaching and learning process. This kind of paradigm shifting requires profound change in societal assumptions and beliefs about the purpose and goals of schools and schooling.

What follows is a discussion of just such a paradigmatic shift that is happening in the economic, political, and educational arenas as the nation moves from an industry-based reality to a technology-based reality. We offer this discussion to help the reader understand the root of the problem of poor mathematics dispositions and noncomprehension. To discover the etiology of these growing societal phenomena we should examine the economic, political, and educational realities of the past that have brought us to our present.

## Economic Realities Then and Now

The economic philosophy of an industrial age, with its hierarchical models of power and its regard for humans as resources for industry, dictated a definition of "mathematics literacy" that was more algorithmic than holistic. This definition has changed dramatically since the "shopkeeper arithmetic" of the smokestack era. Being literate then meant being able to break problems and processes down into smaller parts, perform repetitive routines, think algorithmically, and compute efficiently. These skills had direct applications in an economy that served up goods by means of conveyor belts. Mass marketing, mass communication, and mass education were all desirable for the conforming masses and their learning-to-conform offspring. Those failing to conform could be dismissed, suspended, or forced out. No individual choice was encouraged, no individual needs had to be addressed, no environments had to be tailored for specialists because everyone's behavior was to be contoured to fit the "bolted-down desks" that ordered their learning environments.

With the coming of PCs (personal computers), PINs (personal identification numbers for banking), and pay-per-view television, society today and in the future will have to accommodate to differences, accented by choice. The "Have It Your Way" jingle of television advertisement is characteristic of a new era that is more dependent on spreadsheet sensibilities than on register-tape reckonings. Business leaders are looking to employ workers who can do "what if" posing, "what about" conjecturing, and "what for" explaining. Employers are insisting that their employees be creative problem solvers who are not upset by the "special orders" of an insistent public.

## Political Realities Then and Now

For all the pronouncements of equality and "justice for all," the political philosophy of the U.S. industrial era persisted in investing few with power and fewer with privilege. As a nation, we have struggled with the idea of "for all." We have professed it in our founding documents, but for political, economic, social, and educational reasons, we have made exceptions to who would be included in *all*.

Not until Public Law 94-142 was enacted were educators forced to think of *all* as *everyone without exception.*

The nation's laws have tried to redress this political doublespeak. The Thirteenth, Fourteenth, and Fifteenth Constitutional Amendments provided for the participation of black males in the political structure. The Nineteenth Amendment followed, which made women part of the political mosaic. Various civil rights laws were enacted to correct the economic, social, and educational exclusion of ethnic minorities and women. Not until Public Law 94-142 was enacted were educators forced to think of *all* as *everyone without exception.* This law, predicated on the individual's right to due process, provided that all persons had a right to free public education in the "least restrictive educational environment."

## Educational Realities Then and Now

Prevailing educational philosophies of the industrial era, predicated on the intellectual notions of Thorndike, Skinner, and Binet, were atomistic and persisted in seeing many humans as "defective" who could not benefit from "regular" instruction. Education that reflected the political and economic philosophies of the era found conformity and hierarchical control to be common classroom commodities.

Industrial-age teachers rarely sought or received support in adapting instructional programs to individual needs. Technology-age teachers have available to them materials and support services capable of making their classrooms more responsive to the individual needs of all students. Conveyor-belt-era teachers were required to teach in their individual classroom cubicles and teach compartmentalized disciplines, with "regular" children isolated further from students with "special needs." Microchip-age teachers (including special education teachers) are encouraged to operate in barrier-free learning environments where experiences, subjects, and disciplines are arranged to allow connections to exist and be discovered. Finally, smoke-stack-era teachers engaged in norm-referenced diagnoses of educational, emotional, and physical problems and prognosticated about probable achievement. Computer-age teachers with data-management technologies can analyze the strengths and needs of learners and make decisions that particularize instruction.

Factory-styled learning environments of yesteryear are rapidly disappearing. Diminished is the insistence that students (in "regular" or "special" education) sit in assigned seats in well-ordered rows for work from the same page, in identical texts, with attention expected for the same duration. Educators are calling for "differentiated instruction" (Wilmot and Thornton 1989) and "developmentally appropriate instruction" (Baroody and Hume 1991) to address the needs of all learners. These shifts in environmental conditions for learning mathematics and in instructional techniques will prepare us to survive the technological onslaught of the next century.

Educators are calling for "differentiated instruction" and "developmentally appropriate instruction" to address the needs of all learners.

## A SOCIETY'S STRENGTH

To eliminate the ills of mathematics illiteracy and power deprivation, beliefs about what causes dysfunctional behavior must change and instructional practices must be altered radically. Changing both will require an additional shift away from blaming the victim to looking for systemic causes to the problem of dysfunctionality. As we cast about to find ways to immunize society against mathematics illiteracy, we look to the "mathematically well" for clues to mathematical good health for all. We ask, What are the attributes that prevent mathematics illiteracy? What are the cultural factors, if any, that counter it? What are the behavioral patterns that forestall mathematical misunderstandings?

As we asked these questions, we found that the mathematically powerful construct and reconstruct reality. They represent the world symbolically. They reason from raw data. They examine critically, isolate commonalities, and choose from options with certainty. To be mathematically powerful is not to be like the "red-faced gentleman" caught up in his sums when visited by the Little Prince (de Saint-Exupéry 1943). Mathematical power is not merely calculation. It is the observing of patterns and order—in number, chance, forms, algorithms, and change. It is testing conjecture by observation, simulation, and experimentation to discover truths. It is estimating results with the knowledge that "mathematics offers science both a foundation of truth and standard of certainty" (National Research Council 1989, p. 31). To be mathematically powerful is to employ distinctive modes of thought: modeling, optimizing, logically analyzing, inferring from data, symbolizing, and abstracting. It is to possess the power to better understand the information-laden world in which we live (National Research Council 1989).

When asked, "What is mathematics to you?" the mathematically powerful answer that "mathematics is the invisible culture of our age" (National Research Council 1989, p. 32). Mathematics and statistical ideas are embedded in, and are the fabric of, the environment of technology. Our lives and our work are influenced by these ideas. Practically, mathematics is a vehicle to improve our basic standard of living. Civically, it helps us understand public policy. Professionally, it is the tool of all trades. In our leisure, it provides enjoyment in the form of logical games. Culturally, it is a major intellectual tradition appreciated "as much for its beauty as for its power" (National Research Council 1989, p. 33).

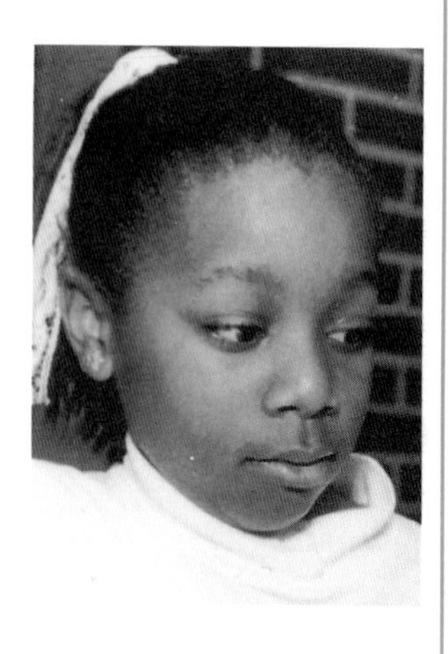

Mathematics is a legacy of human culture—produced by all human cultures—that must be passed on to future generations. Mathematics, they remind us, will be implicated when these questions are asked and answered: What is the cost of allowing any person or group of persons to be mathematically illiterate in a technological world? If we fail to pass on this legacy to future generations, what happens to our standard of living, our public policies, our leisure time, our culture? What resources are required to ensure that mathematical power can be accessed by all members of society, regardless of capacity, ethnic background, or gender?

## A SOCIETY SURVIVES

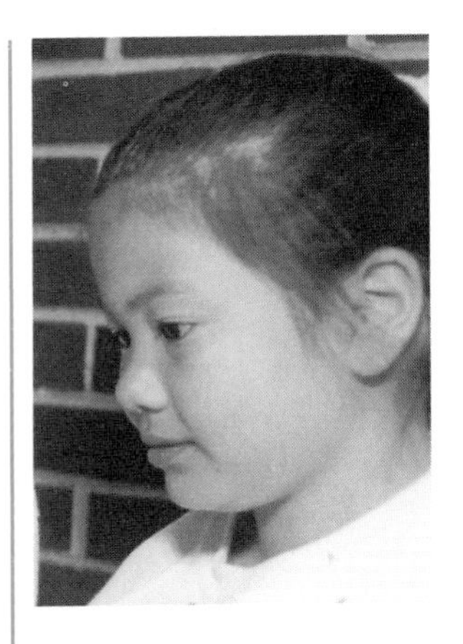

A society's survival is dependent on its ability to cultivate mathematics power in all its citizens and to ensure that a legacy of misunderstanding, apprehension, and fear is not left for any of its offspring. What antidotes exist for the negative dispositions and lack of comprehension exhibited by Walter, Carmen, Amy, and Jesse? The National Council of Teachers of Mathematics has set forth a vision and recommendations for school mathematics reform (NCTM 1989, 1991) that should give us some clues to how to produce a nation of mathematically literate and mathematically powerful individuals.

The genesis of the NCTM reform recommendations can be found in the writings of Lev Vygotsky (1962), a Russian psychologist who suggests that learning is a social phenomenon, even when the language and behavior of the learner seems egocentric, and of Jean Piaget (1954), a Swiss psychologist and educator who gave expression to a constructivist theory of learning. The reform movement owes a debt to "whole language" proponents Kenneth Goodman (1992) and Frank Smith (1986), who argue for contextualizing all literacy experiences, and to the community of special educators who articulated the need to "particularize" education for all learners; who proffered collaboration models of educational assessment, planning, and facilitating in adaptive, inclusive environments; and who challenged "regular" educators to expand their notion of "special" to include all learners.

Strengthened with constructivism and social-linguistic principles of teaching and learning, and buttressed by judicial pronouncements and federal laws, mathematics educators and special educators can collaborate in redefining an appropriate program in mathematics education for all students. During this redefining process, both groups should feel confident that they can generate worthwhile mathematics tasks, organize for discourse, arrange learning environments, and analyze efforts because these actions are rooted in a sound educational theory that is derived from a philosophy of empowerment and inclusion.

What follows is a synthesis of the major thrusts in school mathematics reform, advanced in various NCTM publications. When superimposed on the major provisions of special education legislation, they serve as parameters for redefining "appropriate mathematics instruction for all." (For emphasis, the "major thrusts" of NCTM are italicized and the "major provisions" of the special education legislation are in bold type and were abstracted from the writings of Reynolds and Birch (1988) in their text *Adaptive Mainstreaming*.) The writing and research of noted mathematics and special educators provide corroboration for juxtaposing these NCTM and special education provisions:

### Zero Reject

*All students should be afforded the opportunity of a rich mathematical curriculum.*

Walter, Amy, Carmen, and Jesse are among the many children who, because of poor performances in mathematics classes, get mislabeled "learning disabled" when in fact they are "curriculum disabled," "educationally deprived" victims of inappropriate mathematics curriculum and instruction. Baroody and Hume (1991), echoing the sentiments and the positions held by the National Association for the Education of Young Children (1986a, 1986b), suggest that when mathematics instruction is "developmentally appropriate," "meaningful and purposeful," "active and engaging," then all children will have opportunities to construct their own mathematical understandings and the vicious cycle of defeat, fear, math anxiety, and math avoidance can be broken.

For all children, including those classified as learning disabled, mathematics instruction should (Baroody and Hume 1991, pp. 55–56)—

- promote a broad range of mathematics concepts;
- not be limited to computation;
- be comprehensive, including such topics as geometry and fractions;
- actively involve students in doing mathematics that has a real purpose;
- encourage and build on children's strength: their informal knowledge;
- encourage students to justify, to discuss, and to compare ideas and strategies.

The National Research Council (1989, p. 2) advocates opportunity for all and rejection of none when it writes, "It is time to act to insure that all Americans benefit from the power of mathematics."

### Nondiscriminatory Evaluation

*Evaluation should inform instruction, be ongoing, derive from multiple sources, and reflect multiple perspectives.*

In the years prior to the passage of Public Law 94-142, special education children and their families knew all too well the harm caused by evaluative data collected to sort, seriate, and subjugate. With the passing of PL 94-142, lawmakers have tried to ensure against discriminatory evaluation practices. In the working draft of the NCTM (1993) *Assessment Standards for School Mathematics*, the Council also takes up the banner of nondiscriminatory evaluation.

The Association of State Supervisors of Mathematics (ASSM), realizing the uses and potential for abuses of evaluative data, argues against tests of the seductive norm-referenced variety and in favor of criterion-referenced measures (ASSM 1992, p. 66):

> Most, if not all, currently available norm-referenced tests are not aligned with proposed school mathematics curricula and are detrimental to the implementation of mathematics programs in harmony with the aims of the reform movement. Such tests should not be used for either individual-pupil or total-program evaluation.

Arguing against discrimination in evaluation practices and echoing ASSM members, Wilmot and Thornton (1989, p. 221) write this:

> Achieving minimal competency scores in written tests clearly falls short of the broader goal in the mathematics education of learning-handicapped students—the ability to *think* in quantitative situations. It is critical that they understand what they are doing and why they are doing it. Otherwise, inferences or transfer to new situations cannot be made.

By establishing collaborative working relationships in evaluation activities and by insisting on multiple instruments analyzed from multiple perspectives, special educators and mathematics educators can help eliminate most forms of test bias or discrimination.

## Individualized Eduational Plans (IEP)

*Mathematics teachers should create learning environments in which each learner can construct his or her own understandings about mathematics.*

The National Research Council (NRC) (1989) asserts that "each individual's knowledge of mathematics is uniquely personal" (p. 2). If this assertion were put into practice, mathematics educators and special educators would embrace the differentiated instruction models recommended by Wilmot and Thornton (1989) and the constructivist teaching models proffered by DeVries and Kohlberg (1987). Both of these instructional models require responding in different ways to different learners on the basis of their individual needs, interests, and abilities. Differentiated instruction is essential as students are encouraged to construct their own understandings of mathematics. If classroom instruction is differentiated, work can be tailored for the "less able" and the "more able" to suit individual needs (Wilmot and Thornton 1989). Individualized education plans can be used to accomplish the goals of differentiated instruction and constructivist teaching.

Davis, Maher, and Noddings (1990, p. 3) believe that particularizing instruction is necessitated insofar as—

> learners have to construct their own knowledge—individually and collectively. Each learner has a tool kit of conceptions and skills with which he or she must construct knowledge to solve problems presented by the environment. The role of the community—other learners and teacher—is to provide the setting, pose the challenges, and offer the support that will encourage mathematical construction.

Given these more recent writings on how mathematics is learned and how mathematics could be taught, Spicker and McLeskey (1981) wrote prophetically more than ten years ago that "the time may well come when establishing annual goals and short-term objectives will be recognized as sound education practice not only for the handicapped but for all children" (p. 5). Through their national organizations, mathematics educators have affirmed that indeed that time is now.

## Least Restrictive Environment

*Mathematics should be recontextualized and decompartmentalized.*

Placing children in the least restrictive environment in a mathematics classroom involves taking advantage, to the maximum extent possible, of cooperative, heterogeneous learning groups. Wilmot and Thornton (1989) call for this type of interactive learning, which respects individual differences while providing children "opportunities to recognize and value their interdependence on one another" (p. 215). Although their particular needs at times require targeted attention, gifted and learning-handicapped children alike can profit from interaction in heterogeneous learning groups.

On the value of cooperative-learning groups for children with special needs, Baroody and Hume (1991) offer that, contrary to the expectations in traditional instruction of attentiveness to teacher and independent, quiet completion of assigned tasks, research supports a livelier notion of learning.

Placing children in the least restrictive environment in a mathematics classroom involves taking advantage, to the maximum extent possible, of cooperative, heterogeneous learning groups.

It seems important for children to both think about and justify their solutions to problems. "A useful tool for encouraging both is cooperative learning groups" (p. 56).

The least restrictive environments that can support the goal of thinking for all students include classroom flexibility, celebration of diversity, interaction in cooperative learning groups, and teachers who organize for individual and group needs. In these more adaptive and more inclusive environments, learners construct better and more complete understandings of mathematics.

### Parent Participation

*Parents and members of the community at large should be integral to the teaching-learning-evaluating process.*

The role of parent as first teacher should be used to advantage. Children do not come to school as empty vessels to be filled but rather as individuals with many prior experiences. Building on both the formal and informal knowledge children bring to learning finds increasing support among educators. Building on knowledge gained in the home and the community is more than desirable; it may be crucial to the success of children in understanding, remembering, and applying school mathematics.

Mathematics educators and special educators agree that instead of just discussing children's performance with their parents during assessment periods, teachers should make parents their partners in the daily activities of the classroom and school. The six types of family-community-school collaborations developed by Joyce Epstein (Brandt 1989) describe the interactions that mathematics educators and special educators suggest for parent and community involvement. The more substantive collaboration suggested in upper levels of involvement reflect the suggestions of both groups of educators. At level 4, parents become more active by assisting and monitoring mathematics learning at home; at level 5, they take part in governance, decision making, and advocacy; and at level 6, parents begin to work in collaborative ways to help ensure that all children get the proper support they need for learning.

## A SOCIETY SUCCEEDS

A society can claim success in eradicating the malady of mathematics illiteracy if and only if all its progeny are able to develop to their fullest potential. If its offspring can become employable workers, wisely choosing consumers, and autonomously thinking citizens who can be contributors in the super symbolic quantitative world they will inherit, then society can say, "Victory is ours!"

The curriculum and pedagogy based on the mass marketing and certainties of the past must now give way to a curriculum and pedagogy that can address the niche-marketing issues and uncertainties of the future. Teachers will need to continue to create inclusive, adaptive least restrictive environments where all learners, particularly those who are disenfranchised and those labeled "disabled," can take respected places in the communities of tomorrow.

Teachers will need to continue to create inclusive, adaptive least restrictive environments where all learners, particularly those that are disenfranchised and those labeled "disabled," can take respected places in the communities of tomorrow.

Learners in these communities will no longer lament their distance from mathematics' powers. To replace their anguishing cries, this should be society's song (Elliott 1976, p. 478):

> These mathematical ideas will be useful
> In our "machine-run" society.
> But utility must be coupled with aesthetics,
> For these ideas possess "math artistry."
> The thoughts contained here are priceless,
> The philosophy, I hope, is quite clear.
> Beauty and Utility are here for the asking;
> Let **[all]** children question without fear.

## REFERENCES


Association of State Supervisors of Mathematics. "Position Statements." *Arithmetic Teacher* 39 (February 1992): 66.

Baroody, Arthur J. *Children's Mathematics Thinking: A Developmental Framework for Preschool, Primary, and Special Education Teachers.* New York: Teachers College Press, 1987.

Baroody, Arthur J., and Janice Hume. "Meaningful Mathematics Instruction: The Case of Fractions." *Remedial and Special Education* 12 (May/June 1991): 54–68.

Brandt, Ron. "On Parents and Schools: A Conversation with Joyce Epstein." *Educational Leadership* 47 (October 1989): 24–27.

Davis, Robert B., Carolyn A. Maher, and Nel Noddings. "Introduction: Constructivist Views on the Teaching and Learning of Mathematics." In *Constructivist Views on the Teaching and Learning of Mathematics, Journal for Research in Mathematics Education* Monograph No. 4, pp. 1–3. Reston, Va.: National Council of Teachers of Mathematics, 1990.

de Saint-Exupéry, Antoine. *The Little Prince.* New York: Harcourt Brace and World, 1943.

DeVries, Rheta, and Lawrence Kohlberg. *Constructivist Early Education: Overview and Comparison with Other Programs.* Washington, D.C.: National Association for the Education of Young Children, 1987.

Elliott, Portia C. "Vocabulary of Mathematics." *Mathematics Teacher* 69 (October 1976): 478.

Goodman, Kenneth. "I Didn't Found Whole Language." *Reading Teacher* 46 (November 1992): 188–99.

Marozas, Donald, and Deborah C. May. *Issues and Practices in Special Education.* New York: Longman, 1988.

National Association for the Education of Young Children. "Position Statement on Developmentally Appropriate Practice in Early Childhood Programs Serving Children from Birth through Age 8." *Young Children* 41 (September 1986a): 4–19.

———. "Position Statement of Developmentally Appropriate Practice in Programs for 4- and 5-Year-Olds." *Young Children* 41 (September 1986b): 20–29.

National Council of Teachers of Mathematics. *Assessment Standards for School Mathematics.* Working draft. Reston, Va.: The Council, 1993.

———. *Professional Standards for Teaching Mathematics.* Reston, Va.: The Council, 1991.

National Research Council. Mathematical Sciences Education Board. *Everybody Counts: A Report to the Nation on the Future of Mathematics Education.* Washington, D.C.: National Academy Press, 1989.

Piaget, Jean. *The Construction of Reality in the Child.* 1937. New York: Basic Books, 1954.

Reynolds, Maynard C., and Jack W. Birch. *Adaptive Mainstreaming.* White Plains, N.Y.: Longman, 1988.

Sailor, Wayne. "Special Education in the Restructured School." *Remedial and Special Education* 12 (November/December 1991): 8–22.

Secada, Walter G. "The Challenges of a Changing World for Mathematics Education." In *Teaching and Learning Mathematics in the 1990s,* 1990 Yearbook of the National Council of Teachers of Mathematics, edited by Thomas J. Cooney, pp. 135–43. Reston, Va.: The Council, 1990.

Smith, Frank. *Insult to Intelligence.* New York: Arbor House, 1986.

Spicker, Howard H., and James McLeskey. "Exceptional Children in Changing Times." In *The Mathematical Education of Exceptional Children and Youth: An Interdisciplinary Approach,* edited by Vincent J. Glennon, 1–22. Reston, Va.: National Council of Teachers of Mathematics, 1981.

Steen, Lynn Arthur. "Mathematics for All Americans." In *Teaching and Learning Mathematics in the 1990s,* 1990 Yearbook of the National Council of Teachers of Mathematics, edited by Thomas J. Cooney, pp. 130–34. Reston, Va.: The Council, 1990.

Vygotsky, Lev. *Thought and Language.* 1934. Cambridge, Mass.: MIT Press, 1962.

Wilmot, Barbara, and Carol A. Thornton. "Mathematics Teaching and Learning: Meeting the Needs of Special Learners." In *New Directions for Elementary School Mathematics,* 1989 Yearbook of the National Council of Teachers of Mathematics, edited by Paul R. Trafton and Albert P. Shulte, pp. 212–22. Reston, Va.: The Council, 1989.

**Portia Elliott** is a professor of mathematics education, School of Education, University of Massachusetts at Amherst. She teaches mathematics methods courses and is interested in how volition influences cognition and the affective aspects of mathematics learning. She was chair of the Editorial Panel of *The Arithmetic Teacher: Mathematics Education through the Middle Grades.*

**Cynthia Garnett** is an associate professor of early childhood education, Department of Early Childhood and Family Studies, Kean College of New Jersey, Union, New Jersey. She was formerly a learning disabilities specialist at the International Center for the Disabled, New York, New York. She is interested in literacy development in early childhood populations and the role of culture in all aspects of learning.

2

# A Changing Curriculum for a Changing Age

*Paul R. Trafton*
*Alison S. Claus*

***Are we teaching mathematics so that all students will be empowered to use it flexibly, insightfully, and productively?***

***What does it mean to be mathematically literate in a technological age?***

***Are we teaching the mathematics that all students will need in the future?***

THE dawn of the information age has accelerated efforts around the world to rethink and restructure mathematics programs from kindergarten through graduate school. Fundamental questions are driving change in the mathematics that students learn, the organization of the curriculum, the ways in which students learn mathematics, the roles of teachers in teaching mathematics, and the climate and activities of mathematics classrooms.

Change requires clear direction and consensus about its nature. The National Council of Teachers of Mathematics (NCTM) has published a landmark document, *Curriculum and Evaluation Standards for School Mathematics* (1989), that presents a coherent vision for school mathematics and establishes standards to guide its realization.

The publication, whose message is extended by the *Professional Standards for Teaching Mathematics* (NCTM 1991), reflects the consensus of mathematics teachers at all levels and is supported by professionals in related mathematical sciences, sciences, and education. It highlights characteristics of quality mathematics programs for all students, regardless of economic status, gender, ethnic or racial background, disability, or special learning needs. The two documents provide the supporting framework for all teachers, including those who teach students with special needs, to give *every* student a window of opportunity. They are making change possible by influencing educational policies and energizing creative teachers and curriculum developers.

The *Curriculum and Evaluation Standards* (NCTM 1989) addresses school mathematics at *all* levels for *all* students. It is a broad and comprehensive publication whose message is for *all* teachers and *all* classrooms.

## CHARACTERISTICS OF CURRICULUM CHANGE

The substance and message of the *Curriculum and Evaluation Standards* are not easily captured in capsule form. It is important, nonetheless, to be aware of, and reflect on, its central themes, major recommendations, and key ideas, which transcend type of learner, type of classroom, and type of school. We will also reflect on its particular implications for children with disabilities and other special needs. This section also serves as the foundation for succeeding chapters and the exciting vignettes of teaching and learning presented throughout chapters 8–14.

Two pervasive themes of the document introduce the discussion. The first is mathematics for *all* students. Because mathematics is increasingly affecting all citizens in their personal, vocational, and civic lives in such dramatically different ways, all students need to learn relevant, useful mathematics in order to function successfully in today's world. The necessity for change to affect all students is clear in the following discussion of societal goals for school mathematics (NCTM 1989, p. 5).

> Today's society expects schools to insure that all students have an opportunity to become mathematically literate, are capable of extending their learning, have an equal opportunity to learn, and become informed citizens capable of understanding issues in a technological society.

The second theme is mathematical sense making. A central thesis of change is that all students must become confident "doers" of mathematics. This means that they will be capable and resourceful problem solvers, will develop the ability to communicate and reason mathematically, and will value mathematics as worthwhile and essential. These goals can be accomplished only if mathematics makes sense to them and if they believe in their ability to make sense of it. A central aspect of sense making in mathematics is possessing a strong understanding of mathematical ideas, relationships, and applications. The idea that mathematics must make sense to all individuals is more than a lofty goal, for current research and innovative classroom practice offer strong evidence that, indeed, all students can accomplish this goal and as a result, learn substantially more mathematics.

The idea that mathematics must make sense to all individuals is more than a lofty goal, for current research and innovative classroom practice offer strong evidence that all students can accomplish this goal and learn substantially more mathematics.

The foregoing themes serve as the cornerstone for further examination of change. A discussion follows of three key aspects of change in the content and focus of the curriculum.

### Furnishing a Broad and Balanced Curriculum for All Students

Technology is dramatically changing mathematics; as a result, the mathematics that students learn also must change. It has been said (*Leading Mathematics Education into the Twenty-first Century* [NCTM et al. n.d., p. C-5]) that some mathematics has become more important because technology *requires* it; some mathematics has become less important because technology *replaces* it; and some mathematics has become possible because technology *allows* it.

The current curriculum, with its "shopkeeper arithmetic" focus, repetitive emphasis on arithmetic and algebraic skills, and orientation toward minimal mathematics for most individuals and advanced mathematics for a few, is inadequate for the current and future needs of all students. The mathematics content must reflect current uses of mathematics. A wide range of content has a highly significant role in the workplace, other fields, and daily lives. As a result, the curricular content needs to be substantially broader. An examination of the titles of the standards (see fig. 2.1) reveals the emphasis on a broad range of content.

Children will become confident "doers" of mathematics only if mathematics makes sense to them and if they believe in their ability to make sense of it.

Long-neglected areas of mathematics, such as data analysis, probability, functions, and geometry, have a prominent place, and familiar content, such as arithmetic and algebra, has a substantially different focus. Emerging areas of mathematics are reflected within standards and in separate standards devoted to them, as in the instance of discrete mathematics.

The *Curriculum and Evaluation Standards* also calls for a balanced curriculum in which all content has a significant role and receives ongoing attention. At the K–8 level, data analysis, patterns and functions, number sense, measurement, geometry, probability, and algebra have a significant role, receiving as much emphasis as operations and computation. The inclusion of these topics is a dramatic shift from current programs, in which much content receives cursory treatment, often at the end of the year or as a "nice extra" to the traditional curriculum. The idea of a balanced curriculum is also

CURRICULUM STANDARDS FOR SCHOOL MATHEMATICS

| Standards for Grades K–4 | Standards for Grades 5–8 | Standards for Grades 9–12 |
|---|---|---|
| Mathematics as Problem Solving | Mathematics as Problem Solving | Mathematics as Problem Solving |
| Mathematics as Communication | Mathematics as Communication | Mathematics as Communication |
| Mathematics as Reasoning | Mathematics as Reasoning | Mathematics as Reasoning |
| Mathematical Connections | Mathematical Connections | Mathematical Connections |
| Estimation | Number and Number Relationships | Algebra |
| Number Sense and Numeration | Number Systems and Number Theory | Functions |
| Concepts of Whole Number Operations | Computation and Estimation | Geometry from a Synthetic Perspective |
| Whole Number Computation | Patterns and Functions | Geometry from an Algebraic Perspective |
| Geometry and Spatial Sense | Algebra | Trigonometry |
| Measurement | Statistics | Statistics |
| Statistics and Probability | Probability | Probability |
| Fractions and Decimals | Geometry | Discrete Mathematics |
| Patterns and Relationships | Measurement | Conceptual Underpinnings of Calculus |
| | | Mathematical Structure |

Fig. 2.1. Synopsis of curriculum standards from *Curriculum and Evaluation Standards for School Mathematics* (NCTM 1989)

A broader, better-balanced curriculum that values sense-making offers many advantages for all students, particularly those with special needs.

reflected in the recommendation for a core mathematics curriculum for all students in grades 9–12. The core curriculum is intended to furnish a common body of mathematics, including algebra, geometry, statistics, functions, and trigonometry, that is accessible to all but differentiated by depth and breadth in the way in which content is treated and by the applications included.

A broader, better-balanced curriculum that values sense-making offers many advantages for all students, particularly those with special needs. It prepares them for life in an information age, for students are no longer competent if they have studied only computation; it keeps windows of opportunity open throughout their lives by giving them the strong, diverse preparation that allows them to study additional mathematics later; it supplies successful mathematical experiences, for some students who struggle with numbers and operations are very successful when working with geometry or data analysis; it allows them to study interesting mathematics; it makes them aware of the usefulness of mathematics; and it causes them to believe that they can learn mathematics.

### Connecting What Students Learn with How They Learn It

The *Curriculum and Evaluation Standards* makes clear the inextricable relationship between knowledge and how it is acquired. The document indicates that the purpose for knowing mathematics is to use it to solve problems, to represent ideas in various ways, to discern patterns, to make sense of new situations, to make and justify conjectures, and to communicate one's ideas to others. As Lappan and Schram (1989, p. 20) note,

> the coin of the realm in the twenty-first century will be ideas. It will no longer be sufficient for students to enter the working world with only disconnected rules, theorems, and techniques stored in their mathematical heads. What will be valued in business and industry is being able to think and reason mathematically and to bring the power of mathematics to bear on a problem that needs a solution. The computational aspects of the solution can often be done by a computer, but a human must reason through the situation to decide what techniques need to be applied to solve the problem.

Thus, throughout the vision for the mathematics curriculum, discussions of content are interwoven with the context and processes by which content knowledge is acquired. The purpose of the first four standards is to make connections among context, content, and process. These exciting standards, which can be viewed as "umbrella" standards, highlight what all students need to experience in learning mathematics and what they need to be able to do with their knowledge. They also present a dynamic and comprehensive view of what it means to know and do mathematics. Figure 2.2 presents one example of each of the four standards (the complete statement of each standard at each level is included in appendix A).

The problem-solving standards emphasize the central role of problem solving in mathematics. They call for the inclusion of many types of problems and the use of various strategies in solving them. Of even greater significance is the notion of embedding the mathematics to be learned in rich problem settings.

The communication and reasoning standards highlight the means by which

**Standard 1: Mathematics as Problem Solving**

In grades 5–8, the mathematics curriculum should include numerous and varied experiences with problem solving as a method of inquiry and application so that students can—

- use problem-solving approaches to investigate and understand mathematical content;
- formulate problems from situations within and outside mathematics;
- develop and apply a variety of strategies to solve problems, with emphasis on multistep and nonroutine problems;
- verify and interpret results with respect to the original problem situations;
- generalize solutions and strategies to new problem situation;
- acquire confidence in using mathematics meaningfully.

**Standard 2: Mathematics as Communication**

In grades 9–12, the mathematics curriculum should include the continued development of language and symbolism to communicate mathematical ideas so that all students can—

- reflect upon and clarify their thinking about mathematical ideas and relationships;
- formulate mathematical definitions and express generalizations discovered through investigations;
- express mathematical ideas orally and in writing;
- read written presentations of mathematics with understanding;
- ask clarifying and extending questions related to mathematics they have read or heard about;
- appreciate the economy, power, and elegance of mathematical notation and its role in the development of mathematical ideas.

**Standard 3: Mathematics as Reasoning**

In grades K–4, the study of mathematics should emphasize reasoning so that students can—

- draw logical conclusions about mathematics;
- use models, known facts, properties, and relationships to explain their thinking;
- justify their answers and solution processes;
- use patterns and relationships to analyze mathematical situations;
- believe that mathematics makes sense.

**Standard 4: Mathematical Connections**

In grades 5–8, the mathematics curriculum should include the investigation of mathematical connections so that students can—

- see mathematics as an integrated whole;
- explore problems and describe results using graphical, numerical, physical, algebraic, and verbal mathematical models or representations;
- use a mathematical idea to further their understanding of other mathematical ideas;
- apply mathematical thinking and modeling to solve problems that arise in other disciplines, such as art, music, psychology, science, and business;
- value the role of mathematics in our culture and society.

Fig. 2.2. Used with permission from the *Curriculum and Evaluation Standards for School Mathematics*, copyright 1989 by the National Council of Teachers of Mathematics

students make sense of mathematics—through using models, through expressing ideas verbally and in writing, and through using several approaches to attain and verify solutions. Marilyn Burns (1992, tape 2) captures the power of communication and reasoning:

> Communicating in math lessons, both verbally and in writing, helps children construct their own understanding of mathematical ideas. When children talk with each other about mathematics, they have to clarify their thinking, justify their ideas, and listen to other points of view. Having students write requires that they reflect on their ideas, and helps them cement and extend their thinking. Not only does communicating support children's learning, it provides teachers insight into what their students understand.

The standards on mathematical connections address the need for deliberately making connections among mathematical topics, between mathematics and other curriculum areas, and between mathematics and its multiple applications.

Although these standards are not explicitly content standards, they offer insight into the purposes for learning content, the ways in which students come to understand content, the processes by which mathematics programs can help all students be successful, and the pathway toward helping children become mathematically powerful. They also provide a context for understanding and interpreting content recommendations.

A second connection between content and how it is learned is found in the statements of the content standards (see NCTM [1989]). Young children, for example,

- construct number meanings through real-world experiences and the use of physical materials (p. 38);
- formulate and solve problems that involve collecting and analyzing data (p. 54);
- investigate and predict the results of combining, subdividing, and changing shapes (p. 48).

Middle-grades students—

- develop, analyze, and explain procedures for computation and techniques for estimation (p. 94);
- analyze functional relationships to explain how a change in one quantity results in a change in another (p. 98);
- model situations by devising and carrying out experiments or simulations to determine probabilities (p. 109).

High school students—

- represent situations that involve variable quantities with expressions, equations, inequalities, and matrices (p. 150);
- design a statistical experiment to study a problem, conduct the experiment, and interpret and communicate the outcomes (p. 167);
- determine maximum and minimum points of a graph and interpret the results in problem situations (p. 180).

Building close relationships between content and how it is encountered helps all students make connections between their intuitive knowledge and mathematical knowledge; it causes them to build knowledge for themselves by active mental involvement; it gives them additional time to think things through; it allows them to learn from others by hearing their thinking; and it causes them to feel successful through contributing to group efforts.

## Emphasizing Mathematical Concepts and Relationships

The extraordinary power and usefulness of mathematics derives from its substantive ideas rather than its routines and procedures. The deep understanding of concepts, relationships, and generalizations enables individuals to interpret phenomena and solve theoretical and applied problems. These ideas also allow us to see the connectedness of mathematics and thus become aware of its rich unifying themes. Yet the historic focus on symbol manipulation has kept most students from valuing the subject or becoming powerful users of it. Computer technology has created an opportunity and a sense of urgency to establish the conceptual knowledge base that enables all individuals to keep career opportunities open throughout their lives.

The *Curriculum and Evaluation Standards* thus strongly emphasizes conceptual mathematical knowledge. One aspect of this emphasis is the attention given to number sense—intuitions about numbers and number relationships—and spatial sense—insights and intuitions about shapes.

Students with good number sense about fractions, for example, are aware that 2/5 is a little less than 1/2, a little more than 1/3, and the same as 4/10. This knowledge enables them to compare 2/5 and 1/2 or 2/5 and 5/8 without relying on traditional rules that often make little sense to them. Similarly, by graphing algebraic relationships, students can visualize and better under-

stand what the symbols represent and what it means to solve an equation or a system of equations or inequalities. Current technology makes this approach accessible to all students. Number sense is powerful for children with learning disabilities. It allows a child who, for example, has a memory deficit to generate facts and computational procedures by using relationships among numbers.

Not only is spatial sense essential for all aspects of learning geometry, it is also integral to understanding ideas of numbers and operations. It, too, is particularly useful to children with special needs. The use of models and pictures helps a child who is deaf or has other auditory deficits by stimulating visual strengths. The child with good visual strengths but auditory deficits can compensate for difficulties with symbolic work by using models or writing to overcome difficulties. Teaching with visual models brings understanding and insight to symbolic work.

It is the rich fabric of ideas that needs to receive primary emphasis today in all classrooms for all students, not the rules and rituals of computational mathematics. As noted in *Everybody Counts* (National Research Council 1989, p. 44), "a mathematics curriculum that emphasizes computation and rules is like a writing curriculum that emphasizes grammar and spelling; both put the cart before the horse." Procedural learning is not unimportant; rather, it needs to be placed in its proper perspective.

It is the rich fabric of ideas that needs to receive primary emphasis today in all classrooms for all students, not the rules and rituals of computational mathematics.

An additional reason to emphasize concepts and relationships is their fundamental role in helping students learn. Although we tend to focus on computational deficiencies, research clearly indicates that students' problems are rooted in their lack of understanding of basic ideas. For example, the inability to add fractions is frequently a result of not understanding fundamental fraction concepts. Similarly, not knowing whether to use $2l + 2w$ or $lw$ to find the area of a rectangle indicates a lack of understanding of perimeter and area.

Furthermore, findings from cognitive psychology indicate that concepts achieve power when they have rich associations and are interconnected with other ideas. The mind is not an enormous storage cabinet of discrete pieces of information. A more appropriate analogy is that of a giant Tinkertoy construction consisting of nodes and all possible connections among them (Peterson, Fennema, and Carpenter 1988/89).

Contemporary curricula are devoting substantial time to helping students build representations of concepts and connect mathematical symbols with these representations. They recognize that many ideas, such as the concept of a variable, often pose significant problems for students, and they thus provide many types of experiences to ensure that students build a solid understanding. They allow children to explore the addition of decimals with number strips and base-ten blocks prior to focusing on rules about lining up decimal points. They introduce ideas in the context of problems and everyday events, allowing children to use their own language and ways of thinking to explore the ideas prior to introducing standard language and symbols. Such practices help all children learn mathematics and are especially beneficial to those children who need help the most if they are to be successful learners.

*Rethinking skills*

The commitment and time required for developing mathematics conceptually necessitate a decrease in the current priority on learning skills and memorizing definitions and formulas. Mastery of procedural mathematics is often equated by the public with "knowing" mathematics and is viewed as the mark of a mathematically competent person. However, it is no longer the primary focus in the workplace today. There, workers have reference materials, computers on their desks, and opportunities for interaction with others.

A decreased emphasis on computation concerns some individuals because they interpret it to mean that all skills will be abandoned and envision fifth-grade students counting on their fingers to solve $8 + 3$ or $5 \times 6$. This scenario is not what is intended. Yet at the same time, it is inappropriate to require children to become highly skillful with multiplying two large numbers when a computer can find the product of a million "really large" numbers in a few seconds. It is similarly inappropriate to require students to rationalize denominators in algebra, because this skill is not needed when computers and graphing calculators are used.

Certainly, students need some mathematical facts and procedures at their fingertips, for procedural knowledge can facilitate learning and problem solving. To say, however, that children will "know" their basic facts does not mean that they will learn them by using the unsupported and inappropriate practices of the past. Teachers find it fascinating to see how accurate young children are when solving problems that they care about and how well they acquire basic facts when allowed to think them through in their own ways.

Decreasing the emphasis on rules and routines involves carefully considering what skills are of sufficient value to require mastery and, as we discuss in

The commitment and time required for developing mathematics conceptually necessitate a decrease in the current priority on learning skills and memorizing definitions and formulas.

the following section, requires extended deliberation and letting go of long-held beliefs. Reconsidering procedural mathematics also involves developing a new perspective on how this knowledge is acquired. The *Curriculum and Evaluation Standards*, supported by research and extensive classroom experience, argues that conventional mathematical skills can and should—

- grow out of problems and applied situations, so that students learn that computation is carried out for a reason;
- be developed by using a variety of techniques, including those generated by students, for finding an answer;
- be closely linked with, and develop from, fundamental mathematical ideas;
- develop over time and in a variety of contexts;
- be viewed as only one aspect of a total program.

## Summary

This section has explored the content and dynamics of curricular change. It has attempted to capture a vision of change that must be played out across the mathematics curriculum, as illustrated in chapters 8–14 of this book, and has highlighted what needs to be valued in mathematics and mathematics classrooms.

Special educators also need to see the potential of a broad-based curriculum built around sense making, for it can lead to powerful changes in their students' learning.

Special educators also need to see the potential of a broad-based curriculum built around sense making, for it can lead to powerful changes in their students' learning. A curriculum that focuses on making sense of number, operations, data, and space allows a child who has deficiencies in one modality to use strengths in another to compensate. The use of models and pictures to illustrate such concepts as area and perimeter may help some children by stimulating visual strengths. The use of writing to explain thinking helps a child who has an auditory memory deficit "speak" her thoughts to herself. The child with visual deficits may integrate his understanding of fractions by explaining his reasoning to others.

This curriculum may also require us to rethink and redefine our ideas about being gifted in mathematics.

This curriculum may also require us to rethink and redefine our ideas about being gifted in mathematics. Children who learn to compute quickly and easily are often identified as gifted; yet many of them do not have good conceptual understanding, and they struggle when solving challenging problems that require mathematical intuition and insight. We will become more aware of the mathematical strengths of creative problem solvers as we devote more time to solving problems and justifying solutions. We will become more aware of children who have spatial strengths as they have more opportunities to visualize and manipulate figures. When our teaching and curriculum change, many new and exciting mathematical talents of children will become apparent and will need to be considered.

These are exciting times, for we are rejecting an inappropriate, static, and outmoded curriculum and are creating for *all* students—

- a dynamic curriculum,
- a comprehensive curriculum,

- an integrated curriculum,
- an applied curriculum,
- a problem-driven curriculum,
- a balanced curriculum,
- an active-learning curriculum,
- a curriculum that allows access by removing computational barriers,
- a curriculum that meets their present and future needs.

## THE CHALLENGES OF CURRICULAR CHANGE

As we have seen, the *Curriculum and Evaluation Standards* portrays a very different vision of mathematics programs, as well as a very different view of what needs to be learned and how it needs to be learned. The revolutionary nature of this change is captured in the *Professional Standards for Teaching Mathematics* (NCTM 1991, p. 1):

> To reach the goal of developing mathematical power for all students requires the creation of a curriculum and an environment, in which teaching and learning are to occur, that are very different from much of current practice.

Thus, change entails far more than using manipulatives, adding a problem-solving lesson at the end of a chapter, teaching the statistics chapter earlier in the year, or otherwise tinkering with the current curriculum.

Implementing substantive educational change is complex, for it involves rethinking long-established practices. In mathematics, the process is more difficult because our view of mathematics results from our own experiences with it. Change also involves confronting deeply held beliefs that make it difficult to perceive that things could be different. It challenges accepted views of what teachers do as they teach mathematics; it involves accepting uncertainties associated with using less-structured teaching practices; and it entails dealing with anxieties about current school policies, standard testing practices, and parental acceptance. The issues of change become even more pronounced when working with students whose special needs add complexity to teaching, decision making, and program planning.

This section examines some existing beliefs and practices that need to be confronted. They are deeply embedded in our thinking because of long-standing societal views of school mathematics and learning. The purpose of the discussion is to help all educators understand more fully the task that we face. Many teachers find it easy to give mental assent to arguments for change and overall characteristics of change but still face tensions when putting ideals into practice.

Change involves confronting deeply held beliefs that make it difficult to perceive that things could be different.

The process of implementing change involves examining barriers, beginning the process of thinking through them, and moving beyond them. We are learning that understanding, accepting, and implementing change is a slow process, involving a progressive letting go of existing beliefs and practices.

### Beliefs about Mathematics

Mathematics is commonly viewed as a static, unchanging body of facts and procedures; a symbolic, rule-driven, abstract, and deductive subject; and a prototypical example of a linear, hierarchical, and sequential discipline. These beliefs stand in stark contrast to the contemporary view of a dynamic, growing, and fluid field that involves exploring and experimenting with ideas—a discipline that is intimately intertwined with a host of applications and that is growing rapidly in response to new needs and uses. One teacher beautifully captured the liberating and exciting result of "rethinking" mathematics when she stated, "I used to think that math was chapters in books; now I see that it is everywhere and part of everything we do."

### Beliefs about What Mathematics Is Really Important

Our view of mathematics shapes our thinking about the content that is important for students to learn. We tend to believe that the mathematics we learned is the mathematics that students need to know today. This notion includes the "shopkeeper arithmetic" of an industrial age, the formal mathematics that has been viewed as necessary to prepare students for the next course, and the mathematics that leads to studying physics, chemistry, and engineering.

Our view of mathematics shapes our thinking about the content that is important for students to learn. We tend to believe that the mathematics we learned is the mathematics that students need to know today.

We can easily accept the importance of "nice" mathematics, such as data analysis, problem solving, and informal geometry. We agree that it is interesting, and even useful, mathematics. The problem arises when "nice" mathematics conflicts with "essential" mathematics, for curricular change requires making decisions about the mathematics we need to de-emphasize or delete in order to achieve new goals, to establish new priorities, to plan the sequence of the curriculum, and to rethink essential outcomes.

These decisions often involve traditional computational skills. It just doesn't seem right to cut back (or even eliminate) expectations for long division and mixed-number computation, to accept that factoring trinomials and rationalizing denominators are not essential to "knowing" algebra, or not to expect all second-grade children to become skillful with paper-and-pencil subtraction.

No easy shortcuts can be found to viewing curriculum content and priorities differently. We change as a result of thinking ideas through for ourselves. This process can involve reflecting on arguments for change, becoming more familiar with the details of change, and interacting with colleagues. Perhaps the most powerful stimuli to change are gaining new awareness of our students' capabilities and strengths and seeing their positive responses and new perceptions of themselves. Positive experiences with four-function and graphing calculators, for example, often cause us to rethink views about traditional skills.

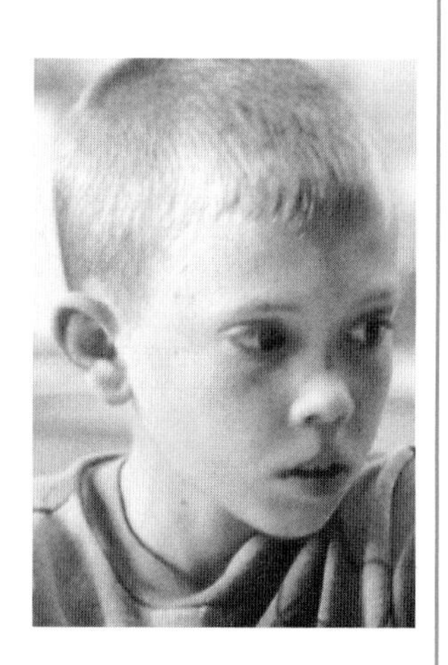

### Beliefs about Children's Capabilities

Traditional practice has led us to expect limited understanding and minimal learning by most students, not just those with learning problems. Society accepts the premise that only a few individuals are capable of being mathematically literate, whereas the rest possess an inherited "disability." Our belief about students' capabilities has been reinforced by a curriculum that has focused on computation, one of the most complex aspects of learning mathematics, particularly when it is taught rotely and out of context.

We find it difficult to believe that the vast majority of children possess significant mathematical strengths. These strengths may be revealed in solving problems, in inventing alternative ways of computing, in spatial reasoning, or in understanding fundamental ideas of algebra or probability. They often are revealed in students' contributions to group work or rich whole-class discussions of ideas and problem solutions.

If we are honest with ourselves, we might admit that teaching students with learning problems can be comforting because it seems that the agenda is more limited: just help them understand what multiplication means, learn what a fraction is, or master some addition facts. Conversely, believing that *all* students can learn a significant amount of mathematics entails accepting responsibility for offering them a rich, challenging curriculum. One of the more exciting outcomes of doing so is seeing the breakthroughs they make and the delightful strengths they possess despite their problems.

Believing that *all* students can learn a significant amount of mathematics entails accepting responsibility for offering them a rich, challenging curriculum.

### Beliefs about How Students Learn

Our behavioral tradition has resulted in well-established beliefs about teaching and learning: state precise objectives, identify the components of a task, teach pieces individually, give clear explanations, reinforce each component, identify errors in the components of a task, reinforce weak subskills, and evaluate each task individually. Related beliefs that are invalid include such views as these:

- The basics must be learned before engaging in higher-order thinking, conceptual learning, or problem solving.
- Understanding follows skill.
- Skills and topics are ordered hierarchically.
- The mastery of elemental skills must precede the learning of more advanced ones.

We have been encouraged to view teaching and learning from this perspective, and traditional diagnostic teaching and evaluation practices reflect these beliefs.

Experience in classes and recent evidence about learning severely challenge these beliefs and offer compelling evidence to the contrary. This evidence suggests that students learn by integrating new ideas with existing knowledge, intuitive understandings, and informal experiences. They learn by establishing links among individual pieces of knowledge. Learning can begin with problem-solving experiences that have specific understandings and skills embedded in them. Thus, learning is neither linear, hierarchical, nor sequential.

It is not difficult to believe that mathematics can be learned differently—at least for some aspects of the curriculum. Yet we find it difficult to let go of traditional beliefs in the instance of such skills as learning basic facts or simplifying fractions. Then we are surprised, and sometimes shocked, at the sudden shift in our thinking. These "really important" things, we find ourselves saying, can be learned only through dogged determination and repetitive drill. This tension is real and difficult to resolve; again, it is often reduced only by experiencing for ourselves that new approaches produce more powerful learning in all areas.

### Beliefs about Programs for Children with Special Needs

Our practices are rooted in our beliefs. These practices influence decisions about programs for children with special needs. As our beliefs change, so do our practices and programs. Making appropriate program decisions is not easy, for many issues need careful consideration. Nonetheless, we need to address aspects of programs that are no longer appropriate, such as those listed below. Other chapters contain additional discussion of several of them.

- Programs that organize the curriculum in a linear, hierarchical fashion according to the perceived level of complexity
- Programs that use behavioral objectives and narrow mathematical criteria to diagnose, plan (e.g., many IEPs), and assess
- Programs that remove "remedial" and "gifted" children with special needs from ongoing class activities
- Remedial programs that focus on "basic skills"
- Programs for the gifted that only accelerate children's progress through advanced skills or are built around activities unrelated to each other or to classroom instruction
- Programs that use computation-oriented tests (that is, most achievement and diagnostic tests) to make decisions about students' placement and the curriculum they should follow
- Programs that inflexibly track children on the basis of ability as determined by tests or a hierarchical curriculum

### Summary

Confronting beliefs is the first step toward changing them, and, in turn, changing the curriculum, changing our teaching, and changing our expectations.

Current beliefs influence our thinking about children and programs. Yet they are not insurmountable barriers to change. Change is still attainable and has tremendous power for helping all children realize their mathematical potential and attain the rewards of being mathematically literate. Confronting beliefs is the first step toward changing them and, in turn, changing the curriculum, changing our teaching, and changing our expectations.

## IMPLEMENTING CURRICULAR CHANGE—A TEACHER'S EXPERIENCE

Can change occur? In complex schools and classrooms? In spite of our traditional views of mathematics and learning? With a wide variety of students, ranging from those who possess special mathematical talents to those for whom we have traditionally held minimal expectations?

These questions are reasonable, for translating the vision of change into everyday practice is complex. This book presents numerous vignettes of what mathematics teaching that addresses *all* students can be. In this section we look at teachers and what occurs when they embrace change with the conviction that it is right for *all* students. We will examine a teacher's perspective of change by presenting her experiences, reflections, tensions, and joys as she rethinks her teaching and curriculum. We will focus on the teacher rather than on teaching.

The following is a view of a "standards rich" classroom as seen through the eyes of one teacher in a primary-grade mainstreamed classroom. This view is supplemented with a few experiences of other teachers who are undergoing similar experiences. The experiences, however, are not unique to the teaching setting or age level, and teachers at other levels and in other settings are urged to make appropriate translations. The story also helps us all see what mathematics classrooms *can* be when the challenge of change is met.

Ms. C teaches twenty-two second-grade children. The group is diverse, with a wide range of abilities, backgrounds, and special needs. Among her students are Zach, who is bright and immediately responds to a challenge of any kind; Andrew, who is bright but hesitant to take any kind of risk; Jessica, who is new to the school and comes from a family for whom English is a second language; Mary, who has been diagnosed as having an attention deficit disorder; and Amanda, who has a visual-motor deficit. Mary receives help from the remedial reading teacher and a volunteer mathematics tutor while Amanda works with the learning resource teacher.

Ms. C is expected to "cover" the goals and content in the district's curriculum guide. She knows that it is also important to the district for her students to score well on the standardized test that is administered each spring. She hopes that her school board and administrators soon will become more enlightened about new forms of assessment that better show what her students are learning. Although Ms. C realizes that she must work within these parameters, she has not stopped her commitment to change her teaching and program.

Ms. C has learned over the years that rote activities and standard textbook pages do not really help her students understand and learn mathematics, nor do they help the children attain the important goals of the change in which she strongly believes. She has attended several workshops and professional meetings and meets regularly with other teachers with similar beliefs. As a result, Ms. C is aware of many new teaching strategies and knows she will receive support as she implements changes.

In particular, Ms. C now believes that manipulative materials are valuable tools for making mathematics come alive for her students, for making abstract mathematical notions concrete, and for stimulating rich reasoning and problem solving. So she has stocked her room with buckets of connecting cubes, containers of links, a large box of base-ten blocks, and several sets of pattern blocks. She also has a calculator for each student and a large supply of paper on which her students can record their experiences and explanations.

Ms. C also knows that problems that encourage various solution strategies are powerful for promoting mathematical growth, so she incorporates many kinds of problems, including classroom situations that she creates or draws from a variety of resources, and problems posed by her students. In particular, she is impressed with the potential of connecting literature and mathematics. As a result, she frequently uses children's books as a source for problems or to create a context for mathematical ideas.

In recent weeks she has used several children's books, including *Alexander Who Used to Be Rich Last Sunday* (Viorst 1979), *Seventeen Kings and Forty-two Elephants* (Mahy 1987), *Counting on Frank* (Clement 1991), and *Amos and Boris* (Steig 1971). She finds that the stories lead to wonderful discussions about the mathematical situations in them. Sometimes her students rewrite parts of the story to fit their own situation. They often create problems related to the book. She still remembers how exciting it was to watch her students trying to figure out how many of the forty-two elephants each of the seventeen kings should take care of, if the work is shared equally. She laughs as she thinks about their struggle over what to do with the seventeenth king after agreeing that each king would care for two elephants and realizing that sixteen kings could each care for another one-half elephant. Initially, they thought of just eliminating the extra king; they finally decided to make him a head king, which means he shouldn't have to do as much of this kind of menial work. She also remembers their efforts in trying to figure out how tall they would be if they had grown at the same rate as a tree, as suggested in *Counting on Frank*.

Ms. C emphasizes having her students share solutions to problems and

highlights the thinking strategies they use to make sense of the problems. This shared knowledge encourages in children's minds the notion that mathematics is a sense-making endeavor and that thinking, exploring, sharing, and explaining are important parts of doing mathematics. By listening carefully to their discussions, she is able to take advantage of mathematical questions and various representations of situations that arise in their work.

She is impressed with how much insight she gains into her students' thinking and understanding through this process and by the ability of all of her students to do powerful mathematical thinking. Andrew is beginning to use his ability—he gets involved in problems and is beginning to take risks; Jessica almost always finds a solution and is gradually growing in her ability to express her thinking; and Mary is one of the class's most creative thinkers (now that she is not limited by traditional computation) and often shares her unique insights with the whole class. Ms. C is continually reminded of how much more all her students are capable of doing than she had previously believed.

Sharing solutions to problems encourages in children's minds the notion that mathematics is a sense-making endeavor and that thinking, exploring, sharing, and explaining are important parts of doing mathematics.

Ms. C is constantly looking for ways of connecting mathematics with the real world. In December she posed a problem about how many boxes of candy canes Santa would need to bring if each child were to get three candy canes. The principal recently asked the children to consider the kind of playground equipment they would find ideal. So Ms. C is having her students design and build their ideal equipment, and as they are doing so, she is helping them review and extend their knowledge of two- and three-dimensional figures, including names and properties. She is pleased that she is finding ways of using everyday contexts for mathematics.

Ms. C is aware that writing, as well as oral communication, is important in helping all her students learn. Thus she tries to incorporate more of it. Many of the children still struggle with writing and can discuss ideas and solutions in greater depth than they can write about them; nonetheless, she encourages them to write about their thinking to the best of their abilities. She often has them share their written work and encourages all children, those who primarily use pictures as well as those who can write more extensively. She is particularly pleased with the connections that she is finding between mathematics and language arts and with how each reinforces many of the goals of the other.

As Ms. C thinks back over her mathematics teaching this year, she is pleased with what she has accomplished and, more important, with how much her students have accomplished. She is still surprised at what they are capable of doing when she lets them take ownership of ideas and problems. Earlier in the year, she was concerned about her less proficient students but is impressed now with their growth and with the power that develops as the result of all children's participating in a "community of learners" who are building knowledge through solving problems and engaging in ongoing discourse.

Throughout the year, Ms. C and the Chapter 1 teacher, who supervises the volunteers, have met to set goals for Mary that can be carried out more effectively in a one-on-one situation where she is not distracted by the sights and noises of other children. She is pleased that Mary has made great progress with relationships among numbers through playing games using base-ten blocks with a volunteer.

Ms. C also feels a strong partnership with the learning resource teacher who works with Amanda. Amanda's visual-motor problems cause her to have a lot of trouble in reading and using tables. When Ms. C had the class write problems using a restaurant menu, a type of table, she and the resource teacher planned activities for Amanda in the resource room that would allow her to make sense of this work by creating her own table using her own items and prices. The special education teacher then spent several periods in the classroom supporting Amanda as she wrote her own problems and helping other children who had difficulty with the activity.

Ms. C values the insights that team planning gives her and finds it helpful to have another knowledgeable teacher in the classroom. The special education teacher tells Ms. C that team planning and teaching are productive for her also, since they allow her to incorporate learning strategies for Amanda in a classroom context. They also give her a better sense of how Amanda is functioning within the classroom and how she is progressing in relationship to her classmates.

Ms. C is becoming aware of many other things as well. She is aware that teaching takes a great deal more reflection on her part, but she enjoys that. She struggles at times to find appropriate, challenging problems and activities. She is still trying to broaden her curriculum, particularly by finding ways to expand her work with graphing, geometry, and measurement. She would also like to find more activities that make connections among mathematical topics. She has learned that *why* she is teaching something and *how* she teaches it are at least as important as *what* she teaches. She has become aware of the importance of knowing what her students are thinking and why they are thinking that way; she is surprised by how much more she now

knows about the performance of her students, although she realizes that she still needs to know more about alternative ways of assessing thinking and learning.

Ms. C is pleasantly surprised that it has been relatively easy to break her dependence on the textbook. She is varying the sequence of the curriculum, often because traditional content and skills emerge from solving problems. She is surprised that traditional subtraction computation has already arisen and developed naturally. She still can't quite believe that! More and more, she is questioning some of the content in the curriculum and is considering omitting some of it.

Along with the importance of *reflection*, Ms. C is more aware of the necessity of *risk taking*, although she is also very much aware that she must accept the anxiety that accompanies it. She knows that most of the parents are pleased and judges that their children are learning a great deal. Yet the other day she heard a more traditional colleague tell a parent that it was necessary to teach the "basics" before the children could do higher-order thinking. She knows she doesn't believe that anymore, but still the thought crossed her mind, "What if he is right?" She also worries at times about how her students will perform on traditional tests. At the same time, however, she finds that when she occasionally uses standard worksheets, her students do very well on them and find them easy.

She still finds it a bit hard to let go and put more control in her students' hands, that is, let them make sense of things in their own way. She has made progress in having her students work collaboratively but realizes that she still has much to learn about making group work productive. She is also learning always to make many kinds of manipulative materials available, because a great variety of material is helpful; for one child, connecting cubes

She realizes how much more professional she is feeling—more like a teacher than a technician.

are most helpful, another pair of children prefer links, and others would rather make drawings.

Ms. C also is finding it more comfortable to have calculators available at all times, realizing that her students use them sensibly when other computational approaches won't help. That certainly was taking a *big* risk! Another aspect of teaching that is still somewhat new to her is observing her students as they work and using that knowledge in large-group discussions and in her planning. Still, it challenges her traditional beliefs about what it means to "teach," and she reminds herself that this, too, is teaching.

Occasionally Ms. C is very tired at the end of the day. Sometimes an activity didn't go as well as she had planned, and she must think about how she can revise it. She must find ways to get Andrew and Jessica more involved and decide with whom they will work well. At the same time, it is such a relief not to have to take a large stack of papers home with her, and paper-grading time has become thinking and planning time.

She wouldn't think of returning to her old ways, but *if only* her efforts were valued more by some colleagues and administrators, *if only* she didn't have a "staffing" before school, *if only* they hadn't planned a faculty meeting tomorrow on test results, and *if only* she had more time to reflect and plan. Still, she is pleased. She has come so far, her students have come so far, and there is so much more to learn. Her old beliefs have changed more quickly than she thought they would. She realizes how much more professional she is feeling—more like a teacher than a technician. Yes, it has been worth it! The challenge of change is being met and the results far outweigh the struggles.

## Summary

This chapter has examined the direction of curriculum change, with particular emphasis on its *values*, *vision*, and *challenges*. We must reconsider what we value, not for the few but for the many, as we face the realities of the information age. As our values change, we seek a new vision built on a broad and balanced curriculum for all students. The vision is exciting, for it holds great potential for helping individuals meet the challenges of this new era. This vision can remain abstract or, as in the example of Ms. C, can become reality. The challenges of change must be confronted individually and collectively. These challenges must also be overcome if we are to align our vision with our values. Change results from embracing *what can be* rather than accepting *what is*.

### REFERENCES

Burns, Marilyn. *Mathematics: Teaching for Understanding*. White Plains, N.Y.: Cuisenaire Co. of America, 1992. Set of 3 videotapes.

Clement, Rod. *Counting on Frank*. Milwaukee: Gareth Stevens Children's Books, 1991.

Lappan, Glenda, and Pamela W. Schram. "Communication and Reasoning: Critical Dimensions of Sense Making in Mathematics." In *New Directions for Elementary School Mathematics*, 1989 Yearbook of the National Council of Teachers of Mathematics, edited by Paul R. Trafton and Albert P. Shulte, 14–30. Reston, Va.: The Council, 1989.

Mahy, Margaret. *Seventeen Kings and Forty-two Elephants.* Illustrated by Patricia MacCarthy. New York: Dial Books for Young Readers, 1987.

National Council of Teachers of Mathematics. *Curriculum and Evaluation Standards for School Mathematics.* Reston, Va.: The Council, 1989.

———. *Professional Standards for Teaching Mathematics.* Reston, Va.: The Council, 1991.

National Council of Teachers of Mathematics, ASSM, CPAM, MSEB, and NCSM. *Leading Mathematics Education into the Twenty-first Century.* N.d., n.p.

National Research Council. Mathematical Sciences Education Board. *Everybody Counts: A Report to the Nation on the Future of Mathematics Education.* Washington, D.C.: National Academy Press, 1989.

Peterson, Penelope L., Elizabeth Fennema, and Thomas Carpenter. "Using Knowledge of How Students Think about Mathematics." *Educational Leadership* 46 (December 1988/January 1989): 42–46.

Steig, William. *Amos and Boris.* New York: Farrar, Straus and Giroux, 1971.

Viorst, Judith. *Alexander Who Used to Be Rich Last Sunday.* Illustrated by Ray Cruz. New York: Atheneum Publishers, 1979.

**Paul Trafton** is a professor of mathematics and fellow in the Regent's Center for Early Developmental Education at the University of Northern Iowa. He is currently involved in a staff development program for primary-grade teachers that is built on a collaborative, shared-expertise model. He chaired the K–4 writing group of the *Curriculum and Evaluation Standards for School Mathematics.*

**Alison Claus** is coordinator of enrichment programs for Lincolnshire–Prairie View School District 103 in Lincolnshire, Illinois. She has taught mathematics to gifted and remedial students at every grade level, K–8. She is also an adjunct faculty member at National-Louis University in Evanston, Illinois, where she teaches several graduate courses in mathematics education, including one for special educators.

3

# Rethinking the Teaching and Learning of Mathematics

*William R. Speer*
*Daniel J. Brahier*

IF WE are to truly embrace the vision and promise of the *Curriculum and Evaluation Standards for School Mathematics* (NCTM 1989) and the *Professional Standards for Teaching Mathematics* (NCTM 1991), we must revitalize mathematics programs and rethink teaching and learning for the benefit of *all* students. In order to capture the flavor of the kind of classrooms that are envisioned, consider the different approaches to problem solving and negotiated learnings that are occurring in the following settings.

In a second-grade classroom, the teacher has presented the following problem:

> Gustavo has saved \$5 toward a game he wants to buy. Including tax, the game costs \$23. How much money does Gustavo still need?

Thomas and Jan decided to use Unifix cubes for dollars to help them solve the problem. They laid out two 10-trains and three extra ones; broke up one train so they could take five units, leaving one 10-train and eight cubes; and declared the amount needed to be \$18.

The teacher invited others to explain how they solved the problem. Emily said that she and her partner had used base-ten blocks. They placed five units on the table and continued to add to this pile until they got twenty-three. "Then we counted to see how many had to be added on to make a total of twenty-three. It took eighteen." She showed the number sentence she had written for the problem: 5 + 18 = 23.

Alisyn explained that she had made two separate piles of Unifix cubes, one with twenty-three cubes and one with five cubes. "Then I compared the number of cubes in each pile and found that one pile had eighteen more than the other."

Juanita said, "I just solved the problem in my head. I subtracted three from twenty-three—that's twenty—and then subtracted two more. That's eighteen."

An environment like that above, in which the teacher expects and encourages children to approach problems in different ways, is characteristic of a teaching-learning setting that strives to accommodate all students. In such a setting, the students form connections among ways of knowing and showing. Collaborative interaction provides the necessary support structure for students with special learning needs. The time devoted to problem solving and concept development is replete with opportunities to share ideas, to challenge thinking, and to negotiate learning paths that are personally relevant to individual students as they construct important understandings about mathematics.

Another example, this one at the secondary level, highlights this point. In an algebra class, the teacher challenged the students with the following question:

> If my employer pays me by the mile to travel, how many miles will I have to drive each month to meet my expenses?

Students in the class began immediately to determine what kind of information they needed to solve the problem. They decided that the make of the car was not relevant, but the average number of miles per gallon was necessary. Similarly, the brand of gasoline was not useful, but the price for a gallon of the fuel was needed to solve the problem.

An environment in which the teacher expects and encourages students to approach problems in different ways is characteristic of an instructional setting that strives to accommodate *all* students.

The class determined the price of the gasoline and, with help, the amount of a typical car payment and the costs of maintenance and insurance. The teacher suggested that a typical travel reimbursement might be 27¢ a mile, and they determined that the fixed costs of owning the car might average $300 a month.

Jose said, "I divided \$300 by 27¢ per mile on my calculator and found out that you would have to drive 1111 miles in the month to break even." Anne countered, "But if you drive 1111 miles, you will have to buy gas to run the car. Since the car gets 25 miles per gallon, you can estimate that the car would burn about 50 gallons of gas to drive that distance. So you will need to drive more miles to pay for the gas."

Tim smiled and said, "If you drive more miles to pay for the gas, it will take even more gas to break even!" The algebra class appeared to have come to the conclusion that the major issues and a process had been identified and that they could continue to "play out" the scenario until an approximation for the answer was reached. Then Mei entered the discussion.

Looking up from her notebook, Mei said, "I did this a different way. Since you know that the car gets 25 miles per gallon, that gas costs \$1.10 per gallon, and that the fixed costs are \$300 per month, I wrote an expression for the monthly cost of the car as $1.10\ (m/25) + \$300$, where $m$ represents the number of miles. Since the mileage reimbursement is 27¢ per mile, you would get $0.27m$ as your payment each month. I put them equal to each other to form an equation that I could solve: $1.10(m/25) + 300 = 0.27m$."

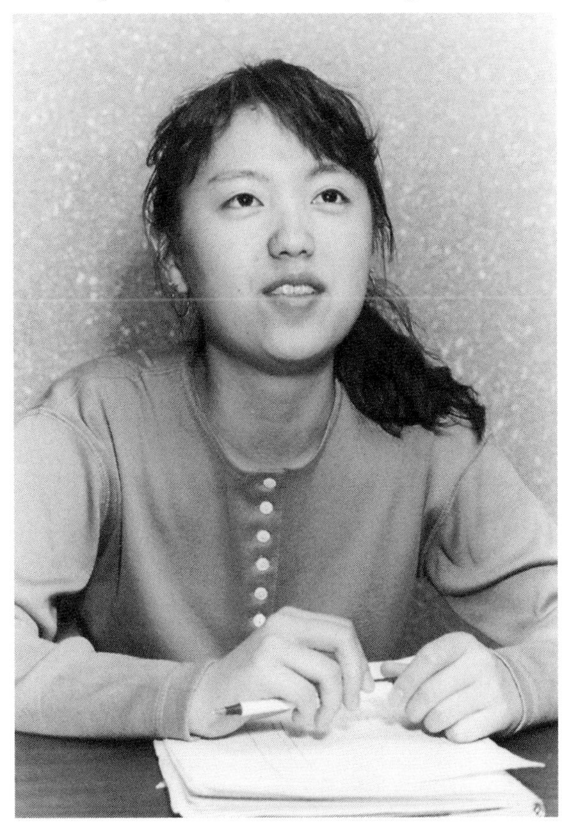

After some discussion, the class reached a consensus that Mei's approach made sense. Then the teacher asked, "Could anyone use Mei's information to solve the problem on a graphing calculator?" The class was familiar with simultaneous equations, and they soon decided that the two expressions could be entered as separate lines whose point of intersection would represent the solution to the problem.

## REFLECTIONS ON "DOING" MATHEMATICS

Classroom interactions related to the "game" and "mileage" problems above illustrate the diversity in thinking that might occur when students are encouraged to approach problems from their own perspectives. Such a classroom harbors a climate conducive to sharing ways of thinking and of building ideas, together with a chance for *all* students to be involved at their own level of understanding.

Students do not learn mathematics just by "doing"—they learn by *thinking about* what they are doing.

For Juanita, who chose to find 23 – 5 mentally, the manipulatives may even have distracted her from her thought process. For Mei, it made more sense to translate the mileage problem into an open sentence than to go through a lengthy guess-and-check process. The problems also serve as examples of rich mathematical tasks that lend themselves to conjecturing, sharing solutions, and attaching or extending meaning to what individual students already know about mathematics.

The problems and the different solutions that were presented or negotiated also illustrate what is really meant by "doing" mathematics. Math is not a spectator sport. Active participation is required for learning. However, students do not learn mathematics just by *doing*—they learn by *thinking about* what they are doing. In the minds of many educators today, doing mathematics is closely linked to active learning modes that invite thinking about, reflecting on, and processing mathematical information in problem contexts: exploring, reasoning, conjecturing, inventing, justifying, connecting, and problem solving. When students construct mathematical knowledge from active participation in meaningful, thought-provoking experiences that build naturally from their own understandings, they are much more likely to retain and use what they have learned.

In a world in which information is growing at an alarming rate, it is not sufficient for students to have memorized a few selected basic facts and algorithms for adding, subtracting, multiplying, and dividing. The goals of school mathematics programs have changed—from procedural and rule-oriented to exploratory and inferential thrusts. The emphasis has shifted *away from* only computational proficiency to problem solving based on understanding and application of mathematical concepts or principles.

Teachers are being challenged to spend valuable instructional time nurturing thinking and "pulling math out" of students instead of "pushing math in."

Teachers are being challenged to spend valuable instructional time nurturing thinking and "pulling math out" of students instead of "pushing math in." Rather than *teaching mathematics **to** students*, today's emphasis is on offering learning opportunities that enable students to actively construct their view of mathematics. The desired perspective is not hinged just on semantics. The new role for teachers as facilitators is—

- to offer experiences that enable students to make sense of mathematics;
- to ask questions and set rich problem tasks that tap and extend personal understandings;
- to establish a climate that respects and employs both individual work and collaborative problem solving;
- to expect and promote a high level of mathematical thinking and related interaction among students;
- to nurture important connections among what students think, say, do, and write as they solve mathematical problems;
- to ensure ready access to calculators (or other technology) and to a variety of manipulatives that students can use for modeling and solving problems;
- to challenge students to apply mathematics as a tool for reasoning and problem solving in situations that stem from, or relate to, their own experiences.

This perspective on doing mathematics is most clearly reflected in the classroom environment and in the day-to-day actions of both the teacher and the students in this environment.

For example, a fourth-grade class is reviewing multiplication ideas. On the basis of her assessment of students' thinking, the teacher decides that although knowledge of basic multiplication facts is relevant and important, her students also need experiences related to estimation and problem solving. Further, she concludes that explicit attention to, and support in, these difficult areas is particularly crucial for the special education students in her class. She poses the following problem:

> Today's challenge involves only six digits: 4, 5, 6, 7, 8, and 9. Your task is to create the multiplication problem with two three-digit numbers that has the largest possible product. Every digit should be used, and no digit may be used twice.

After working for a few minutes with a partner on the problem, George explained, "We decided that the numbers in the hundreds place would have to be 8 and 9, and then we tried all the other combinations on my calculator until we got the highest number."

Susan commented, "We didn't use a calculator. We decided we needed 70 to multiply the 900 and 60 to multiply the 800 to get the largest answer. We just thought about it in our heads."

Another example of doing mathematics involves a group of secondary school students using graphing calculators. On previous days, the students had used the calculators to create, interpret, modify, and compare functions and their graphs. They had also explored the difference between the slopes of such lines as $y = 2x + 3$, $y = 3x + 3$, and $y = 1/2x + 3$, and they welcomed the fact that they didn't have to spend great amounts of time making tables of values and graphing the functions by hand.

Today the teacher decided that his students were ready to form and test conjectures about $y$-intercepts, the effect of a negative-valued slope, and the relationships of the equations of parallel and perpendicular lines. He predicted that because these ideas could be explored with just a few calculator entries, his students would be able to explore a greater number of different ideas, be more involved in discussion, and leave the session with good conceptual understandings.

Explorations like these on a graphing calculator characterized a shift in emphasis in his teaching that was affecting all his students, including those with special learning needs. This new focus offered increased opportunities for learning and led to paths of understanding, applying, and using mathematics in ways that previously had not been accessible without the technology.

## BUILDING ON STUDENTS' KNOWLEDGE

All students should be afforded opportunities to construct their understanding of mathematical concepts and ideas in meaningful ways that fit with what they already know. In a sense, prior knowledge is a "hook" onto which new learnings and problem-solving strategies are hung. Because each student's background experiences and perspectives are different, there are probably as many hooks as there are students in a classroom. Therefore, to assume that one way of teaching or learning is "the best" or "the only" way does not make much sense.

Consider the shape in figure 3.1. What does one see when viewing it? Some will claim that they "see" a cube, a three-dimensional object with *six* congruent square faces. Still others may see only a large *six*-sided plane figure (a hexagon) with a "Y" in it. If we use this figure to talk about cubes, can we be sure which aspect of *sixness* has caught a student's attention?

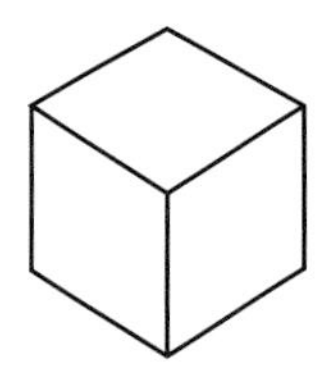

Fig. 3.1. What aspect of sixness catches a student's attention?

As a further example, consider three people entering a shopping mall. The first person is a jogger and views the building as an excellent location for exercising. The second person is a teenager who imagines the good times that will be spent in the mall with friends. The last person is the owner of a small business and looks at the building as an entrepreneurial opportunity. Three people view the same building, but each sees it from the perspective of his or her own personality and experiences.

The challenge for teachers is to stay attuned to the different perspectives, knowledge, and backgrounds of their students as the basis for ongoing planning and instructional decision making. Consider, for example, these opportunities in the daily lesson:

> The challenge for teachers is to stay attuned to the different perspectives, knowledge, and backgrounds of their students as the basis for instructional decision making.

- *At the beginning* of a mathematics lesson, many teachers are beginning to step back and ask questions or frame problem tasks to help them determine what individual students already know about a concept or topic.
- *Within a lesson*, they ask students to state predictions; observe students as they work; probe for different explanations and approaches; and strive to create a supportive mathematical climate for conjecturing, inventing, discussing, justifying thinking, and refining approaches.

- *At the end of a lesson,* these teachers might ask the students themselves to summarize "big mathematical ideas" that have surfaced.

It is essential to listen to and respect students' ideas.

This approach furnishes a further opportunity for teachers to observe, analyze, and reevaluate the learning that has taken place as a basis for determining the "next steps" to extending and challenging the mathematical growth of all students.

Teachers who incorporate approaches like those listed above believe it is essential to listen to, respect, build on, or connect with students' ideas. They want students to construct, extend, or modify their thinking and to develop and exert their own strategies for sense making in mathematics rather than depend on the authority of the teacher or textbook as a guide to what is "right." They expect and encourage students to work cooperatively to share, create, debate, modify, and develop—in active, personally meaningful ways—important mathematical ideas and ways of problem solving. In their teacher-as-facilitator role, they study the scope and sequence for which they are responsible, identify the big content ideas, and consider appropriate options for addressing those major thrusts. They plan focused lessons and activities with enough flexibility and open-endedness to accommodate individual needs and interests and sometimes shift plans within a lesson in response to students' questions or comments.

An example of an unexpected shift in plans occurred in an eighth-grade lesson on roots. The teacher asked the class to estimate the fourth root of 107. The students decided that since $3^4 = 81$ and $4^4 = 256$, the root would be between 3 and 4 but much closer to 3. The teacher then challenged the class to find an approximation with their calculators. After a moment of thought, Matt suggested the following sequence of calculator strokes:

$$107 \; \boxed{\wedge} \; 4 \; \boxed{x^{-1}}$$

The teacher inquired, "Can you explain why you think that will work?"

Matt responded, "Well, instead of raising 107 to the fourth power, we are really doing the reverse of that. So pressing the $x^{-1}$ key takes a fourth root instead of raising to a fourth power."

The teacher then asked, "So what is the reverse of squaring a number?" It did not take much discussion for the class to agree that it involved "taking a square root."

Ironically, nothing in the teacher's lesson plan for that day called for discussing inverses, but Matt's response created an unplanned teachable moment that provided an excellent opportunity to pursue the topic. So the teacher decided to extend the discussion on inverses, leaving the lesson originally planned for another day. The class had chosen their own path, and it made more sense to follow the lead they had taken.

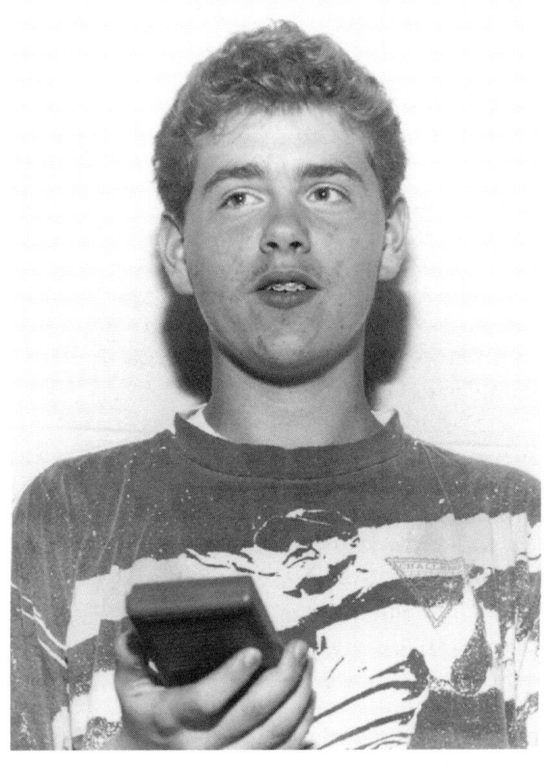

Clements and Battista (1990, p. 34) have spoken of the classroom in which "knowledge is actively created or invented by the child, not passively received from the environment" and where students "create new mathematical knowledge by reflecting on their ... actions." As was pointed out in the last section of this chapter, students do not learn mathematics simply by doing; they learn by thinking about and discussing what they are doing.

Reflection—essential to personal growth in mathematics—is stimulated by questions and problem tasks that cultivate the seeds of understanding already present. These questions and problem tasks, then, offer students the opportunity to build, or instill the confidence that they are capable of building, bridges from their own knowledge and language to the language and context of the mathematics being studied. Being able to approach concepts from personal perspectives gives students a chance to contribute, to feel involved, and to grow more powerful in the mathematics they are learning. A student who views learning as a passive event has much less at stake than one who views learning as an opportunity to contribute and even direct the course of a lesson.

Being able to approach concepts from personal perspectives gives students a chance to contribute, to feel involved, and to grow more powerful in the mathematics they are learning.

If a student has seldom been asked to think about a concept or has never been asked to take a different look at an idea, then some hesitation may be expected—perhaps even some outright reluctance to do so. When students say, "Can't you just show me how?" they are projecting their perception of what they think is important. Right or not, such students have inferred that what is valued in this situation is the answer, not the thinking, understanding, or process for deriving the answer.

This inference might stem from all the standardized testing to which they have been exposed in which only the answer mattered. This inference might be traced to a wealth of textbook exercises in which the desired result seemed to be *the* correct answer listed in the back of the book or given by the teacher. This inference might come from parents who have stressed answers over meaning. It doesn't matter. If the perception exists, it is going to surface in some manner no matter what its genesis might be.

Consistent with recommended shifts in curricular emphasis and in both teaching and learning styles, the teacher's new challenge is to get students to say, "Can we discover why?" instead of, "Can't you show me how?" If the "why" is missing from the experience, then the "how" becomes an isolated and fragmented bit of knowledge that seldom, if ever, is transferred to a related but different situation.

## IMPLICATIONS FOR TEACHING AND LEARNING

Approaches that appeal to *all* students of mathematics nurture the building of intuitions or sense making and encourage students to attach their own idiosyncratic meanings to the content. Such approaches will be most readily employed by establishing a student-centered classroom environment and by acknowledging, as well as accepting, the loss of some control of the direction of a lesson. To do so requires that the teacher be flexible and willing to take some risks.

In the example previously mentioned, in which the class was asked to find the fourth root of 107, flexibility was the key to meeting the needs of the students. The lesson plan was not like a book, in which the print was fixed and unchangeable. Instead, on the basis of students' questions and performance in the lesson, the lesson was refocused. Even though a general plan and an outline had been prepared, changes were made as dictated by "audience reaction." Such changes may also occur after the fact, as patterns on homework or tests suggest the need to approach a topic from another perspective.

Some educators feel comfortable with a fairly rigid lesson plan that allows

for some flexibility but is not malleable enough to permit a total change of direction. Structured teachers frequently have difficulty letting go, since in doing so they perceive too great a risk to their control over the lesson sequence. They may fear that students will attach some meaning or make a connection for which the teacher is unprepared. Yet in this constructive context, students are most likely to find themselves "doing" mathematics and formulating personal understandings through exploration, conjecture, and validation of their thinking.

Within any classroom, there is likely to be a range of intellectual capabilities, interests, and learning needs, as well as an uneven landscape of concept formation, attitudes, and learning habits. Some students will have acquired mathematical skills through memorization; others, by example and visualization. Part of the class may have a firm grasp of concepts but be weak on skill application. The rest of the class may have the opposite problem.

Some individuals will demonstrate a facility for learning new ideas and will readily explore new topics with self-confidence. Other students will be hesitant about new concepts, be reluctant to pose questions, and exhibit misgivings about their aptitude for mathematics. The role of the teacher, then, is to allow for, encourage, and assist in explorations and problem-posing and problem-solving ventures that appeal to the myriad of backgrounds, interests, and needs of the students in the class.

To adjust to the new recommendations for teaching and learning mathematics that have been summarized in this chapter and that affect *all* students, teachers today are challenged to build their mathematics programs around thoughtfully designed rich problem tasks that address important content goals and that, consistent with the discussion above—

- build on students' existing knowledge or prior experiences;
- involve students actively, reflectively doing mathematics.

A comprehensive analysis of these tasks should also include an examination of their nature, the logistics required to complete them, the types of desired student interaction, the discourse that might occur as a result of experiencing the tasks, and possible extensions of the tasks to related explorations. The following discussion considers each of these implications for nurturing approaches to teaching and learning mathematics for the benefit of all students.

*Engage students in rich, meaningful mathematical tasks and problem explorations.*

Students are innately interested in learning new things. By taking advantage of their present knowledge, expanding interests, and growing cognitive abilities, teachers can actively involve students in rich problems and tasks that—

- are mathematically significant;
- require students to *think* rather than rely on memorized procedures;
- invite hypothesizing and generalizing;
- lead to other problem questions or extensions;
- allow students to *learn* in the process of completing the task or solving the problem;
- allow for several acceptable answers or solution approaches. (*Note:* Although helpful to all students, this open-ended feature is a key to stimulating creative thinking among talented students and for engendering confidence among students with learning difficulties or disabilities.)

Additionally, such experiences often give students opportunities to learn how to organize information, identify patterns, and apply deductive and inductive thinking. The effectiveness of a problem task, then, is not judged on the basis of students' involvement alone. Rather, it is judged on how well it elicits investigation, exploration, and thoughtful justification and discussion of solutions and on how well it promotes mathematical reasoning, student ownership, and making connections.

The following problem is illustrative of a focused yet open exploration that was chosen by a teacher because of its potential for promoting mathematical reasoning. The teacher directed his students to fold a sheet of paper in half and then in half again (see fig. 3.2). He asked, "If you cut off the double-folded corner at a 45-degree angle, what shape do you think will appear when the paper is unfolded? Tell your partner. What shape do you think would appear if a 60-degree cut were made instead? Tell your partner what you think now. Try it to see. Decide with your partner who will make the 45-degree cut and who will make the 60-degree cut."

Presenting open-ended tasks or problems that do not restrict students to a single correct answer or solution approach is a key to stimulating creative thinking in talented students and for engendering confidence among students with learning difficulties or disabilities.

The problem is simple yet rich in several ways:

- It challenges students to *think* in order to predict or hypothesize what might occur. They cannot rely on any memorized strategy. Specifically, the problem calls for spatial visualization, since students must imagine that a square is formed if the paper is cut at a 45-degree angle and that a general rhombus is formed if the paper is cut at any other angle.

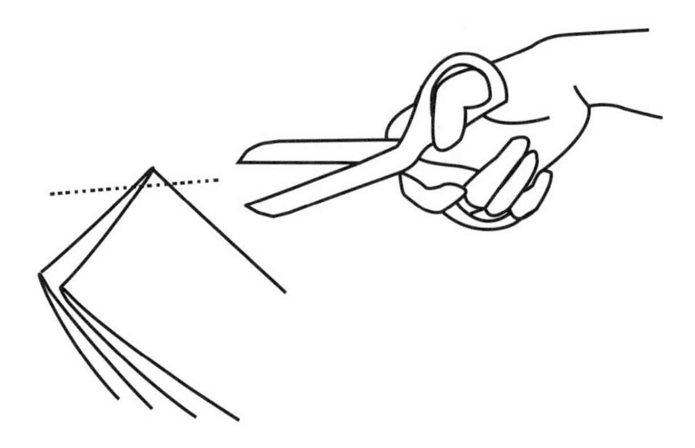

Fig. 3.2. What shapes result when a double-folded corner is cut at various angles?

- It requires students to analyze properties of geometric shapes; hence, it involves sound and significant mathematics.
- The task is open-ended in that it invites making any number of different cuts while working to solve the problem.
- As different students report their results or as student teams make many different cuts, their original hypotheses are either confirmed or rebutted, serving as a basis for formulating a generalized statement.
- Related questions may flow naturally from this task; for example, "Will a rhombus always result from the cut?" "Is it possible to reformulate the problem so that a rectangle or other specific shape is obtained from the cutting?"
- As part of the problem solving, students have the opportunity to consider and compare properties of a square and of a generalized rhombus and possibly *learn* that a square is a special kind of rhombus.
- Most important, the task invites class discussion, the sharing of findings, and student ownership of the properties that are discovered.

Additionally, the exploration might invite students to draw several conclusions about the nature of the diagonals of the rhombus. Traditional geometry programs explore theorems that highlight how the diagonals bisect one another, intersect at right angles, form four congruent right triangles, and so on. A paper-folding activity such as this can help students to conjecture and devise their own theorems, with informal hands-on proofs, including the derivation of a formula for finding the area of the rhombus by looking at the triangles. Meaningful mathematical tasks like this one transform the traditional "I'll give you the properties, and you formally prove them" thrust into one in which "you tell me what properties you have discovered and why they seem to be true."

*Select problem tasks that build on and naturally extend students' thinking and knowledge.*

For any task to have meaning and offer an opportunity for success, students should be able to relate it to mathematical understandings and skills

they already possess. Careful analyses of students' knowledge and the goals of instruction enable teachers to select appropriate problem tasks that help connect what students know and can do with what is to be learned.

Problem tasks that a teacher judges to be appropriate at a given time for building on, extending, or stretching students' understandings can be drawn from many sources. Problems might arise spontaneously from daily transactions in the classroom, from problems posed by children, and from children's books. These or problems from other sources might be selected to introduce a topic or to highlight a piece of mathematics. Some of the best problems arise naturally from students' comments and interests and are addressed because the teacher recognizes their value in nurturing important mathematical understandings or processes that are aligned with instructional goals.

A first-grade class, for example, was curious to know how many cans of food have been collected when each of twelve students has brought in three cans for a food drive. A junior high school class that consisted of fourteen boys and ten girls was interested in finding the probability of the occurrence of this ratio among twenty-four children born in the town in the same year. A high school class was interested in knowing how much money a student could accumulate in the bank if $1000 were invested every year at 7 percent interest until retirement at age 65. These situations furnished natural opportunities to address important instructional goals while building on students' curiosity and addressing issues of relevance to individuals in the class.

Students mature in their ability to think mathematically through frequent and challenging encounters with appropriate problem tasks like these. When students are engaged in meaningful problems, they often ask, "Why?" "What would happen if …?" and other important questions that stimulate reflection

When students are engaged in meaningful problems, they begin to view mathematics as ideas that they explore rather than as a collection of algorithms and skills.

and learning. Students, then, begin to view mathematics as ideas that they explore rather than as a collection of algorithms and skills. Consequently, students raise their own questions that can further the learning of the class.

For example, a student who has just learned to use substitution to solve a system of equations may ask, "What will happen when you try to use substitution but the lines are parallel?" The teacher's response could be, "Let's find out. Can someone give me two examples of equations of parallel lines?" The students can then explore the situation to answer their own questions. A fourth-grade student with a calculator having a broken addition key may lead the class into a discussion of how to find the sum of two numbers, such as 187 and 376, without using the addition key.

Organizing instruction in such a way that children can develop their own understanding of significant mathematical ideas requires problem tasks that build on or extend students' thinking. Only when students "understand" can the goals of mathematics instruction be achieved and will students recognize and use the power of mathematics.

*Allow enough time for students to address problem tasks by using their own strategies.*

Children solve problems in various ways and from various perspectives. The fact that two students arrive at the "correct answer" to a problem does not mean that they solved the problem in the same way or that they have the same understanding of the concepts underlying the problem.

One student, for example, may solve a problem by recalling a memorized strategy for adding two fractions. Besides "knowing how," a second student with deeper intuitions and stronger number sense may understand *why* addition is appropriate in this situation, *why* the computational procedure works, and whether the solution obtained is reasonable. While remaining open and flexible to different correct strategies, a teacher is challenged to be attuned to the range of solution strategies used by various students and to encourage them to think through a problem rather than rely on memorized procedures.

Students need considerable time to accommodate sound understandings to their existing knowledge as a basis for learning and communicating mathematically.

Further, because students often take considerable time to develop sound understandings and work out meaningful strategies for approaching problems, sufficient time must be allocated for completing the tasks that are posed. One outcome of this approach is that students are encouraged to be independent and responsible for their learning. Agreement is growing among educators that learning is best nurtured when students are given the time to explore significant problems, to produce strategies that fit their individual ways of doing things, and to interact productively with classmates and the teacher. This latter point is addressed more fully below.

*Give students time to discuss, share, and justify their processes of solving problems.*

Although some inferences about students' thinking can be made by a careful analysis of their written work, the most effective ways to assess students' understanding are to observe students as they work and, more important, to listen to them explain their solution approaches. "Sharing time" opportunities can be embedded in both cooperative-group and whole-class discussions.

In one secondary school mathematics class, for example, a teacher presented the class with the following system of equations: $y = 2x - 3$ and $3x + 4y = 7$. He directed the students to form groups and identify an effective method for solving the system. He knew that the class had learned to solve systems of equations by graphing, substitution, addition and subtraction, and matrices. He was interested in knowing *why* a student would choose one method over the other and asked each table of students to prepare a report including their preference and reasoning.

As the students interacted within their groups to reach a consensus for their group report and as they presented, argued, and justified their statements, the teacher observed and listened so as to informally assess the students' thinking as the basis for further questioning and instructional planning. The cooperative-group setting became a forum for thoughtful justification and discussion and for promoting mathematical reasoning and enhancing students' learning. As the students shared their thinking and strived for consensus, the more advanced students learned the concepts in a new way as they were challenged to reexplain them to others and all students gained from hearing the way that others thought and reasoned.

The interchange of ideas that occurs in small-group settings typically is extended to whole-class discussions, often as a way of bringing closure to a problem task. As the following classroom episode illustrates, providing adequate time for such discussions has rich payoffs.

As the students present, argue, and justify their conclusions, teachers can informally assess the students' thinking as the basis for refocusing instructional plans.

The teacher posed this problem to her class:

> "Suppose that I have 34 feet of rope and wish to cut it into two pieces so that one piece is 8 feet longer than the other. How long would the two pieces have to be?"

The students discussed the problem at their tables until each table had an answer. During sharing time, James said, "We just subtracted the 8 from 34 to get 26 and then divided the 26 by 2. We got 13 feet, to which we added the original 8 feet. So the pieces must be 13 and 21 feet."

Mary Jo spoke for her group: "We ended up the same, but we worked it out a different way. We just divided 34 by 2 to get 17 feet. Then we figured that one piece would be 4 feet longer than that, and one would be 4 feet shorter. So the lengths must be 13 and 21 feet." Finally, John stated, "We just used guess-and-check. We knew it couldn't be 17 and 17, so we tried 16 and 18, and so on, until we found the right combination."

The teacher might easily have accepted the first response as "the correct answer" and then moved on to the next question. However, she thought that the students' best interests were served by allowing adequate time for sharing solutions and justifying their thinking. She emphasized problem solving, not simply the answer. After the class had shared their various solution strategies, the teacher asked students to decide whether any method seemed to be more efficient than others and which strategy they would use to solve a similar problem in the future.

Although the teacher may have wanted to keep the discussion moving more rapidly, she was aware that rushing to "get through" a planned agenda would not foster understanding. Students often learn alternative—and sometimes

more efficient—strategies for problem solving by listening to those of others. The teacher's question about whether or not students would use the same process the next time may cause some students to evaluate which method would work best for them. The following examples suggest further instances in which teachers structured time for sharing and discussion, to the benefit of all.

Students in one middle school class were asked to determine the "average" height of members of the class. In analyzing the data, one group discovered that the tallest student's measurement actually caused the arithmetic mean to have a value that represented a height greater than that of everyone else in the class! Consequently, these students listened with interest when other groups reported why they chose the median or mode to describe the data. A rich discussion on the different ways of interpreting "average" resulted.

In a primary-grades classroom, the teacher presented the class with a sequence of pattern blocks on the overhead projector (see fig. 3.3) and asked the students to determine what the next block in the sequence might be and why. The children were given time to discuss their responses at their tables before they reported to the class. The teacher realized that more than one "correct" answer was acceptable (e.g., a red block because of the color pattern, or a square because of a pattern in the number of sides) and encouraged students to voice their opinions, even if they appeared to be different from the answers that others in the class had given.

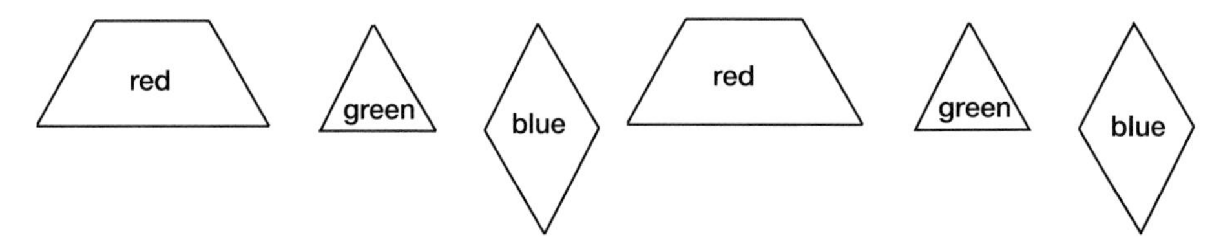


Fig. 3.3. What is the next block in the sequence?

*Take advantage of "sharing times" to focus on key mathematical ideas, to make connections, and to refocus instructional planning.*

Instead of much direct instruction at the beginning of a lesson, teachers today are beginning to involve students in problem tasks as the focus of lessons and to interact or react during "sharing times" to correct misconceptions, bridge conceptual gaps, help articulate or extend important mathematical understandings, and generally help students organize learning or make appropriate connections. Assignments flowing from these sharing times can reinforce the key mathematical ideas discovered during a session, reinforce basic skills that appear to be lacking, or focus on applications.

During such a sharing session, a student in the third grade asked the teacher about the square-root key on the calculator. The teacher asked the student to form a square by using nine color tiles on the overhead projector and then asked for the dimensions of the square. When asked to punch the square root of 9 into the calculator, the student was surprised to find the number to be the same as the unit length of a side of the square. Students

repeated the process for squares formed by using sixteen and twenty-five tiles and then defined the square root. The teacher—to prompt further discussion—asked the class what would happen if they wanted to form a square with twenty tiles and explored the idea that square roots are not always integral.

Similarly, during a sharing time at the junior high school level, Stephen noticed that when he punched in the square root of –16, his calculator displayed an error message, and he asked why this should happen. The teacher captured the opportunity to explore the issue of numbers that are undefined in the set of real numbers, and students were asked to explain why this value could not exist.

Often, such questions as these that emerge during sharing times prompt clarifications, reteaching, or new explorations of mathematical ideas. The ideas are student generated and therefore more meaningful to individuals in the class. Allowing ideas to come naturally from students makes mathematics look less artificial than it appears when presented in tidy units. For example, the student who states, "I have a quick way to get the answer when you multiply two two-digit numbers" may prompt a discussion of other methods that different students have discovered and allow the teacher to explore mental computation without treating it as a separate topic.

The teacher's role is to elicit, engage, and challenge students' thinking in relation to important mathematical ideas or processes.

Sharing times have the potential of welding the students into a community of learners, with the teacher as coach. As the National Council of Teachers of Mathematics suggests in its *Professional Standards for Teaching Mathematics* (NCTM 1991, p. 45), sharing times provide opportunities for students to—

- listen to, respond to, and question one another and the teacher;
- initiate questions and problems;
- make conjectures and present solutions;
- try logically to justify to one another the approaches taken;

- summarize important ideas or procedures that have been learned or used.

Further, the *Professional Standards for Teaching Mathematics* (p. 35) highlights how these sessions allow the teacher to—

- pose questions that elicit, engage, and challenge students' thinking;
- listen carefully to students' ideas;
- ask students to clarify and justify their statements;
- decide which ideas to bring up in depth among those that surface;
- decide when and how to attach mathematical notation and language to students' ideas;
- decide when to provide information, when to clarify an issue, when to model, when to lead, and when to let a student struggle with a difficulty;
- decide when and how to encourage each student to participate.

Perhaps most important, sharing sessions also afford opportunities for teachers to obtain a sense of what students are thinking and can do as the basis for making adjustments in instructional plans to better meet the needs of all learners.

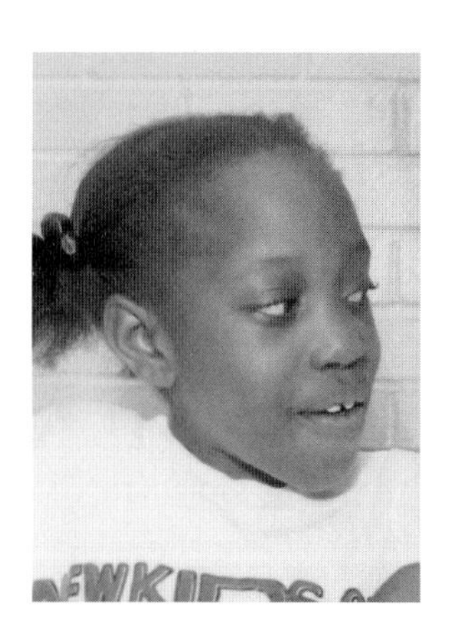

## NEW BEGINNINGS

What is in the minds of the students matters, and on this foundation new concepts are built. Students have reached their current levels of mathematical learning through different routes by developing the understandings they possess, the attitudes they have acquired, and the modes of thinking on which they rely. The study of mathematics can be a stimulating endeavor incorporating a student-centered approach by which *all* students build from their individual experiences and expand their abilities to think mathematically. A breath of fresh air through a window of opportunity! Surely, to think mathematically in problem settings and to reflect on that thinking is to do mathematics!

### BIBLIOGRAPHY


Clements, Douglas H., and Michael T. Battista. "Research into Practice: Constructivist Learning and Teaching." *Arithmetic Teacher* 38 (September 1990): 34–35.

Committee on the Mathematical Education of Teachers. *A Call for Change: Recommendations for the Mathematical Preparation of Teachers of Mathematics.* Washington, D.C.: Mathematical Association of America, 1991.

Davis, Robert B. "One Point of View: How Many Ways Can You 'Understand'?" *Arithmetic Teacher* 34 (October 1986): 3.

Driscoll, Mark J. "Learning Elementary School Mathematics: Individual Styles and Individual Needs." In *Research within Reach: Elementary School Mathematics*, edited by Mark Driscoll, pp. 84–90. Reston, Va.: CEMREL, National Institute of Education, and National Council of Teachers of Mathematics, 1981.

Johnson, Howard C. "How Can the Curriculum and Evaluation Standards Be Realized for All Students?" *School Science and Mathematics* 90 (October 1990): 527–43.

National Council of Supervisors of Mathematics. "Essential Mathematics for the Twenty-first Century: The Position of the National Council of Supervisors of Mathematics." *Arithmetic Teacher* 37 (September 1989): 44–46.

National Council of Teachers of Mathematics. *Curriculum and Evaluation Standards for School Mathematics.* Reston, Va.: The Council, 1989.

———. *Professional Standards for Teaching Mathematics.* Reston, Va.: The Council, 1991.

National Research Council. Mathematical Sciences Education Board. *Everybody Counts: A Report to the Nation on the Future of Mathematics Education.* Washington, D.C.: National Academy Press, 1989.

———. *Reshaping School Mathematics: A Philosophy and Framework for Curriculum.* Washington, D.C.: National Academy Press, 1990.

U.S. Department of Labor. *What Work Requires of Schools: A SCANS Report for America 2000.* Washington, D.C.: U.S. Government Printing Office, 1991.

**William Speer** is a professor of mathematics education in the College of Education and Allied Professions at Bowling Green State University (BGSU), Bowling Green, Ohio. He teaches mathematics methods courses and is director of the BGSU Mathematics Clinic for Children. His research interests include diagnosis and the identification of intervention procedures.

**Daniel Brahier** is a mathematics and science consultant in the Catholic Youth and School Services office for the Diocese of Toledo, Ohio. His primary responsibilities include curriculum development and in-service opportunities for teachers. In addition, he is a part-time instructor of mathematics methods courses and teaches one eighth-grade mathematics class each day.

# 4

# Issues of Identification

*Ruth E. Downs*
*Janice L. Matthew*
*Marilyn L. McKinney*

***What are the defining characteristics of students with special needs?***

***How can the screening, prereferral, and referral processes be made more effective?***

***What should be the function and structure of school prereferral teams?***

***Do inequities exist in the process for identifying special-needs students?***

THE ongoing process of developing effective, equitable strategies for identifying students with special needs and designing programs that respond appropriately to their individual needs is a constant challenge. Raising broad, basic questions such as those above is an integral part of the process. The perceptions that influence our answers to such questions greatly affect the design of any identification and placement process. As a basis for reflection and action, this chapter addresses issues related to each of these questions in turn.

The first major section of this chapter reviews the definitions of different categories of students served by special education programs, including those for gifted and talented students. A knowledge of the parameters defined by law for serving exceptional individuals is the first important consideration in identifying special populations.

Issues that are central to the screening, referral, and identification of students for special programs are addressed in the second major section. Problems are posed and solutions are discussed that may assist school professionals in appropriately placing special students. The function and structure of school prereferral teams is also addressed.

The third section focuses on equity issues related to identification and placement into special education programs. Specific situations and instances are given that clearly demonstrate a need for a change in the present procedures for identifying and placing students into special programs.

Finally, goals for rethinking screening, prereferral, and referral in the identification and placement of students with special needs are presented. These goals challenge teachers, administrators, and the entire educational community to respond.

## STUDENTS WITH SPECIAL NEEDS: DEFINING CHARACTERISTICS

In recent years, a great deal of attention has been devoted to the referral and identification of students with special needs and their placement into special education programs, including those for gifted students. *Special education* is defined as instruction that is specially designed to meet the unique needs of a student who is disabled or mentally gifted. It consists of adaptations or modifications to the regular-education curriculum, instructional environments, methods, or materials. Such specialized interventions are designed to meet the unique needs and abilities of exceptional students and go beyond the services and the program that a student would receive as part of regular education (Pennsylvania Department of Education 1991).

Specially designed instruction meets the unique needs of special education students.

A noticeable shift has occurred toward accepting and accommodating both a greater number and a greater range of individual differences in the regular-education classroom. Appendix B summarizes the legal definitions of each of the categories of special education. These definitions cannot begin to describe the unique social, physical, psychological, and educational strengths and weaknesses displayed by an individual with disabilities or giftedness, but they serve as a practical framework for defining special education populations.

A continuum of options must be considered when determining special education programming.

## Students with Disabilities

In 1975, Congress passed the Education of the Handicapped Act (EHA), which is more commonly known as Public Law (P.L.) 94-142. This law contained two profound mandates regarding special education students. The first requirement was that disabled students be placed in the "least restrictive" educational environment in which they can learn successfully. This provision recognized the need to provide equal educational opportunities for all students within an environment where exceptional students are educated with their nonexceptional peers to the maximum extent appropriate. Hence, a continuum of program and service options must be considered when determining appropriate special education programming.

The second mandate affirmed that such students have a right to appropriate instructional programs to meet their unique needs. Appropriate instructional programs are based on a student's Individualized Education Program (IEP). An IEP is a written document that identifies specific, specially designed instruction and services needed by an exceptional student.

Congress amended the EHA in 1990 by reauthorizing the programs and by changing the name to the Individuals with Disabilities Education Act (IDEA). Often referred to as P.L. 101-476, IDEA reflected the preference for the use of the term *disabled* over *handicapped*. IDEA also added the categories of "autism" and "traumatic brain injury" to the list of disabilities that may qualify students for special education programs (Individuals with Disabilities Education Act 1991).

IDEA defines students with disabilities as children with mental retardation; hearing impairments, including deafness; speech or language impairments; visual impairments, including blindness; serious emotional disturbances; orthopedic impairments; autism; traumatic brain injury; other health impairments; or specific learning disabilities. These students must also demonstrate a need for special education services or programs as documented in school referral information.

In a *legally binding* memorandum that clarified the policy (Davila, Williams, and MacDonald 1991), children with attention deficit disorders (ADD) or attention deficit hyperactivity disorders (ADHD) were declared eligible for special education and related services under the "other health impairment" category of IDEA in instances in which the ADD (or its variations) was a chronic or acute health problem resulting in limited alertness adversely affecting educational progress.

This policy memorandum further protected the rights of ADD and ADHD students to special placement by citing Section 504 of the Rehabilitation Act of 1973, which ensures that a free, appropriate public education—including special education or adaptations in regular education programs—is provided for each qualified handicapped child.

More detailed definitions given to the different types of disabilities by the public laws or by policy memorandums are summarized in appendix B.

## Gifted and Talented Students

The literature on identification and definitions of gifted and talented students is extensive (MacRae and Lupart 1991). Even among the leaders and theorists in the field of gifted education, however, little consensus can be found on what constitutes giftedness. Perhaps in their defense, Davis and Rimm (1989) point out that no one theory-based definition of *gifted and talented* will fit all programs and circumstances. Services for the gifted and talented are not included in the federal laws or guidelines for the disabled. The most commonly cited definitions are included in Appendix B.

A different concept of giftedness has been proposed by Renzulli (1978). In his frequently quoted definition, *giftedness* is an interaction of above-average intellectual ability, high levels of commitment to tasks (motivation), and high levels of creativity (see fig. 4.1). Gifted and talented children are those possessing or capable of developing this set of traits and applying them to any potentially valuable area. His three-ring model has several advantages (Kitano and Kirby 1986). It focuses on a combination of traits that characterize many children who have led creative, productive lives and extends the idea of giftedness to include many areas valued by society.

Davis and Rimm (1989) recommend Renzulli's use of the three-ring definition in conjunction with his revolving-door identification model (RDIM). In this model, a more inclusive talent pool (up to 25% of the students in some schools) could be selected according to established criteria. Those students showing or developing creativity and motivation may self-select for independent projects in a resource room. This model assumes that a large number of students manifest gifted behavior under certain conditions and at different times. The RDIM serves as a mechanism for allowing students to participate in independent projects as their needs dictate (Davis and Rimm 1989).

Several models of identification have been built around the idea of multiple talents, as suggested by Gardner (1985), who proposed seven different domains of intelligence: linguistic, musical, mathematical-logical, visual-spatial, bodily kinesthetic, interpersonal (ability to understand and interact with others), and intrapersonal (ability to know oneself and have a developed sense of identity). This theory of multiple intelligences supports a broad definition of giftedness that includes individuals who are socially, personally, and motorically gifted (Kitano and Kirby 1986).

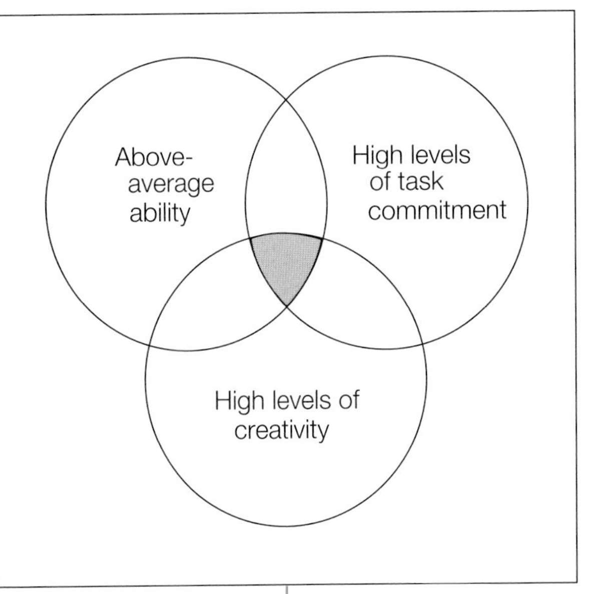


Fig. 4.1. Renzulli's three-ring conception of giftedness

Legal definitions of giftedness exist but tend to be vague and vary from state to state. Some states specify gifted *and* talented and include the visual and performing arts, whereas other states recognize only superior academic achievement as giftedness. Some states use *gifted* and *talented* interchangeably, and others address only *gifted* and ignore the term *talented* in their definition.

Some authorities distinguish between *gifted* and *talented* by separating the intellectual and academic from the artistic and creative. Some define as gifted those children who are highly intelligent in all areas and as talented those who have high ability in just one area. In some states, extensive services are mandated for gifted or talented students, whereas in others these services are considered discretionary (Adderholdt-Elliot et al. 1991).

Identification procedures vary not only from state to state but also within states. This situation raises the possibility that because of variations between and within states, a child who qualifies as gifted in one system may not necessarily qualify as gifted in another (Kitano and Kirby 1986). Some states or school districts, for example, may require a particular cutoff score on an individual IQ test or a minimum SAT score to qualify for a gifted program, whereas others may not.

### Challenges Associated with Identification

Clearly, the identification of students for placement in special programs is affected by many issues, particularly these:

- Although federal law defines areas of disability, individual states may interpret the definitions slightly differently according to their populations.
- No one definition of *gifted and talented* is accepted by everyone.
- The specific definition of *gifted* accepted by a state or school district or school will determine the selection instruments and procedures (and hence the population served)—and often the program design and its delivery.
- The definition that a school selects for *gifted* could potentially exclude economically disadvantaged students from various geocultural groups and disabled, underachieving, or female students.

For each population of special students on the continuum from disabled to gifted, the issues affecting identification deserve thoughtful analysis. The challenge is to bring any disparities or inequities into focus and accelerate appropriate action so that necessary and appropriate programmatic changes can be created and implemented in a timely fashion.

## SCREENING, PREREFERRAL, AND REFERRAL ISSUES

In the past decade, the growth in students identified as disabled has been astronomical.

Recent government figures indicate that there are more than eight million children with disabilities in the United States (McCormick 1990). This statistic, compared with that in a 1988 government report that stated that four and one-half million children were identified as handicapped, illustrates the astronomical growth in the number of students identified as disabled and in need of special education programs.

The high numbers in these reports have caused widespread concern that large numbers of children are being inappropriately placed in special education programs. It is suggested that a major contributing factor may be an inflexible, regular-education system that may not be modifying the basic curriculum to meet the range of needs of students in a regular-education classroom. According to Sartain (1989, p. 31), "graded schools, organized since 1840 so that students of one age are all expected to learn and achieve at exactly the same level, are certain to cause some students to feel so highly frustrated that they either become very emotionally disturbed or very aggressive."

As most teachers are aware, within a classroom, particularly in the middle grades, students can be within a range of five grade levels. A typical fifth-grade class is likely to have students achieving at the third-, fourth-, fifth-, sixth-, and seventh-grade levels.

What options are available to the regular-education teacher, other than a special education referral, when faced with a class that contains this type of variation in skill levels? Listed below are several suggestions:

1. Make an effort to learn the developmental level at which individual students are working. Be sensitive to what students think and understand about different topics in the curriculum as the basis for instructional decision making.

2. Know the scopes and sequences of the curricula that are being taught. This knowledge enables the teacher to modify the curricula and pacing to better match the developmental levels and thinking of special-needs or high-achieving students.

3. Seek assistance from the principal, colleagues, and instructional specialists or supervisors concerning additional strategies for modifying the curricula or for alternative curricular materials.

4. Evaluate the needs of the students at either end of the range in order to determine if their needs are so extensive that they require modifications or interventions beyond what can feasibly be accomplished by regular education. Students whose needs cannot be met *may* be appropriate referrals for special education programs.

5. Center instruction on problem tasks that involve the collaboration of students with partners or in mixed-ability groups. Whenever possible, integrate open-ended questions or tasks that tend to better accommodate a broad range of responses.

6. Systematically plan times for pulling lower- or higher-level student groups aside to attend better to their specific needs. Doing so will allow for reteaching or special assistance for slower students while also providing a forum for interacting on special projects or promoting accelerated exchange among more capable students.

To reduce the number of special program placements, there is much interest in implementing selected interventions in the regular-education program before the actual referral to special education. One study suggests that "these interventions might ameliorate the problems many students experience, thus negating the need for them to go through the formal referral process" (Lloyd et al. 1988, p. 44).

The regular-classroom teacher is key to the success of the prereferral intervention process. The teacher's responsibility is to assess the student's strengths and weaknesses, understandings, and skill levels in order to develop alternative instructional interventions and to monitor the effectiveness of those interventions. Classroom interventions are derived from the results of classroom assessment data. These data can be collected in a number of ways:

The regular-classroom teacher is key to the success of prereferral intervention.

- Daily work samples
- Chapter and unit tests

- Standardized test scores
- Observations of a student's performance on selected tasks
- Anecdotal comments based on the observation or analysis of a student's written or oral responses
- Portfolio entries
- Teacher-made checklists and assessment devices, some of which might summarize response or performance patterns over time

Once the teacher has determined the understanding and skill levels of a student in different significant areas, the next step in the prereferral process is to develop appropriate intervention strategies that reflect the needs identified through the assessment. When the interventions are implemented, the progress or lack of progress should be monitored. Students should be taught at their instructional level for a time to determine the degree of success of the interventions.

Many schools have chosen to establish prereferral teams whose purpose is to assist the teacher in collecting assessment data on students. In addition, the team collectively develops appropriate intervention strategies that can be implemented in the classroom setting. The members of the prereferral team might include the following personnel: the school psychologist, guidance counselors or social workers, the building principal, the mathematics or reading specialist, or a speech and language therapist.

Children should advance to the referral stage only when the following questions have been answered effectively (Wolf 1990):

Many schools have established prereferral teams to collect and analyze assessment data.

- Do you have academic concerns about the student?
- Is the curriculum effective for this student? (*Note:* Be alert to multicultural bias.)
- Is the evidence clear that the student is not learning or not challenged?

- Is there evidence of systematic efforts to identify the source of difficulty and to take corrective action?
- Do concerns persist despite efforts to identify and alleviate them?
- Have other program alternatives been identified and tried?
- Do concerns continue despite the application of systematic, quality intervention alternatives?

If it is determined that a student requires special education services, then appropriate combinations of regular-education modifications and special education interventions to meet the particular student's needs should be considered. Only when a student's needs are so significant that they cannot be met in a combination of regular and special education should a total special education program be considered (Lewis and Doorlag 1991).

In summary, schools should be oriented toward collecting and analyzing data that will be used for instructional planning as opposed to simply collecting data to justify a label and subsequent placement into a special education program. Specific instructional objectives should be developed along with intervention strategies to meet a student's needs. Progress in these objectives should be monitored over time to determine the following points:

1. Whether the interventions are successful
2. Whether alternative interventions should be implemented
3. Whether a student should be considered for a special education program

## EQUITY ISSUES

### Inappropriate Placement Decisions for Special Education

The use of the results of standardized tests as the major criterion for identifying special-needs students—both disabled and gifted—may be inappropriate, since some research suggests that these tests are culturally biased. Successful performance on standardized tests depends primarily on cognitive learning style, or the way information is processed (Cohen 1969). Since cognitive learning styles are culturally based (Gilbert and Gay 1985; Shade 1982), certain individuals and groups of individuals are at a distinct advantage or disadvantage when standardized tests are used for identification and referral for special-needs placement. Although cognitive learning styles have been found to be related to achievement (Kogan 1971; Gage and Berliner 1984; Downs 1989), they are unrelated to general ability (Witkin et al. 1971). This point adds validity to the claim of cultural bias in standardized tests.

Traditional standardized tests may be culturally biased and therefore inappropriate for identifying special-needs students.

If we look at cognitive learning styles on a continuum, we find those with a more global cognitive style at one end and those with a more analytic style on the other. (See fig. 4.2.) Students with a global cognitive style tend to view a field of information as a whole and apply that whole body of information directly. They are designated as having a field-dependent cognitive-learning style. At the other end of the continuum are those whose cognitive-learning style is characterized as field independent. These students separate out pieces of information from a larger field of information before applying them.

Although individual student scores may fall anywhere along the continuum, low-achieving students, females, and African American students *as a group* tend toward the field-dependent end of this continuum (Shade 1983; Garner and Cole 1986; Downs 1989).

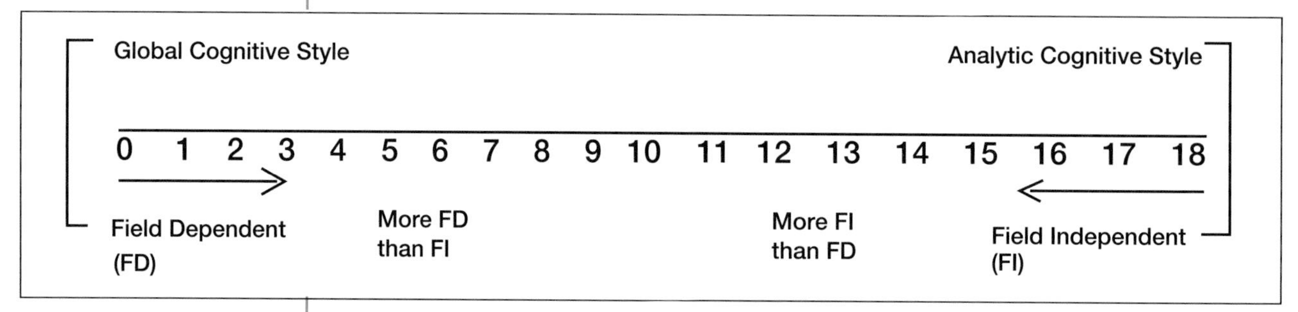


Fig. 4.2. Cognitive-styled continuum. The numerical values represent possible scores on the Group Embedded Figures Test (Witkin et al. 1977), which is a measure of field dependency.

Standardized tests of achievement are made up of both items that assess knowledge of specific concepts and items that measure an increase in a student's ability to extract relevant information embedded in a context. For example, students must extract key points from a written passage to answer reading comprehension questions or formulate an arithmetic problem from a written context. For many years researchers have noted (e.g., Cohen 1969) that achievement tests are weighted somewhat toward logical skills or skills of analytic abstraction.

Since students with a more field-dependent, or global, cognitive style were so designated partly because of their difficulty in selecting and using relevant information embedded in a larger field, it is not surprising that these students have been found to perform poorly on standardized tests. Their poor performance on these tests contributes to identifying them for placement in some special-needs classes and rejecting them for placement in gifted or talented classes.

Certain populations are overrepresented in programs for the disabled and underrepresented in programs for gifted students.

Nationally, and perhaps in your local school district, African Americans and other ethnic groups are overrepresented in programs for the disabled (e.g., seriously emotionally disturbed or mentally retarded) and underrepresented in programs for the gifted and talented. Inappropriate placement decisions based on such criteria as standardized tests may account for these occurrences. Moreover, despite the availability of information concerning nonbiased assessment and models for developing the potential of gifted students, the majority of states continue to employ standardized measures as the major vehicle for the identification of gifted students (Kitano 1991).

In 1979 the Office of Civil Rights (OCR) established a panel to assist them in identifying factors that might account for the nationwide disproportionate representation of African American, Hispanic, Native American, white, and Asian students in special education classes. By analyzing nationwide data from 1978 to 1984 on these geocultural groups' total enrollment in school districts and the percentage of these students enrolled in special education classes, the panel confirmed that the representation of some geocultural groups continues to be disproportionately high in classes for educable mentally retarded, trainable mentally retarded, seriously emotionally disturbed, and learning disabled students and disproportionately low in classes for gifted and talented students. The panel concluded that there is overwhelming evidence that many children in the past, as in the present, have been misidentified—accounting in part for the disproportionate numbers of certain geocultural groups in special education programs.

Similar studies have shown that African American, Hispanic, and Native American children participate in gifted programs at only about one-half their prevalence in the larger society and that Asian Americans participate at twice their representation in the American population (Kitano 1991). If the current identification and referral processes were equitable, it would seem that enrollment in all areas of special education would be proportionate to the total population of the various geocultural groups within a school or school district. Current research suggests that minority groups *should* participate in special programs in proportion to their representation in the school district (Richert 1991, p. 87; Maker and Schiever 1989, p. 235).

Participation by geocultural groups in special education programs should be in proportion to their representation in the school district.

African American males constitute yet another group that is disproportionately represented in special education classes. A study (Grady and Simmons 1990) published by the Prince George's County Public Schools in Upper Marlboro, Maryland, reported that African American males composed 33 percent of the total school population but constituted nearly 47 percent of the students receiving special education services. The committee conducting the study suggested that the overenrollment of African American males in special education classes may be related to the problems males encounter in adapting to a predominantly white and female teaching force (Part 1, p. 22). In an effort to alleviate this pattern of disproportionality, the committee recommended that its school system examine the instruments, procedures, and practices used by teachers and by school psychologists to refer students and place them in special education classes.

One alternative instrument is the System of Multicultural Pluralistic Assessment (SOMPA) (Psychological Corporation 1978). Recent research reports evidence that SOMPA procedures can serve as an alternative standardized method that increases the proportion of minority students in programs for the gifted, particularly in states that use IQ cut-off scores for placement decisions (Matthew et al. 1992).

Culture-fair tests and other assessment intruments are available for use.

## A Case of Inappropriate Denial of Programs for the Gifted

An incident that was shared by a former secondary school mathematics teacher serves to strengthen the argument for revised practices and procedures for placement in special-needs programs, including those for the gifted and talented. An African American male whom we will call Derek displayed exceptional talent in mathematics in the ninth-grade algebra class in which he was enrolled. It became apparent to the teacher that Derek was mathematically gifted. The standardized test scores found in the student's records did not qualify him for consideration for the gifted program.

The teacher decided, however, on the basis of Derek's mathematics performance that he should be referred for special education services. The teacher's major concern was that no former counselors or teachers had noticed or pursued alternative methods to determine that the student was gifted. Only because of this teacher's persistence and willingness to fight the system's red tape and biased perceptions was Derek appropriately evaluated and eventually placed into the gifted program.

With the support and encouragement of his teachers, Derek did very well in the advanced mathematics classes and in his other courses. Because of his excellent Scholastic Aptitude Test (SAT) scores, he received a full scholarship to Colgate University where he majored in computer science.

Derek has had occasional conversations with his former secondary school teacher concerning his progress at Colgate and often has shared how the encouragement and support given in high school helped ensure his success in mathematics. Although Derek completed his college education, he thought his grade-point average would have been higher if a support system similar to that in the high school had been in place.

Teachers need to become advocates for gifted students who do not fit the stereotype of a gifted student.

Statistics showing disproportionate numbers of African American males enrolled in programs for the gifted and talented represent a complexity of issues and problems. However, educators such as this teacher who are willing to become advocates for students who generally are not perceived as gifted represent one practice that may move us toward more equitable referral and identification procedures. The results of the Prince George's study and the personal experience of Derek are examples of another type of bias that

may be present in identifying and referring students for placement in special education programs.

## Other Critical Issues

### *Opportunities for girls*

Traditionally, the rates of referral and identification of girls for gifted programs have been low. A conscious effort should be made to foster giftedness in girls. Often the behaviors of competitiveness, risk taking, and independence that are associated with screening for possible giftedness are not generally encouraged in girls (Kerr 1985). Similarly, young women are not often encouraged to enroll in higher-level mathematics courses in secondary school or to pursue mathematics- and science-related careers.

Low teacher expectations of the mathematics achievement and ability of female students, even as early as the first grade, are usually grounded in myths and false assumptions (Fennema et al. 1990; Shore et al. 1991). These expectations and assumptions in turn influence classroom instruction and decisions that have a negative effect on the mathematics performances of females. Figure 4.3 presents an autonomous learning behavior model proposed by Fennema and Peterson (1985) that may explain gender differences in performance. Several other studies have indicated that counselors have not encouraged, and in some instances have discouraged, young women's study of mathematics and their pursuit of career goals related to mathematics (Fox 1980). The point made here is that systematic bias can prevent students from performing to the height of their potential.

Systematic bias can prevent students from performing to the height of their potential.

Internal belief system
Autonomous learning behaviors
Gender differences in mathematics
External influences (teachers)

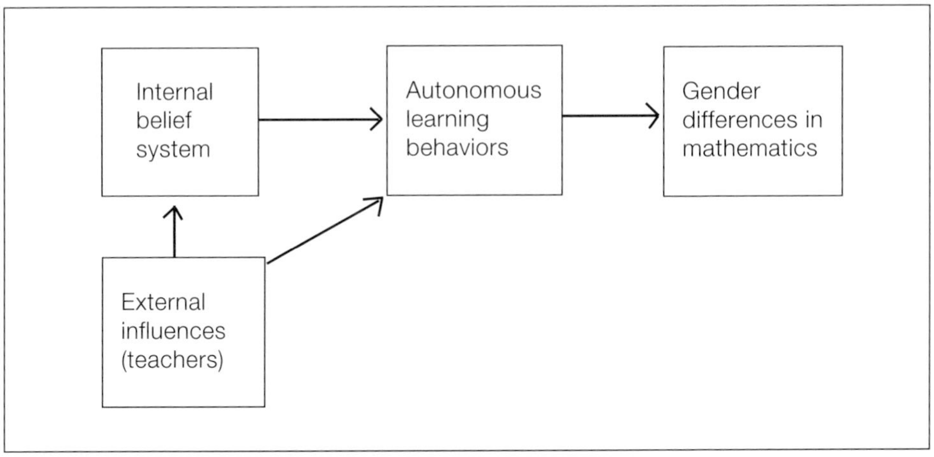

Fig. 4.3. Autonomous learning behavior model

### *Opportunities for economically disadvantaged students*

Economic bias is another problem associated with standardized testing that must be addressed. One instance in which economic bias becomes apparent is in the identification of preschool and kindergarten students for placement into programs for the gifted. Clearly, students who come to school with rich learning experiences are at an advantage when taking standardized tests. Students who have not been given the opportunities for such experiences but have the ability and would perform well on these tests if afforded these opportunities may not be included in gifted programs. Unfortunately, these students are usually those from lower socioeconomic backgrounds.

## Appropriate Placement

Students who are appropriately placed in classes typically served by special education and programs for the gifted should receive all the intervention and resources necessary to meet their educational needs. However, the use of standardized tests for placement into special programs has resulted in many students' being misplaced.

Perhaps contributing to misplacements for students with perceived learning deficits is the fact that teachers may, even on an unconscious level, communicate lower expectations to certain subcultures of students or to those who are perceived to be low achievers or have certain disabilities. Brophy and Good (1984) and Cooper (1983) found that low expectations cause students to perform or adjust their learning so that it coincides or is consistent with the level of expectation communicated by the teacher.

Teachers may, even on an unconscious level, communicate lower expectations to perceived low achievers and special-needs students.

Whether the problem resides in the identification or the referral process or both, steps must be taken to adopt alternative methods of assessment to more equitably and effectively identify students for placement in special programs. *Denial* of appropriate programs and *misplacement* in special education when interventions in regular programs are needed are both issues that challenge the professional community to intervene thoughtfully and timely.

## Students with Unattended Needs

A major area of concern is the students who seem to "fall through the cracks" of our educational system. The following types of students fit into this category:

- Students who are disabled and gifted
- Students who are gifted and underachieving
- Students who have special needs but do not meet the criteria for special education services or programs, including these:
    1. Those who are very bright but do not qualify for a district's gifted program
    2. Those who are low achievers but are not eligible for special education

Many students are disabled and gifted, but their dual exceptionality is not recognized. The problem is twofold:

1. Students may be identified by their disability, and their program goals are based on remediation of the weaker area.

2. Many gifted students have advanced coping strategies that mask their area of disability or weakness. Their weak area depresses their giftedness and the giftedness masks the disability, so the student is never identified as exceptional (Davis and Rimm 1989, pp. 369–86; Whitmore and Maker 1985).

The second category, the underachieving gifted student, has received little attention over the past decade. According to Gallagher (1991), underachieving gifted students possess considerable intellectual potential but are performing in a mediocre, or worse, fashion in the educational setting. This underachievement may be due to a number of factors, such as motivation, low self-esteem, poor study habits, peer-acceptance problems, an unchallenging curriculum, or poor concentration in school (Rimm 1986).

Another group comprises children who are not identified as gifted but still are very bright and are functioning at a very high level when compared with their peers. Generally, there are minimal to no provisions for these students except what regular education provides. It falls to the classroom teacher to furnish additional enrichment and acceleration opportunities for these students. These types of opportunities are presented in the *Professional Standards for Teaching Mathematics* (NCTM 1991) and have been discussed in various chapters in this book.

Students who feel that they "can't," "won't."

Students whose standardized test scores identify them as low achievers but do not qualify them for special education may also be in danger of having real needs unattended. Students who are not in classrooms where supplemental services, such as the federally funded Chapter 1 programs, are in place may not receive the support needed to raise their level of achievement or to develop positive attitudes about their mathematics performance. Students who feel that they "can't," "won't." Students who "won't" are usually left alone to fall through the cracks.

Until we are able to resolve effectively the problems inherent in the identification and referral processes that are currently in vogue, all four of these groups of students may remain hidden in the educational system and continue to perform beneath their true potential. As the regular-education classroom moves toward the accommodation of a greater number and range of differences by incorporating cooperative learning techniques and other instructional and assessment practices, a special focus on detecting and addressing the unattended needs of these special students is appropriate.

## The Inequity of Instruction Driven by Inappropriate Assessment

An important issue affecting not only identification and placement but also the focus of instructional programs is the degree to which standardized tests fail to assess the level of thinking required of students in curricula that are being established by those in the forefront of today's mathematics education community. A major goal of the National Council of Teachers of Mathematics (NCTM) is the development of mathematical power for all students that includes the ability to explore, to conjecture and reason logically, to solve nonroutine problems, to communicate mathematically, and to make mathematical connections (NCTM 1991).

Mathematical power for all students can be realized only when the tests that measure the effectiveness of the mathematics curriculum are in full alignment with this goal. Current standardized tests make reasonable assessments of computational speed but do not assess many important aspects of mathematical reasoning (see, e.g., Resnick and Resnick [1991]). These researchers have stated (p. 47) that none of the standardized tests that were examined during their investigations even approximated the standards established for all students by organizations such as NCTM and the Mathematical Sciences Education Board. On the basis of their analysis, they concluded that students who practiced mathematics in the form found in the standardized tests would never be exposed to the kind of mathematical thinking sought by all who are concerned with reforming mathematics education.

Compounding this problem is the fact that what is tested is usually what is emphasized in the mathematics curriculum. If standardized test scores are the major criterion for placement into special programs and teachers base subsequent instruction and IEPs for these students on this assessment, students will not have the opportunity to have the rich mathematical experiences that might foster the development of mathematical power.

Effective instruction and IEPs should be based on effective assessment.

In recent years, the need for reforming mathematics assessment practices at all levels has been an important issue within the mathematics education community and in other professional circles. The problem is complex and not easily addressed. Leaders in many states and countries are seeking or suggesting alternative assessment practices (see, e.g., Stenmark [1989, 1991]). Chapter 5 also focuses on this important issue and offers useful suggestions for educators involved with special-needs students.

## TODAY'S NEEDS, TOMORROW'S GOALS

This chapter has discussed crucial issues related to the processes involved in identifying and placing students in special programs. Much attention has been given to inequities that exist in the screening, prereferral, and referral of students for placement in special education. These inequities include the following:

- The inappropriate placement of African Americans and other ethnic groups into special education
- The inappropriate denial to specific student populations of programs for the gifted
- The cultural bias and inappropriateness of some standardized measures used as criteria for placement
- Low expectations for students based on gender, race, or ethnicity

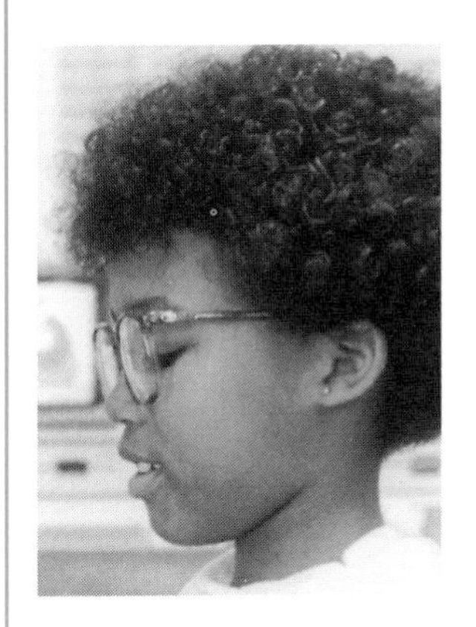

The response to these inequities by thinking educators should call for a move toward more equitable and effective referral, identification, and placement procedures, for example, the following:

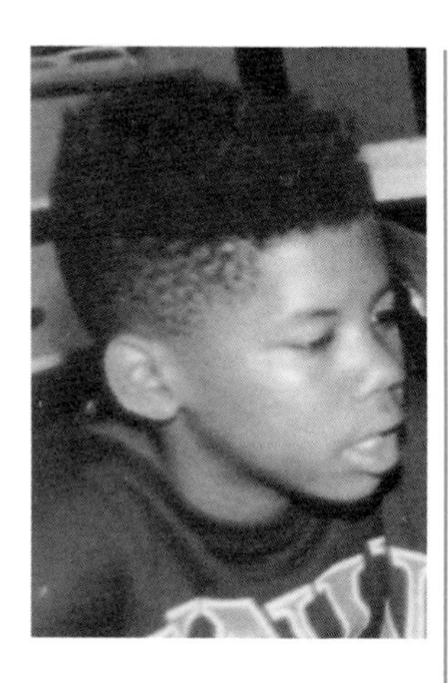

- The use of culture-fair instruments such as the following (both from Psychological Corporation):
  *a*) The System of Multicultural Pluralistic Assessment (SOMPA) (1978)
  *b*) Raven Progressive Matrices—Revised (1984)
- The use of additional methods of screening and prereferral practices discussed in this chapter, which included these:
  *a*) Instructional intervention teams collaborating with the regular classroom teacher to address the special needs of students
  *b*) The monitoring of the intervention over time to determine the results of the intervention
  *c*) The consideration of the least restrictive environment when determining special education program options for students

Another area of concern in this chapter is that some individuals or groups of students are allowed to "fall through the cracks" of our educational system. These students can benefit from special education, but they are not placed into special classes because their standardized test scores are either too high to qualify them for special education programs or too low to qualify them for programs for the gifted. In either instance, it is imperative that teachers make modifications and adjustments in the regular-education program that will meet the needs of these students.

This chapter has highlighted how students' special needs are poorly met or unattended in school mathematics because of inappropriate identification and placement. These *needs* dictate the statement of important *goals* for timely and engaging action: to design more equitable and effective referral, identification, and placement procedures. This direction must be taken in order to open wide the windows of opportunity for *all* students.

## BIBLIOGRAPHY

Adderholdt-Elliot, Miriam, Kate Algozzine, Bob Algozzine, and Kay Haney. "Current State Practices in Educating Students Who Are Gifted and Talented." *Roeper Review* 14 (September 1991): 20–23.

Brophy, Jere, and Thomas Good. *Teacher Behavior and Student Achievement*. Occasional Paper 73. East Lansing, Mich.: Institute for Research on Technology, Michigan State University, 1984. ERIC Document Reproduction Service no. ED 251 422.

Cohen, Rosalie. "Conceptual Styles, Culture Conflict, and Nonverbal Tests of Intelligence." *American Anthropologist* 71 (October 1969): 828–56.

Cooper, Harris M. "A Historical Overview of Teacher Expectation Effects." Paper presented at the annual convention of the American Psychological Association, Anaheim, Calif., August 1993. ERIC Document Reproduction Service no. ED 239 176.

Culross, Rita. "Measurement Issues in the Screening and Selection of the Gifted." *Roeper Review* 12 (December 1989): 76–78.

Davila, Robert R., Michael L. Williams, and John T. MacDonald. *Memorandum on Clarification of Policy to Address the Needs of Children with Attention Deficit Disorders within General and/or Special Education*. Washington, D.C.: U.S. Department of Education, Office of Special Education and Rehabilitative Services, 16 September 1991.

Davis, Gary A., and Sylvia Rimm. *Education of the Gifted and Talented.* 2nd ed. Englewood Cliffs, N.J.: Prentice Hall, 1989.

Downs, Ruth E. "Correlates of Differential Achievement at the Elementary Level in an Urban School District." Ph.D. diss., University of Pittsburgh, 1989.

Fennema, Elizabeth, and Penelope L. Peterson. "Autonomous Learning Behavior: A Possible Explanation of Sex-Related Differences in Mathematics." *Educational Studies in Mathematics* 16 (August 1985): 309–11.

Fennema, Elizabeth, Penelope L. Peterson, Thomas P. Carpenter, and Cheryl A. Lubinski. "Teachers' Attributions and Beliefs about Girls, Boys, and Mathematics." *Educational Studies in Mathematics* 21 (February 1990): 55–69.

Fox, Lynn H. *The Problems of Women and Mathematics: A Report to the Ford Foundation.* Washington, D.C.: Library of Congress, 1980.

Gage, Nathaniel L., and David Berliner. *Educational Psychology.* 3rd ed. Boston: Houghton Mifflin Co., 1984.

Gallagher, James. "Issues in the Education of Gifted Students." In *Handbook of Gifted Education*, edited by Nicholas Colangelo and Gary A. Davis, pp. 14–23. Boston: Allyn & Bacon, 1991.

Gardner, Howard. *Frames of Mind: The Theory of Multiple Intelligences.* New York: Basic Books, 1985.

Garner, William C., and Ernest G. Cole. "The Achievement of Students in Low-SES Settings: An Investigation of the Relationship between Locus of Control and Field Dependence." *Urban Education* 21 (July 1986): 189–206.

Gilbert, Shire E., and Geneva Gay. "Improving the Success in School of Poor Black Children." *Phi Delta Kappan* (October 1985): 133–37.

Grady, Michael, and Warren Simmons. *Black Male Achievement: From Peril to Promise.* A Report on the Superintendent's Advisory Committee on Black Male Achievement. Upper Marlboro, Md.: Prince George's County Public Schools, 1990.

*Individuals with Disabilities Education Act.* U.S. Code. Vol. 20, sec. 1401 (a) (20) (1991).

Kerr, Barbara. *Smart Girls, Gifted Women.* Columbus, Ohio: Ohio Psychology Publishing Co., 1985.

Kitano, Margie K. "A Multicultural Educational Perspective on Serving the Culturally Diverse Gifted." *Journal for the Education of the Gifted* 15 (Fall 1991): 4–19.

Kitano, Margie K., and Darrell F. Kirby. *Gifted Education: A Comprehensive View.* Boston: Little, Brown & Co., 1986.

Kogan, Nathan. "Educational Implications of Cognitive Styles." In *Psychology and Educational Practice*, edited by Gerald S. Lesser, pp. 242–92. Glenview, Ill.: Scott, Foresman & Co., 1971.

Lewis, Rena B., and Donald H. Doorlag. *Teaching Special Students in the Mainstream.* 3rd ed. New York: Macmillan Publishing Co., 1991.

Lloyd, John W., Paula E. Crowley, Frank Kohler, and Philip Strain. "Redefining the Applied Research Agenda: Cooperative Learning, Pre-referral, Teacher Consultation, and Peer Mediated Interventions." *Journal of Learning Disabilities* 21 (February 1988): 43–52.

MacRae, LaDonna, and Judy L. Lupart. "Issues in Identifying Gifted Students." *Roeper Review* 14 (Winter 1991): 53–58.

McCormick, Linda. "The Gifted and Talented." In *Exceptional Children and Youth*, 5th ed., edited by Norris Haring and Linda McCormick. Columbus: Merrill Publishing Co., 1990.

Maker, C. June, and Shirley W. Schiever, eds. *Critical Issues in Gifted Education: Defensible Programs for Cultural and Ethnic Minorities.* Vol. 2. Austin, Tex.: Pro-Ed, 1989.

Matthew, Janice L., Anne K. Golin, Mary W. Moore, and Carol Baker. "Use of SOMPA in Identification of Gifted African-American Children." *Journal for the Education of the Gifted* 15 (Summer 1992): 344–56.

National Council of Teachers of Mathematics. *Professional Standards for Teaching Mathematics.* Reston, Va.: The Council, 1991.

Pennsylvania Department of Education. *Guidelines for Assessment, Evaluation, and the Individualized Education Program.* Harrisburg, Pa.: Department of Education, 1991.

Psychological Corporation. *Raven Progressive Matrices—Revised.* San Antonio, Tex.: The Corporation, 1984.

———. *System of Multicultural Pluralistic Assessment.* San Antonio, Tex.: The Corporation, 1978.

Public Law 94-142. *Education of the Handicapped Act of 1975.* Federal Register 42 (23 August 1977): 163.

Renzulli, Joseph. "What Makes Giftedness? Reexamining a Definition." *Phi Delta Kappan* 60 (November 1978): 180–84, 261.

Resnick, Lauren B., and Daniel P. Resnick. "Assessing the Thinking Curriculum: New Tools for Educational Reform." In *Changing Assessments: Alternative Views of Aptitude, Achievement and Instruction*, edited by Bernard R. Gifford and M. O'Conner. Boston: Kluwer Academic Publishers, 1991.

Richert, Susanne E. "Rampant Problems and Promising Practices in Identification." In *Handbook of Gifted Education*, edited by Nicholas Colangelo and Gary A. Davis. Boston: Allyn & Bacon, 1991.

Rimm, Sylvia. *Underachievement Syndrome: Causes and Cures.* Watertown, Wis.: Apple Publishing Co., 1986.

Sartain, Harry W. *Nonachieving Students at Risk: School, Family, and Community Intervention.* Washington, D.C.: National Education Association, 1989.

Shade, Barbara J. "Afro-American Cognitive Style: a Variable in School Success." *Review of Educational Research* 52 (Summer 1982): 219–44.

———. "Cognitive Strategies as Determinants of School Achievement." *Psychology in the School* 20 (October 1983): 488–93.

Shore, Bruce M., Dewey G. Cornell, Ann Robinson, and Virgil S. Ward. *Recommended Practices in Gifted Education.* New York: Teachers College Press, 1991.

Stenmark, Jean Kerr. *Assessment Alternatives in Mathematics: An Overview of Assessment Techniques That Promote Learning.* Berkeley, Calif.: EQUALS Project, Lawrence Hall of Science, University of California, 1989.

Stenmark, Jean Kerr, ed. *Mathematics Assessment: Myths, Models, Good Questions, and Practical Suggestions.* Reston, Va.: National Council of Teachers of Mathematics, 1992.

Whitmore, Joanne R., and C. June Maker. *Intellectual Giftedness in Disabled Persons.* Rockville, Md.: Aspen Publications, 1985.

Witkin, Herman, Carol Moore, Donald Goodenough, and P. W. Cox. "Field-Dependent and Field-Independent Cognitive Styles and Their Educational Implications." *Review of Educational Research* 47 (Winter 1977): 1–64.

Witkin, Herman, Phillip K. Ottman, Evelyn Raskin, and Stephen A. Karp. *Group Embedded Figures Test.* Palo Alto, Calif.: Consulting Psychologists Press, 1971.

Wolf, J. "The Gifted and Talented." In *Exceptional Children and Youth*, 5th ed., edited by Norris Haring and Linda McCormick, pp. 447–90. Columbus: Merrill Publishing Co., 1990.

**Ruth Downs** is a supervisory instructional specialist in the Pittsburgh Public Schools. She supervises mathematics teachers in the elementary and middle school Chapter 1 Program. She is interested in achievement-gap issues and in raising the mathematics-achievement level of low-achieving students. She was awarded the Dissertation of the Year Award from the University of Pittsburgh's School of Education for work in these areas.

**Janice Matthew** is the supervisor of gifted programs for the Pittsburgh Public Schools. She has supervised and taught special education and basic education for over twenty years. Her recent publication is "Use of SOMPA in Identification of Gifted African-American Children."

**Marilyn McKinney** is the assistant director for the Division for Exceptional Children, Pittsburgh Public Schools. She has taken an active role in the field of special education for twenty years as a teacher and administrator.

5

# Assessing Mathematical Abilities and Learning Approaches

*Maria R. Marolda*
*Patricia S. Davidson*

***Are there particular mathematical profiles that characterize students with specific learning differences?***

***How are mathematical profiles revealed and assessed?***

***How does the understanding of mathematical profiles translate into better instructional opportunities in mathematics for students with special learning needs?***

A PRIMARY consideration in the effective assessment and teaching of mathematics is the recognition that students bring to the learning of mathematics a wide range of abilities and learning approaches. These differences among learners must be recognized, revealed, and addressed in both assessment and teaching efforts. Students' learning profiles are marked by different constellations of relative strengths and relative weaknesses with which they face the world. In the consideration of differences among students, the critical questions become, "When does a 'learning difference' reflect special learning needs?" and "When do special learning needs render a student 'learning disabled'?"

Students characterized "learning disabled" are those whose learning profiles place them at a particular disadvantage in dealing with the typical demands and presentations of the ongoing curriculum. They are often intolerant of a variety of approaches and require quite personalized strategies. Though they may have relative strengths, it may be that these strengths are not being tapped or used effectively. It is necessary, therefore, to have a means to assess and understand more fully the differences in the learning profiles of learning disabled students, to examine what they bring to the circumstance of mathematics, and to generate meaningful mathematical profiles that teachers can use to meet students' needs.

In education, the search for strengths must be included along with the examination of specific deficits and must serve as the focal point of all assessment activities.

It is important to recognize that the diagnostic process in education is quite different from the diagnostic process in medicine and other fields. Whereas the medical diagnostician is looking to uncover "what is wrong" and what the patient "can't do," the educational diagnostician must strive to uncover the student's strengths and what the student "can do." The goal is to find those strengths that can be used to address the weaknesses and difficulties inherent in students' learning approaches. In education, the search for strengths must be included along with the examination of specific deficits and must serve as the focal point of all assessment activities.

Central to the generation of mathematical profiles for any student, and most particularly for those students with special learning needs, is the premise that the content and methods of assessment should acknowledge, recognize, and reflect an underlying cognitive theory. A neuropsychological approach to learning offers a useful framework within which to consider the differences brought to the learning of mathematics by students. Differences in the development of various brain systems lead to different learning profiles, with resultant variations in the interplay of innate talents and vulnerabilities of students at particular stages of cognitive development. A neuropsychological approach is useful because it contributes to the understanding of the current abilities and approaches of students, allows for predictions about future issues they might face in mathematics, and suggests the parameters around which instruction can be customized so that maximum growth is achieved.

The purpose of this chapter is to present the *child/world system* (Bernstein and Waber 1990) as a framework for understanding the mathematical profiles of students with special learning needs as they relate to their cognitive and behavioral aspects, to discuss assessment techniques that will generate comprehensive mathematical profiles, and to consider the implications for classroom instruction of the interface between the child and the world of mathematics.

## A CHILD/WORLD SYSTEM IN MATHEMATICS

Mathematics learning should be considered as a dynamic interplay between what students inherently bring to the circumstances of mathematics and what the circumstances of mathematics require of the students. Considered separately, neither the features of the students alone nor the specifics of the mathematical topics themselves are sufficient explanations for variations in the learning process. The learning process is better understood by building a *child/world system* that characterizes the reciprocal relationship of the developing child and the mathematical world in which that child must function. The construction of a child/world system highlights areas of "match" and "mismatch" between a student's complement of skills and the demands placed on those skills (see fig. 5.1).

Mathematics learning should be considered as a dynamic interplay between what students inherently bring to the circumstances of mathematics and what the circumstances of mathematics require of the students.

The child/world system as applied to the learning of mathematics is consistent with a neuropsychological approach to cognition and serves as an effective and useful framework for understanding the mathematical profiles and learning differences among students. It offers a particular contribution to the

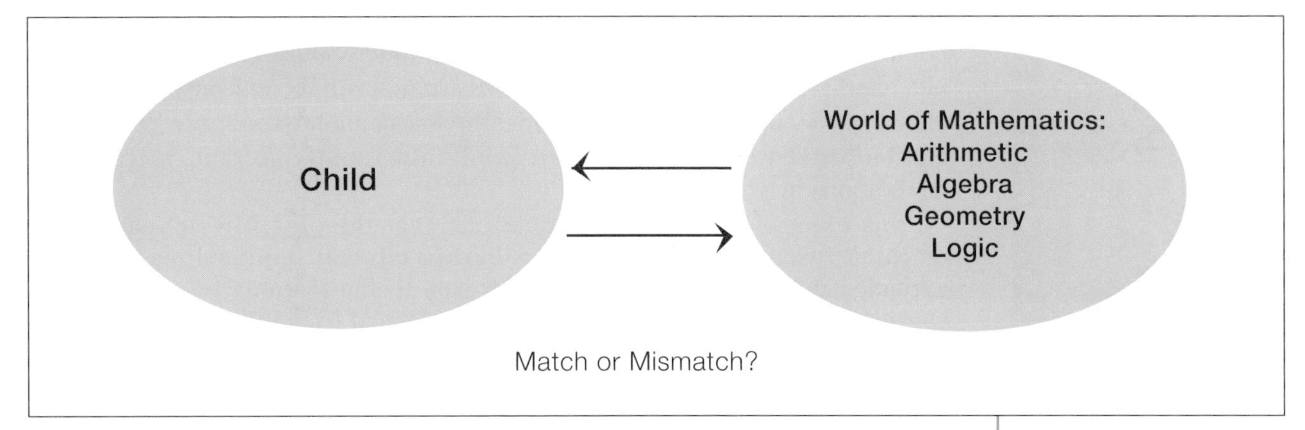


Fig. 5.1. A child/world system for learning mathematics

understanding of students with special learning needs in mathematics. It directs attention to brain/behavior relationships and explores learning by focusing on the interplay of the relative strengths and weaknesses present at different stages of the child's cognitive development. It "builds on the systemic psychology of Anokhin, Vygotsky, and Luria and extends the process approach elaborated by Kaplan.... [I]t emphasizes the dynamic interplay of neural and behavioral systems in development" (Bernstein and Waber 1990, p. 312). Brain/behavior relationships are examined along three axes of the brain: the cortical/subcortical axis, the anterior/posterior axis, and the lateral axis, involving the two cerebral hemispheres and the corpus callosum, the large bundle of fibers connecting the two hemispheres. (Readers should note that the term *axis* is used in the neuropsychological sense, not in the mathematical sense.) Specific behaviors, it would seem, are "virtually wired into the structure of the nervous system" (Gregory 1987, p. 547) and can be described in terms of activities associated with the three axes. Although all systems are required for effective behaviors, the lateral axis is of particular interest to educators because of its consequential implications for intellectual functioning.

Study of the "wiring" of the brain's structure, particularly the lateral axis, is a relatively recent activity. Although research is ongoing and many questions must be investigated, some principles are well established. The left hemisphere of the brain is primarily concerned with processing discrete elements in a series (one at a time), is strongly oriented to details, is analytical, and plays a special role in language. The right hemisphere is concerned with overall perspectives, is spatially oriented, is intuitive, is adept at visual communication (such as recognizing faces and reading body expressions), and plays a major role in the analysis of visual and spatial dimensions of the world. The corpus callosum carries on the important and rapid communication between the two hemispheres and allows the merging of information from both hemispheres. The allocation of responsibilities to specific cerebral hemispheres must be carefully interpreted. Distinctions are made for the purpose of understanding the lateral system itself. In terms of actual behavior, both hemispheres and the corpus callosum must work in concert for effec-

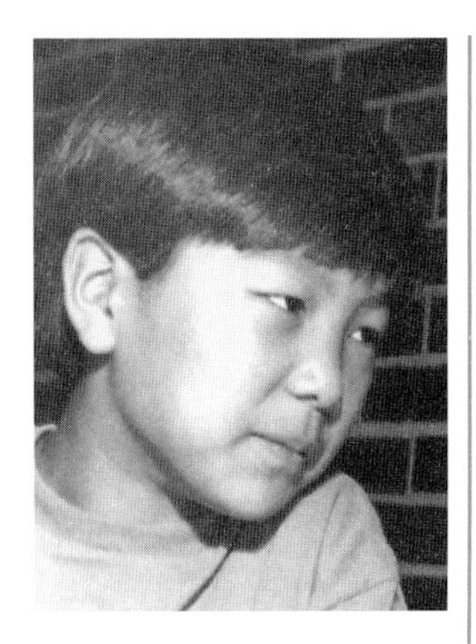

tive results. The main aspect to consider is the relative contribution of each component to a specific activity. With an understanding of the neural and behavioral underpinnings of students' learning profiles, not only can current effectiveness in learning mathematics be better understood, but predictions can be formed about possible areas of vulnerability so that instructional adaptations may be made.

From extensive clinical investigations over the past fifteen years and a detailed research project to corroborate clinical impressions, we have extended the neuropsychological approach to the learning of mathematics. Our approach capitalizes on a neuropsychological framework and becomes a useful paradigm in examining and understanding a child/world system in mathematics. We define two distinct learning styles in mathematics that are highly correlated with hemispheric considerations. In the first (Mathematics Learning Style I), the student is better at activities that require forward processing, such as counting forward and the concepts of addition and multiplication. Such students prefer a "recipe" approach to mathematical procedures, in which they follow a step-by-step sequence, moving forward to a solution. They need to know the "how" before they can deal with the "why" of a specific situation. They seldom estimate, tend to remember parts rather than wholes, and have a strong need for talking themselves through procedures. Once these students have mastered required recipes, they are often very diligent in carrying them out. Although they might arrive at the correct answer, they may remain totally unaware of the underlying principles that give meaning to what they are doing.

The second style of learner (Mathematics Learning Style II) uses skills involving reverse processing, such as counting backward as well as the concepts of subtraction and division, and is quite comfortable using them. These students are often impatient with step-by-step procedures and are likely to make mistakes while executing them. Such students are good at estimating, may spontaneously give a correct answer, resist paper-and-pencil techniques, and are superior at recognizing large-scale patterns. They are especially good at appreciating various spatial relationships and in dealing with geometric situations involving both two-dimensional and three-dimensional configurations.

In designing effective assessment techniques in mathematics for students with special learning needs, we must scrutinize both the *child* and the *world* aspects of the child/world system. First, the *child* must be considered in terms of those characteristics that are most relevant to the child's participation in mathematics, that is, specific mathematical profiles. Then, the *world* of mathematics must be comprehensively defined in terms of the mathematical activities that are most salient to effective understanding. Finally, the interface of the *child* and the *world* must be examined to generate those occasions of "match" and "mismatch" that affect the child's performance. Assessment must provide opportunities for teachers or evaluators (1) to observe the various cognitive predispositions that students bring to the classroom, (2) to evaluate students' specific learning approaches in mathematics, and (3) to determine the students' effectiveness in dealing with the inherent demands of the mathematical topics themselves.

## THE *CHILD* OF THE CHILD/WORLD SYSTEM

To understand the student with special learning needs, a multidimensional view must be taken and a variety of diagnostic parameters (fig. 5.2) must be considered:

- Developmental readiness for dealing with specific mathematical topics
- The preferred models with which mathematical concepts are interpreted
- The preferred processing strategies with which mathematical topics are pursued
- The memory, language, and output or production skills that affect students' ability to participate in mathematical activities
- Salient behavioral postures that affect performance in mathematics

**Diagnostic Parameters**

- Developmental readiness to consider particular mathematical topics
- Mathematical models
  - Preference for discrete (set) models vs. preference for continuous (measurement) models
- Processing strategies
  - Linear, step-by-step procedures vs. global, gestalt procedures
  - Organizational skills
  - Integration skills
- Memory skills
  - Retrieval of arithmetic facts
  - Recall of steps of procedures
  - Visual memory
  - Memory of math data
  - Auditory memory
- Language skills
  - Facility with directions and explanations
  - Facility with verbal requirements of word problems
  - Ability to explain approach
- Output or production skills
  - Pace
  - Precision
  - Graphomotor skills
- Salient behavioral postures
  - Need for physical movement
  - Perseveration
  - Persistence
  - Attentional skills
  - Self-image and self-confidence in mathematics

Fig. 5.2. Diagnostic parameters for understanding the child. From *Mathematics Diagnostic/Prescriptive Inventory* (Davidson and Marolda 1993)

These parameters must extend from a consideration of those cognitive postures that are readily available and easily mobilized to those postures that are elusive and less effectively pursued.

A true understanding of the *child* must consist of a definition of the strengths on which mathematics instruction can be based as well as a description of those areas of vulnerability and weakness that may intrude and inhibit the student's efforts in mathematics.

## Developmental Readiness

A student's mathematical profile should incorporate an appreciation of the developmental maturity of the student at various ages.

The *Curriculum and Evaluation Standards for School Mathematics* (National Council of Teachers of Mathematics 1989) has stated that mathematical goals should be *developmentally appropriate.* "It is clear that children's intellectual, social, and emotional development should guide the kind of mathematical experiences they should have..." (p. 16). Thus, a definition of a student's mathematical profile should incorporate an appreciation of the developmental maturity of the student at various ages. There are many developmental milestones with respect to mathematical readiness for dealing with numerical, spatial, and logical topics. For numerical concepts, the developmental milestones consist of an appreciation of number, the concept of quantity, conservation of quantity, one-to-one correspondence, and principles of set inclusion. For spatial concepts, the construct of space, conservation of length, and conservation of volume must be considered. Finally, for logical thought, developmental milestones include the concepts underlying classification, seriation, associativity, reversibility, and inference. Many of the fundamental developmental milestones that have been described for number, space, and logic are acquired by children between the ages of four and seven (Piaget 1965).

More recent clinical and teaching practice has suggested that the concept of place value might also be developmentally mediated. That is, an appreciation of place value depends on the state of the child more than on specific teaching or learning experiences. If the child is not cognitively "ready" to deal with place value, then the concept of place value cannot be formally or meaningfully developed, despite teaching efforts. The concept of place value seems to be established for most children between the ages of six and eight. The appreciation of the principles of place value is of particular importance, since place-value concepts are necessary prerequisites to the understanding of larger quantities and the pursuit of all procedures involving multidigit calculation.

Therefore, since the formal study of topics would have little meaning for a student not ready to deal with them, it is important to know which developmental milestones are in place. Curriculum expectations must be examined to determine if the curriculum is asking students to do what their cognitive systems are ready to attempt. Since concepts that are developmentally mediated are reliant on the inherent state of the child rather than on the specific presentation of the topic, it is important to know the child's developmental status in order to design appropriate instruction. If developmental readiness is not yet established, instructional time would best be directed to the nurturing of intuitions. If developmental maturation, and thus readiness, has been achieved, direct instructional efforts for the building of specific concepts and procedures are appropriate. For a complete understanding of a student, it is important that assessment efforts define the developmental milestones that have been achieved by the student and distinguish them from broader learning issues.

Several mathematical topics call on the integration skills associated with the corpus callosum, whose development seems to be ongoing between the ages of eight and twelve. These topics include the concept of fractions, operations with fractions, the interpretation of three-dimensional configurations, and the constructs underlying algebra. Since these topics seem to be reliant on the function of the corpus callosum, which is subject to developmental maturity, they too might be developmentally mediated.

Once it has been established that the student has the necessary readiness or mathematical maturity to approach a particular topic, it is important to investigate the best way to develop that topic so that it is most meaningful to the student. "[I]n evaluating a child, it is essential to observe not only the adequacy of the product, but how he or she goes about accomplishing the task. It is only by observing the routes to success or failure on a given task that alternative pathways available to the child can be diagnosed and remediation determined" (Bernstein and Waber 1990, p. 316). The understanding of the *routes and pathways* available to a student should incorporate a consideration of the particular mathematical models that are most meaningful to the student as well as the processing strategies that are most natural and effective for the student. The *routes and pathways* available determine students' effectiveness in dealing with the input, processing, and output aspects of mathematical tasks and must be incorporated in all mathematical assessment.

## Models for Mathematical Concepts

Mathematical situations can be interpreted with *concrete*, *pictorial*, or *symbolic* models. For some students, mathematical situations that are direct and rely on the actual manipulation of physical materials are the most comfortable. Other students prefer situations that involve pictorial interpretations, where they can rely on their visual senses to come to conclusions. Finally, some students are comfortable with abstract or symbolic interpretations, involving verbal or numerical cues. Assessment activities should be offered in all three modalities, and students should be allowed to approach any task in any of these three modalities. Furthermore, responses may be given in any of these same modalities. The teacher or examiner should note the medium a student spontaneously selects as well as the one in which the student appears most comfortable. Once a student has completed a task using a particular modality, the teacher or examiner should direct the student to "try this another way." The teacher can then determine whether alternative modalities are available or whether there is an actual inability to operate in a particular modality for that task (see fig. 5.3).

Mathematical situations can be interpreted with *concrete*, *pictorial*, or *symbolic* models.

## Processing Strategies

Assessment activities must also investigate the *processing strategies* that students most naturally use to approach mathematical situations. Processing strategies can be distinguished into two broad categories: logical, step-by-step procedures versus gestalt, global procedures. Strategies must also be

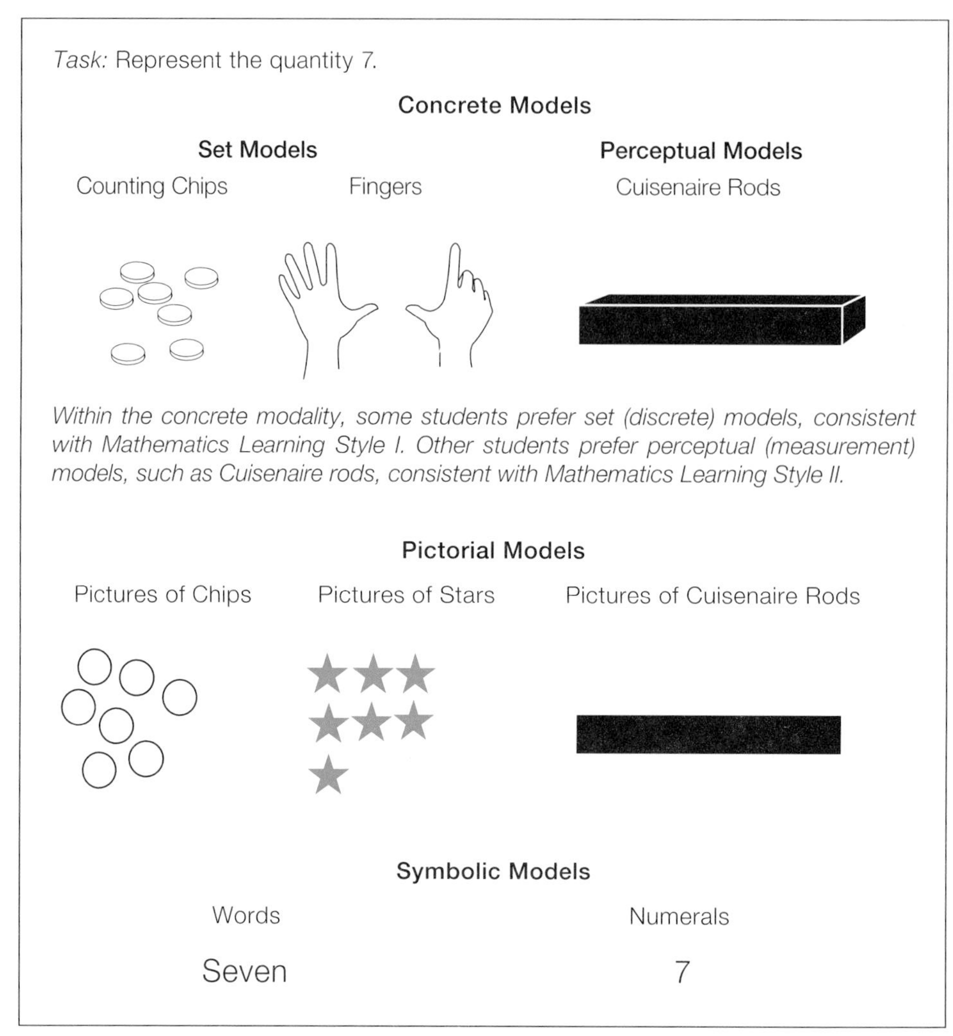


Fig. 5.3. Various models for interpreting quantity

evaluated according to the effectiveness with which organizational and integration demands are met. The *Curriculum and Evaluation Standards for School Mathematics* (National Council of Teachers of Mathematics 1989) has reaffirmed the importance of considering the process as well as solutions in assessment. "Students' strategies and thought processes must be looked at as well as their answers" (National Council of Teachers of Mathematics 1991, p. 9).

Processing strategies can be elicited in any situation where different approaches can be applied to achieve a correct final solution. These situations are available to teachers during classroom instruction or may be planned for specifically designed assessment activities. The approaches that students use often have a great deal of consistency from task to task. Once students have been observed in a variety of situations, the nature of their strategies can be generalized and characterized so that predictions can be made on how they might proceed in other mathematical situations.

The ways in which students process or approach mathematical situations follow two distinct patterns. Some students process situations in a linear fashion, building forward to an exact final solution. Sometimes, they are so focused on the individual elements before them that the overall goal or perspective to which the elements contribute is obscured. This style of processing is often characterized as sequential or step by step. For other students, a careful, building-up approach holds little inherent meaning. Such students prefer to establish a general overview of a situation first and then refine that overview successively until an exact solution emerges. Such students are often prone to imprecision and tend to lack appreciation of all relevant details. This style of processing is often described as global, gestalt, or random. The two styles are shown in figure 5.4.

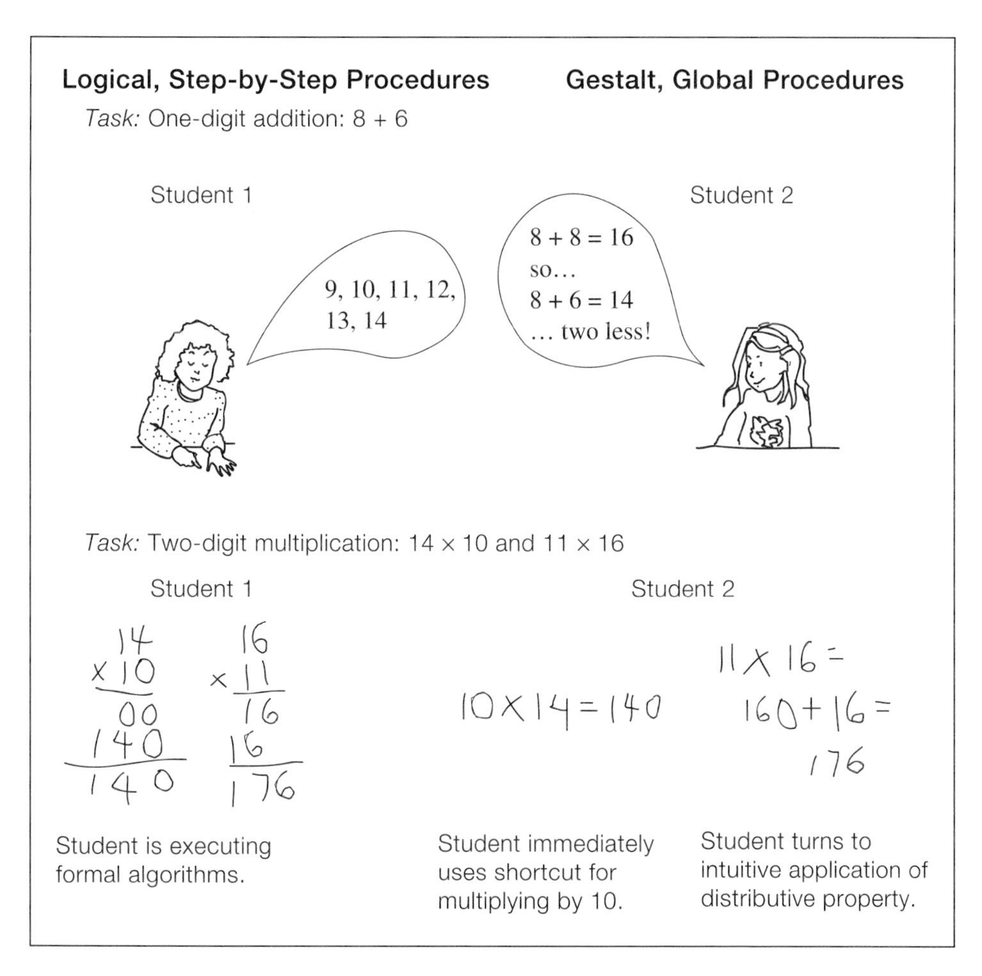


Fig. 5.4. Two different processing strategies in mathematical situations

From the responses made, it is clear that Student 1 is adept at dealing with the specific elements or steps of a situation and proceeds in a careful, building-up approach. Student 1 would be characterized as demonstrating

Mathematics Learning Style I. In contrast, Student 2 is looking at the overall pattern or relationship that prevails. Once the general direction is established, this student uses successive refinements until an exact, final solution is achieved. Student 2 would be characterized as having Mathematics Learning Style II.

The definition, through classroom observations or in assessment activities, of a student's preferred models for mathematical concepts and preferred processing strategies in mathematical situations will enhance the possibility of an effective *match* between the student and the mathematical topic. Those resources that are immediately available and comfortable for the student can be designated as areas of relative strengths. Those areas of strengths may then be used and fashioned to address the areas of weaknesses or inefficiencies that beset the student. "Remediation itself, moreover, is likely to be more effective when directed at strengthening the alternative pathways that are naturally available to the child rather than training the function through the more usual route" (Bernstein and Waber 1990, p. 316). For example, if a student prefers verbal, step-by-step approaches, geometric designs might be carefully described in terms of their various components. If a student prefers global, perceptual interpretations, arithmetic procedures might be introduced through estimation strategies, with successive approximations used to yield an exact solution.

Teachers must use the "operating" processing style and translate all activities into that operating style.

To be full and successful participants in mathematics, students must be able to mobilize both Mathematics Learning Style I and Mathematics Learning Style II. For most students, the issue is to define the processing strategy that offers the best entry into a topic. The student with special learning needs, however, is often limited to one processing style alone and is unable to mobilize the skills and strategies associated with the alternative processing style. For success with students with special learning needs, teachers must use the "operating" processing style and translate all activities into that operating style, building a scaffold that integrates the areas of strengths and weaknesses so that they complement one another.

## Organizational and Integration Skills

Not only is the choice of the type of processing important, it is also crucial to consider the effectiveness of the pursuit of the chosen strategies; in particular, the ability to organize the elements and steps of the process and the ability to integrate multiple aspects or considerations simultaneously must be examined. Some students appreciate all the elements or steps of a procedure but then have difficulty in bringing those elements together in a meaningful way. They may have difficulty organizing all the salient components or in sustaining multiple steps over multiple digits or multiple iterations (see fig. 5.5). Even when students can mobilize and pursue effective organizational strategies, their efforts are sometimes frustrated when multiple aspects or considerations must be incorporated. These students may have difficulty integrating those various considerations or in merging the solutions of the subroutines involved (see fig. 5.6).

**Arithmetic Situation: Addition with Regrouping**

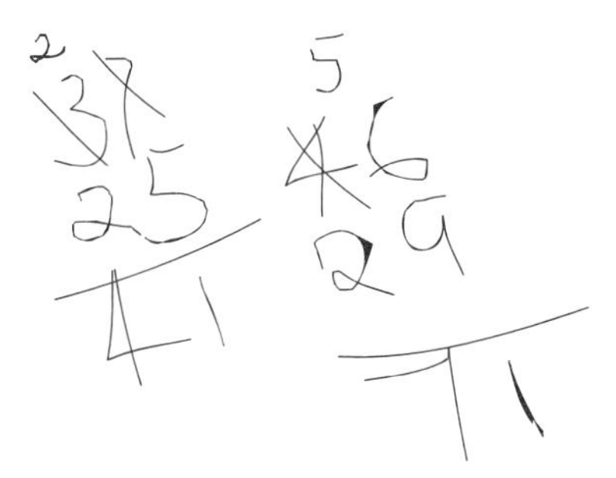

The student has a disorganized, but consistent, approach to both addition problems. In the first example, the student crossed out the 3, indicating a recognition of some kind of regrouping, and then added 7 + 5, but recorded this partial sum by "bringing down the 1" and "carrying the 2." The final step was to add 2 + 2 in the tens place.

Fig. 5.5. Organizational difficulties

**Arithmetic Situation**

*Task:* Draw a picture showing 6/9.

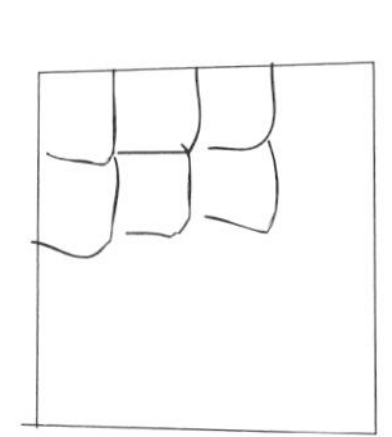

A "whole" is drawn. Six parts are identified. *The dual aspects are not integrated.*

**Geometric Situation**

*Task:* How many cubes are needed to make this building?

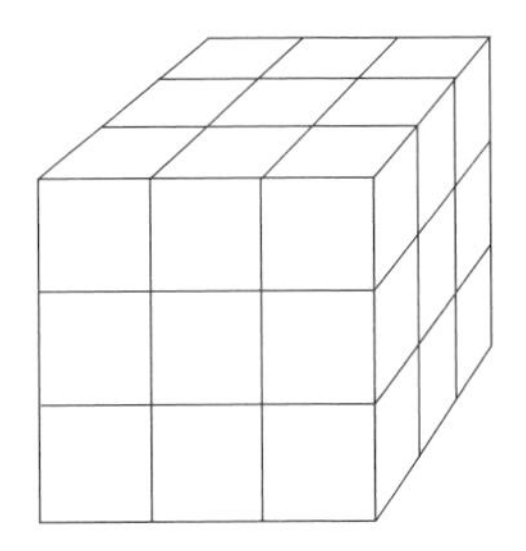

*Solution:* $9 \times 6 = 54$
*Reasoning:* 9 cubes to a side and 6 sides.
*The two- and three-dimensional aspects are not integrated properly.*

Fig. 5.6. Integration difficulties

## Memory Skills

Often students are characterized as having difficulty in mathematics because they "can't remember." The attribution of mathematical difficulties to a global memory deficit is somewhat simplistic. However, at least five memory requirements are important in mathematics: the retrieval of solutions to arithmetic facts, the recall of multistep procedures, the visual memory of geometric configurations, the recall of mathematical data presented in a situation, and the auditory memory of information. Assessment efforts must include the examination of students' various memory skills in mathematics, with differences compared and contrasted.

Cognitive psychologists suggest that memory issues, in general, are very complex (Holmes 1988, p. 186):

> In evaluating a child's recall of materials, the clinician should recognize the various components of the process loosely called memory: registration of the stimulus, encoding, organization, storage, and retrieval.... Learning disabled children, however, are constantly described in the psychological and educational literature as having memory deficits of various types, usually "visual" or "auditory" (short term or otherwise). In almost all cases, the impairment involves either the initial encoding or the effective retrieval of information.

In mathematics, it is also important to consider the distinction between the encoding and retrieval aspects of memory.

In mathematics, it is also important to consider the distinction between the encoding and retrieval aspects of memory. Is the student having difficulty "remembering" the fact or procedure because it was never properly understood and therefore not encoded for storage in memory? Or is the student having difficulty "remembering" the fact or procedure because it cannot be accessed from the student's repertoire of learned skills?

If encoding is the specific memory issue for students, then the remediative focus must be directed toward building better understanding of the concept or process to be encoded. The implicit framework or structure underlying the concept or process must be made explicit and interpreted with meaningful referents or within meaningful contexts. The encoding of concepts or procedures is particularly enhanced when students are active participants in constructing and developing the frameworks and concepts being used.

If retrieval is the specific memory issue for students, then the remediative focus must be directed toward building alternative or more effective accessing routes to a concept or process that has been stored. For example, if students cannot retrieve the solution for $7 \times 9$, an alternative "accessing route" might be as follows: "It's one of the 'hard' ones...it might be 63, 72, 54, 56, 49, 64, 81, or 48 ... $7 \times 9$ must be odd, since an odd number times an odd number is odd ... so, it can't be 72, 54, 56, 64, or 48 ... it is either 49, 63, or 81 ... it must be 63, since the digit in the tens place in a '9 times fact' is one less than the multiplier...." Similarly, for the student who cannot retrieve the name of "triangle," the name might be accessed as follows: "It's a shape ... with THREE sides and THREE angles ... it is a TRI-angle." For those students with retrieval issues, recognition formats may aid in tapping their learned skills. Given many choices, they often can *recognize* the correct response, even when they cannot access or *retrieve* it.

In order to address difficulties in retrieving the steps to multistep procedures, students often turn to verbal mediation techniques or mnemonics to direct their efforts. For example, to access the rule for multiplying a binomial by a binomial, the mnemonic FOIL may be helpful: First terms, Outer terms, Inner terms, Last terms. Other students can visualize the procedure "in their minds" and rely on visual formats to direct their efforts (fig. 5.7).

In dealing with geometric designs, students need to use visual memory skills. Some students may remember a design because they describe the

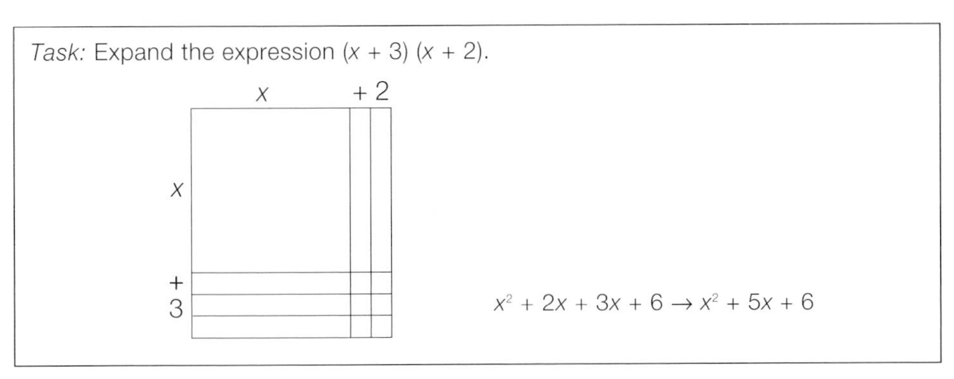


Fig. 5.7. A visualization strategy to retrieve the steps of the binomial × binomial procedure

design verbally and then remember the verbal description; other students "picture" the design in their minds (see fig. 5.8).

*Task:* Using patterning blocks, produce this design using visual memory.

*Student with Learning Style I:* "It has four squares that form a larger square and four triangles that form a larger triangle."

*Student with Learning Style II:* "It looks like a house lying on its side."

Fig. 5.8. Visual memory strategies

Difficulties with visual memory requirements can also manifest themselves in nongeometric situations. Some students have difficulty copying problems from their texts and from the board. For these students, the teacher must minimize "copying" requirements and monitor students' attempts at recording problems given to them.

Students who have difficulty with auditory memory for information often have trouble remembering the directions given in mathematical situations, remembering all the relevant data presented, and remembering all relevant information in word problem situations. These students are helped by writing or recording the pertinent information and by being allowed to refer to their written records as they proceed. Students with apparent memory issues are often confused with students whose primary difficulties are in language, where memory difficulties are secondary to specific language issues.

## Language Issues

Language skills become an issue in mathematics when students are required to deal with lengthy verbal presentations typical of classroom instruction, when they are faced with word problem situations, and when they are asked to explain their solutions or approaches. Concerning the difficulties that may be posed by verbal presentations and classroom explanations, instruction in mathematics may reduce some students' difficulties with language, since the "language" of mathematics is characterized by a limited set of universal symbols and is generally marked by consistent explanations of procedural requirements. Students with auditory comprehension difficulties, sometimes called "receptive language" issues, benefit from this limited vocabulary in their study of mathematics and explanations that are direct, succinctly stated, and carefully presented.

For those students who find the verbal demand of word problems particularly troublesome, assessment techniques must determine if they can choose the correct approach once the situation before them is understood. A further distinction must be made on whether the reading requirement or the basic language comprehension requirement is the source of the student's difficulty. To minimize the burden of reading, word problems can be read to the student. The content of word problems should be presented deliberately and in meaningful "chunks" that are more manageable by the students. If these accommodations for reading requirements are not sufficient, then the information contained in word problems should be presented visually in an effort to minimize the language difficulties.

For those students who find the verbal formulation of responses in mathematical situations overwhelming—sometimes called students with "expressive language" difficulties—the availability of the different modalities in which mathematics can be expressed offers particular support. Those students should be encouraged to turn to concrete or pictorial models to express the mathematics at hand. When mathematics is presented and produced in concrete or pictorial terms, the difficulty with verbal formulation is minimized.

Among the goals for mathematics education, the *Curriculum and Evaluation Standards for School Mathematics* (National Council of Teachers of Mathematics 1989) has emphasized "mathematics as communication" and has encouraged teachers to ask students for explanations or justifications of their approaches and solutions. This new emphasis in mathematics poses a particular challenge for students with verbal formulation difficulties. Though they may have learned the mathematics, they may not be able to retrieve the appropriate words or organize the content they want to express. Communication skills can break down at one or more of the three levels of input, processing, or output. Regarding language issues of students with special learning needs, communication goals in mathematics must be extended to include both verbal and nonverbal modes of communication. Moreover, students with special learning needs often need nonconfrontational circumstances and may require external support in order to participate in the mathematical discussions of the classroom. For the student with special learning needs, the increased emphasis on communication in mathematics should be approached with particular sensitivity by the teacher so that the language issues that are so prevalent in the population of students with special learning needs do not inhibit their participation in mathematics.

## Output or Production Issues

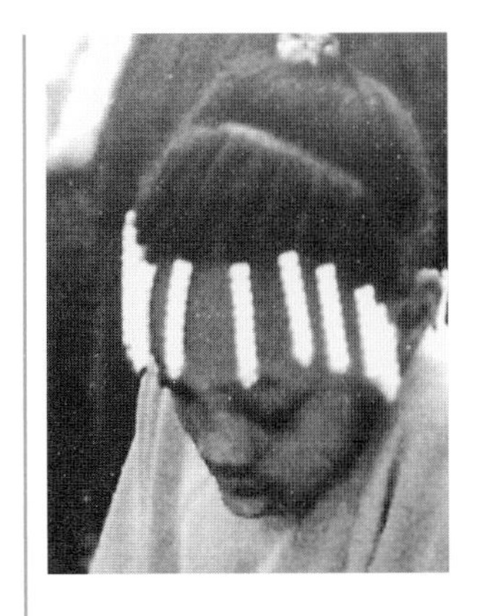

For some students, the issue in mathematics is not the learning of mathematical topics and procedures but rather the ability to "output," or produce, solutions effectively. For these students, issues of pace, precision, and graphomotor skills must be carefully considered. In the search for students' strengths, assessment efforts must ensure that strengths are not obscured by production issues. The goal should be to determine those production conditions in which the students can best perform.

In assessment activities, in either the classroom or a separate setting, students should be encouraged to use the speed with which they are comfortable. Confrontational circumstances should be avoided, and students should be given permission to "take your time." The teacher or examiner must then report if the pace the student assumes in order to produce the best solutions is compatible with the pace expected in a typical classroom setting. If an adjusted pace yields better solutions, the teacher can make accommodations in class in the amount of time allowed or the number of problems assigned.

For other students, difficulty maintaining precision obscures the quality of their learning. Despite correct approaches, there is "slippage" in the solutions to specific arithmetic facts or in dealing with perceptual cues and spatial relationships. Many students, in fact, pursue a specific approach or arithmetic operation correctly but do not maintain exactness in their solutions. These students are often advised that they must avoid "careless" errors. For example, many of these students can produce solutions to one-digit facts when those facts are presented in isolation but cannot maintain that precision within the context of longer calculations. Difficulties maintaining precision may reflect an "overload" phenomenon: students are so focused on executing the steps of the procedure that little energy is left to monitor the precision of the one-digit facts or specific details involved. To counter imprecision, students must develop monitoring and corrective strategies. In mathematical problem-solving activities and in other subject areas where exacting calculations are required, the provision of a calculator may result in the most advantageous and motivating approach for students to achieve correct final solutions.

Other students have difficulty maintaining precision when copying problems or recording problems in written formats. Trouble achieving precision in written problems is often reflective of spatial and perceptual difficulties. To counter imprecision in recording multidigit calculation problems, perceptual accommodations and support should be offered by minimizing the amount of copying and by furnishing graph paper or lined paper turned on its side.

The student's efficiency and automaticity in meeting the graphomotor requirements of written numerals are of particular concern, especially among the population of students with special learning needs. Students with graphomotor problems have difficulty with the motor programming or motor control required to produce written numerals. Their productions may be inconsistent in size and poorly formed. The nonalignment of digits combined with poor form of the digits themselves often leads to incorrect solutions.

Furthermore, for many students multiple erasures and self-corrections are indicative of the additional effort needed to record the problems accurately. For students with graphomotor problems, the extra effort expended to actually write numerals and record numerical formats diminishes the amount of energy available for processing the actual mathematical requirements involved (see fig. 5.9).

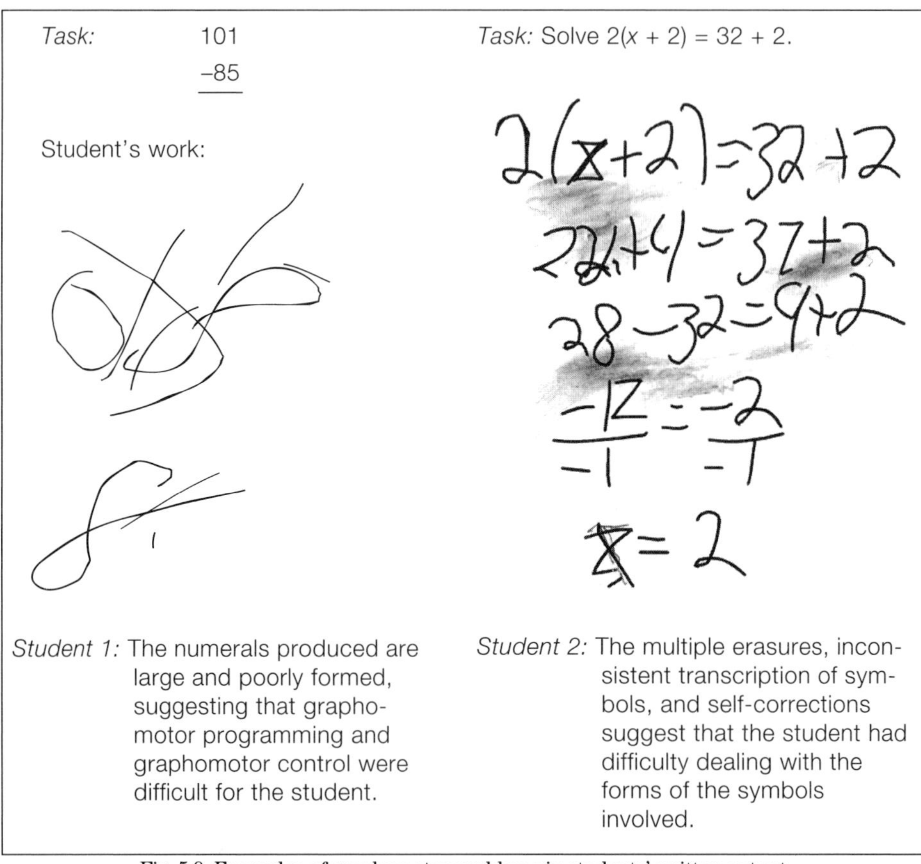


Fig. 5.9. Examples of graphomotor problems in students' written output

In any assessment activity, it is important to distinguish output (production) issues from learning difficulties. Errors should be scrutinized to determine if the approach is correct but the solution is not. If the approach is incorrect, then reteaching is indicated. If the approach is the appropriate one, then students would benefit from additional time to allow them to record the problems, use alternative formats, or apply metacognitive strategies that help them to monitor, pace, and amend their solutions.

## Salient Behavioral Postures

During any assessment activity, not only should an individual's particular cognitive postures in mathematics be observed, but salient, "nonmathematical" postures and behaviors that affect performance in mathematics should be noted as well: need for physical movement, tendencies toward perseveration, lack of persistence, attentional deficits, and a lack of self-confidence with respect to mathematical competence.

Some students feel particularly constrained if they do not have an opportunity to move about when participating in mathematical activities. Sometimes they need to stand as well as sit. Moving around allows them to release excess energy and be more attentive to the mathematical task before them. Seat work and a focus on written work often frustrate them. Mathematical experiences that include manipulative approaches, cooperative grouping, or flexible learning environments facilitate the participation and concentration of these students.

Some students demonstrate tendencies toward perseveration in mathematics, that is, when they switch from one task to another, they inappropriately bring to the new activity elements of the previous activity. For example: "If there are six children, how many legs do they have?" The student correctly responds, "12." "Now, for the same six children, how many toes do they have?" The student incorrectly responds, "120," inappropriately bringing the 12 from the previous problem. Other examples of perseveration are seen when students are presented with a variety of arithmetic examples. Often students with perseveration problems will complete all the examples using addition if the first problem requires addition. This posture suggests that the students have difficulty shifting from one procedure to another. In geometric activities, perseveration can be seen when students incorporate elements of previous designs in the particular design before them. In logical contexts, attribute blocks (Marolda 1993) offer particularly useful situations in which tendencies toward perseveration can be identified. Once perseveration is diagnosed, teachers can interpret error patterns in situations where the student is possibly giving correct answers, but to questions different from the ones presented.

Students might also demonstrate a lack of persistence in mathematical situations where they are asked to sustain a concept or procedure over multiple steps, multiple digits, or multiple iterations. Often they can demonstrate the mathematical skill in a direct application but cannot persist in more complex applications unless specific support is provided.

Recently, the issue of attentional deficits has become prominent in the discussion of students with special learning needs. Students who have difficulty attending to the mathematical tasks at hand are often easily distracted from the tasks or attracted to unrelated aspects of their environment. They may also have difficulty maintaining their focus and efforts over a period of time. Attentional issues can be addressed with instructional as well as medical interventions. Students' performance in mathematics can often be maximized when the students are offered explicit structure, increased opportunities for student-teacher interactions, or a small-group setting within a quiet environment.

Finally, students who have had academic failures in the past often approach mathematics with trepidation and hesitation. They lack confidence in their mathematical competence and frequently need external support and encouragement to initiate a response or attempt a strategy.

It should be determined if nonproductive behavioral postures are specific to the task or the topic. Often students demonstrate counterproductive behavioral postures in arithmetic, but the behavioral postures of those same students are dramatically different in perceptual or concrete situations. For example, whereas students may be restless and distractible during arithmetic activities, where they are under stress from the arithmetic procedures, those same students may be well focused and persistent in geometric activities. Behavioral postures may vary from structured to unstructured circumstances. Some postures that can be appropriately contained or managed in the one-to-one setting may not be available in a larger, group situation. The observation and reporting of salient behavioral postures exhibited during mathematical activities should be included in the building of mathematical profiles, since they reflect a student's availability to instruction and participation in mathematics.



In summary, the descriptions of students' mathematical profiles must be broad and multidimensional. They should include cognitive postures inherently called upon in mathematics as well as more general cognitive dispositions involving memory, language, output or production skills, and salient behavioral postures that affect performance. Now that specific features of mathematical profiles that are useful in understanding the *child* in the child/world system have been presented, the mathematical *world* in which these profiles are revealed and assessed must be considered.

## THE *WORLD* IN THE CHILD/WORLD SYSTEM

The *world* of mathematics with all its diversity and with all its richness must be fully reflected in the assessment process, whether that process is undertaken in the classroom or in a separate setting. In the past, assessment in mathematics has been generally confined to arithmetic, the manipulation of quantity (often in purely symbolic terms), arithmetic concepts (often solely expressed in numerical form), and applications that relied on formal arithmetic procedures. As the range of mathematics that students face in their daily lives as well as in the classroom increases, so must the assessment process include a more broadened and enriched view of mathematics. Moreover, it is only when the range of mathematical activity is sufficiently wide that there is allowance for converging evidence—from different contexts—to substantiate the features of students' mathematical profiles.

As the range of mathematics that students face in their daily lives as well as in the classroom increases, so must the assessment process include a more broadened and enriched view of mathematics.

A full range of mathematical possibilities should be incorporated in assessment activities. Content areas should include a comprehensive view of arithmetic concepts and procedures, a wide variety of geometric and spatial experiences, and various problem-solving activities applied to numerical, perceptual, and logical contexts. (See fig. 5.10.)

### Oral Counting Skills

Oral counting skills are generally not included on standard assessment measures. Yet, it is important to recognize that counting skills serve as foundations for later arithmetic skills and that all arithmetic calculations can be alternatively pursued using counting techniques. Thus, the assessment of the extent of students' counting skills becomes essential in formulating their mathematical profiles. The specific counting skills that are of interest include counting forward, counting on, counting backward, and counting by tens, fives, and twos. The assessment of counting skills offers a measure of the foundations underlying many arithmetic topics and also offers indications of alternative routes available to support many arithmetic topics. For example, being able to count on is a useful strategy that can be used to generate solutions to one-digit addition facts; being able to count backward supports the student facing a number of one-digit subtraction facts. Skip counting by tens, fives, and twos allows students to develop effective strategies with which to produce solutions to many one-digit multiplication facts. Skip-counting skills not only often serve as foundations for multiplication concepts but provide students with useful skills for dealing with the topics of time and money.

**Content of Mathematics Assessment**

- Oral counting skills
- Basic quantitative concepts
- Mathematical symbolism
- Arithmetic operations
  - Whole numbers
  - Fractions
  - Decimals
  - Percents
- Informal geometry
- Informal graphing and data analysis
- Problem solving
  - Real-life applications
  - Numerical and arithmetic patterns
  - Patterns in analogies
  - Patterns in sequences
- Logical reasoning

Fig. 5.10. Specific content areas to be included in mathematics assessment. From *Mathematics Diagnostic/Prescriptive Inventory* (Davidson and Marolda 1993)

## Basic Quantitative Concepts

The assessment of basic quantitative concepts should include an examination of the concepts underlying quantity, money, place value, fractional relationships, and time. Concepts such as one-to-one correspondence, the cardinality of a set, the conservation of number, and set-inclusion principles must be assessed to establish that appropriate developmental milestones are available for the pursuit of more formal work with number and arithmetic. In the assessment of quantitative concepts, a variety of models for quantity should be involved, including concrete models using both sets of objects and lengths, pictorial models using pictures of real-life objects and pictures of manipulative models, and symbolic models using numerals and verbal interpretations. When different modalities are available, opportunities arise that reveal the preferred mathematical models with which students interpret quantity. Students' preferred models for quantity should be defined so that quantitative concepts and subsequent arithmetic activities can be interpreted in the most meaningful manner.

Once fundamental appreciations of quantity are established, the consideration of place value becomes a preeminent concern. Place-value appreciations are very important, since they serve as necessary prerequisites to understanding larger numbers and multidigit computational algorithms. Place-value concepts emerge as a result of developmental readiness and the nurturing of intuitions with relevant concrete experiences. The development of place value starts with an appreciation of the one-to-many relationship (i.e., that a single unit or element can stand for a value greater than one), proceeds to the application of this principle to manipulative models or familiar contexts involving money, and finally progresses to the use of place-value considerations in computational procedures and estimation techniques (fig. 5.11).

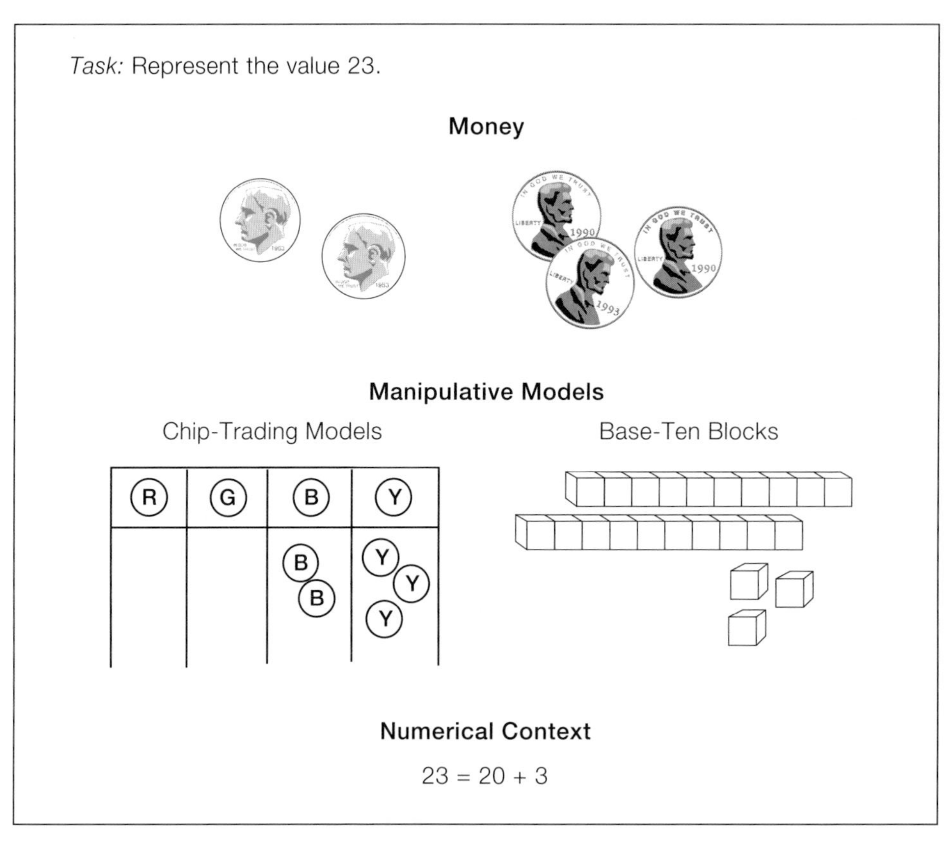


Fig. 5.11. Various interpretations of place value

**Arithmetic Situation: Subtraction with Regrouping**

| 46 | 46 |
|---|---|
| −29 | −29 |
| 23 | 20 |

Fig. 5.12. Error patterns in subtraction with regrouping

Since place-value concepts are necessary prerequisites for all multidigit algorithms, the assessment of students' levels of development of place-value appreciations is a primary interest. For example, the contrast between the two incorrect solutions for the problem "46 – 29 =" in figure 5.12 yields insight into the direction remedial efforts should take.

Both examples suggest that the underlying place-value concepts are not being effectively applied. Whereas the first attempt suggests that the student may have recognized that "nine can't be taken from six," the student resolved that dilemma by merely "taking six from nine," suggesting that position holds little relevance and that appreciation that the operation of subtraction, unlike addition, has a prescribed order has not been established. This student requires further experiences in the basic concept of place value and its role in numerals and numerical operations.

In the second example, however, the student recognized that "nine can't be taken from six because six is not enough." The student then reasoned that "zero" is left and the "rest can't be taken away." This approach suggests that the second student recognizes that the position of the digits is important and cannot be randomly changed. The possibility of regrouping, however, has not yet been consolidated. This student may benefit from instruction focused on regrouping. The student might express the initial quantity, 46, in a concrete modality (with money, trading chips, or base-ten blocks) and then recognize that the quantity 29 must be taken from that collection. In order to "take away" the proper collection, the initial collection must be reinterpreted using regrouping (i.e., trading). After the regrouping approach has been so modeled, the numerical problem can be used as a reminder or record of similar subtraction situations.

Although many tests include an assessment of fraction and decimal procedures, the underlying concept of a fraction is often overlooked. A thorough assessment of the fraction concept is essential in planning instruction in fraction, decimal, and percent topics for students. Students should be asked to interpret fractional models given to them and to represent fractional relationships with models they choose. The models chosen and the ability to shift between various models not only provide a view of the stability of the fraction concept but also offer clues about the preferred model with which the student is most comfortable and would best be taught. Students with Mathematics Learning Style I tend to prefer linear models, whereas students with Mathematics Learning Style II tend to prefer region models. A thorough understanding of a fraction is prerequisite to all fraction operations, and therefore it is important not only to offer a variety of models but also to emphasize those models that are most compatible with a student's preferred learning style (fig. 5.13).

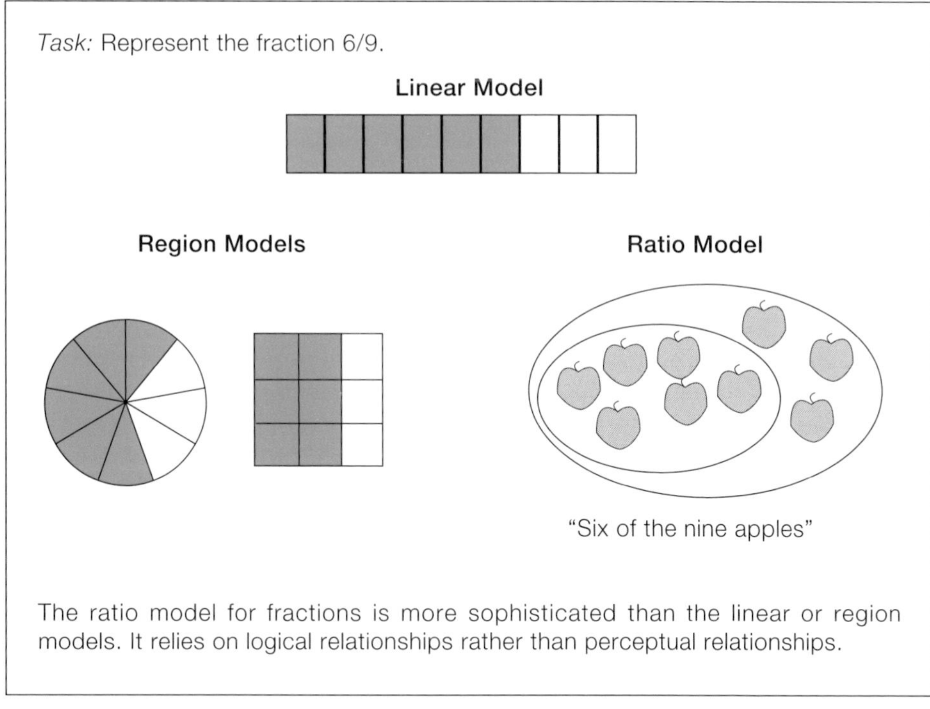


Fig. 5.13. Alternative interpretations of the concept of the fraction 6/9

## Mathematical Symbolism

The use of mathematical symbolism is yet another area that is not comprehensively covered in most assessment instruments. Students' skills in reading and writing numerals must be carefully examined. Difficulty with mathematical symbolism takes many forms in the population of students with special learning needs. The efficiency with which students read numerals is often superior to their proficiency in writing the numerals. The graphomotor requirements in actually producing the written symbols have been discussed earlier in this chapter. Beyond the graphomotor issues of motor programming and motor control, the numerals actually produced must be examined. They can be marked by reversals of digits, reversal in the order of the digits, and difficulties integrating the digits with their respective place values.

Reversals of single digits (fig. 5.14) suggest orientation problems, specifically the organization and orientation of the overall form of the digit. It is not unusual for young children to produce digits that are reversed. Given time and gentle external reminders about the appropriate orientation, these reversals often are spontaneously resolved. Therefore, contrary to some parents' concern, the reversal of digits alone does not indicate a learning disability in mathematics. Except for possible embarrassment to adults, these reversals need not interfere with forward growth in mathematics.

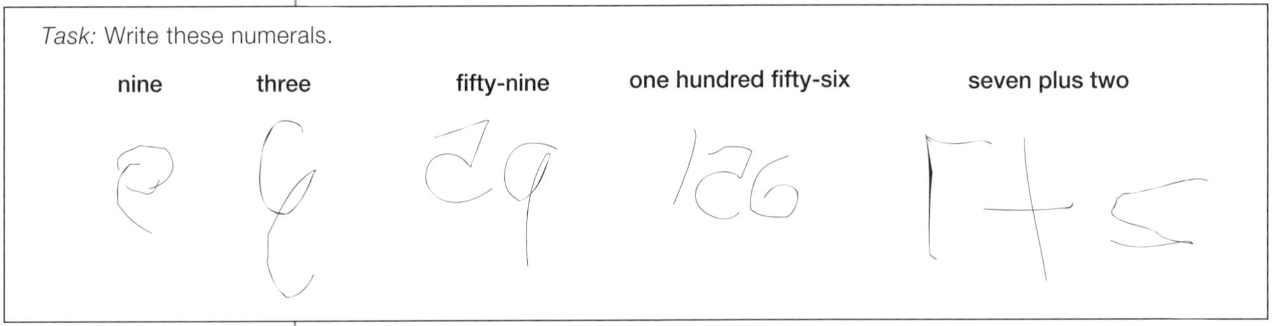

*Task:* Write these numerals.

nine three fifty-nine one hundred fifty-six seven plus two

Fig. 5.14. Sample of students' errors involving reversals of single digits

For some students, reversals are limited only to the sequence of digits in "teen" numerals, where the verbal sequence is misleading, as in "17": we *say* "*seven*teen" with the "7" first but *write* the "7" second, as "17." The reversal of teen numerals is often overcome when teachers point out the inherent idiosyncrasy of these numerals and students adopt compensatory strategies.

For other students, even when the verbal cue is not misleading, reversals in the sequence of digits occur—for example, "forty-seven" written as "74" (fig. 5.15). The reversal in the sequence of any two-digit numeral is often seen in students whose efforts are compromised by sequencing issues. In contrast to

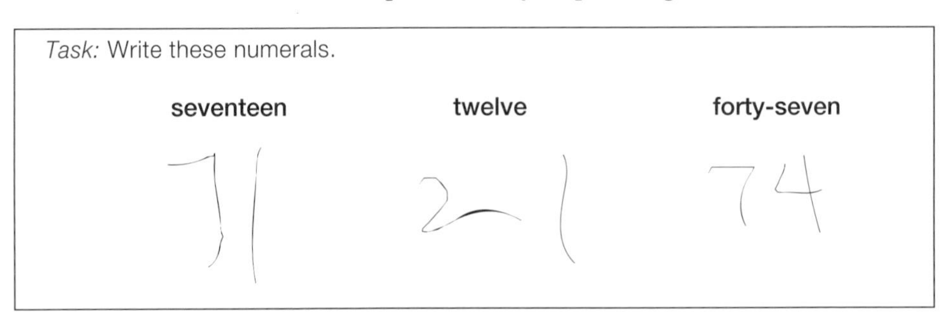


Fig. 5.15. Samples of students' errors involving reversals of the sequence of digits

the reversal of single digits, reversals in the order of digits, when they are either read or written, may interfere with the development of arithmetic topics. For students with these difficulties, special monitoring and corrective strategies must be developed. In time and with repeated practice, students should learn to assume monitoring and corrective strategies themselves.

In writing numerals for larger numbers, students sometimes demonstrate that they understand the place values involved but have difficulty integrating the place-value considerations with the digits in those places. Students who make these errors can sometimes recognize that what they have produced is "too long" and then make a correction. Other students require additional instructional and "rehearsal" opportunities to consolidate the dual aspects involved. (See fig. 5.16.)

*Task:* Write this numeral.

**three thousand two hundred eighty-nine**

Fig. 5.16. Samples of students' errors in integrating digits with their place values

Students' difficulties with mathematical symbolism, particularly in the actual writing of the symbols, are often overemphasized and unnecessarily seen as a barrier to the development of mathematical topics. These difficulties, in fact, sometimes obscure the strengths of students. The lack of skill in dealing efficiently and effectively with mathematical symbolism is not necessarily a reflection of an inability to learn mathematical concepts and procedures. Difficulties with symbolism, however, will affect the ability to produce solutions in the written form. Effectiveness with symbolic requirements in mathematics is an important area for assessment and must be incorporated to determine if students can meet the written demands of the classroom.

## Arithmetic Operations

Arithmetic operations for whole numbers, fractions, decimals, and percents are topics frequently included in mathematical assessments. Typical assessment or testing procedures, however, have focused primarily on the examination of solutions. Such findings can confirm the absence of skills, but they often do not offer information or insights about how or why errors are made. Moreover, results do not directly contribute to the design and implementation of alternative learning experiences for students. It is certainly true that progress to later arithmetic topics requires an understanding of the arithmetic skills that students have already acquired. Thus, it continues to be true that the ability to generate solutions to arithmetic facts and multidigit computation is important to evaluate. The extent of arithmetic skills, as well as estimation skills, however, must be thoroughly probed. The qualitative aspects of arithmetic skills, their flexibility, and their application must also be evaluated. Assessment should evaluate the ease with which students change from one type of problem to another and the degree to which the students can explain or justify their solutions.

Moreover, it is important to assess the process by which the students generate solutions as well as their appreciation of the reasonableness of the solutions they have generated. See, for example, figure 5.17.

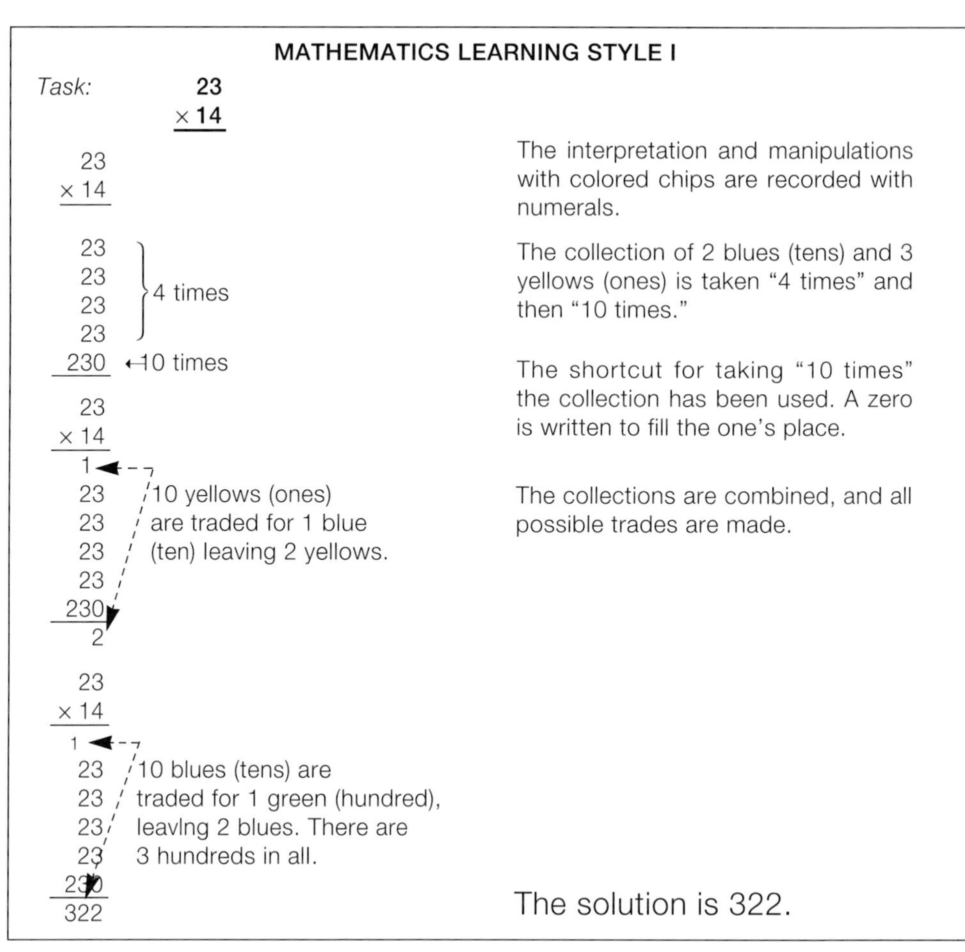


Fig. 5.17. Alternative processing

Sometimes students generate the correct response, but their approach is not the typical one. The approach chosen is often the manifestation, in mathematical terms, of the underlying mode or processing strategies that students prefer. Thus, students should be asked, "How did you get the answer?" The only clue from the student's record of the problem in figure 5.18 that a different approach is being used is the absence of "cross-outs," indicating that regrouping, or "borrowing," was not pursued. The student, in fact, used a global, gestalt approach, typical of Mathematics Learning Style II, and circumvented the need to regroup by proceeding with successive approximations.

Similarly, in fraction, decimal, and percent topics, the qualitative aspects must also be considered. For the student with special learning needs, it is important to assess these topics in real-life applications such as cooking, carpentry, sewing, and money applications as well as in formal algorithms. For the student with special learning needs, these topics may only be available and best elicited in relevant and meaningful contexts.

Finally, students' arithmetic performance must be viewed according to whether the student appreciates the reasonableness of the solution pro-

**MATHEMATICS LEARNING STYLE II**

*Task:* 23 × 14

The product is conceived as the area of a rectangle with dimensions 14 and 23 to be represented by base-ten blocks. The region is then covered with the fewest number of blocks. The four separate rectangular regions that result correspond to the four partial products.

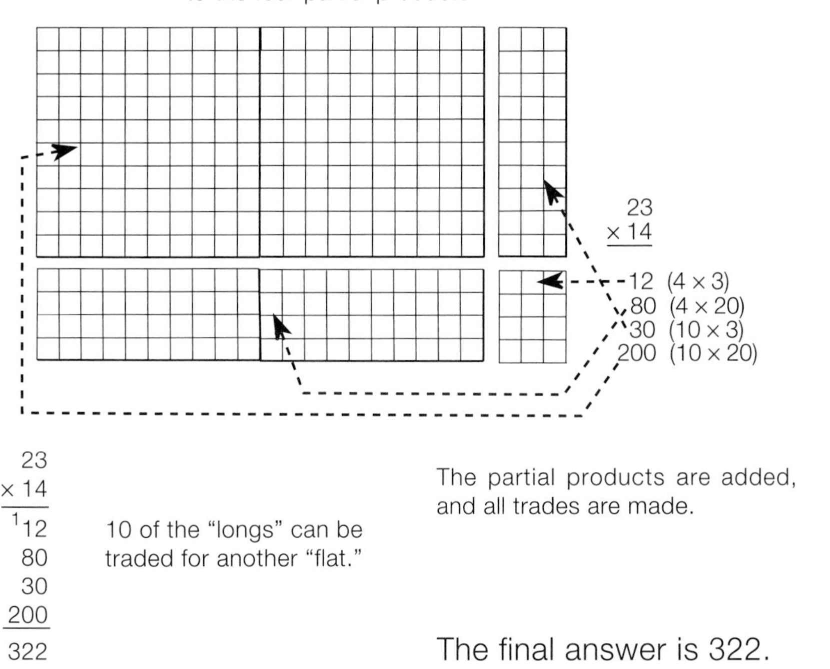


$$\begin{array}{r} 23 \\ \times\ 14 \\ \hline {}^{1}12 \\ 80 \\ 30 \\ 200 \\ \hline 322 \end{array}$$

10 of the "longs" can be traded for another "flat."

The partial products are added, and all trades are made.

The final answer is 322.

approaches to multiplication

duced. For example, when dealing with the problem 1/2 + 1/3, the solution 2/5 is frequently presented. Some students may not recognize that 2/5 is an inappropriate solution, protesting, "That's how I thought I was supposed to do it." Other students recognize immediately that the solution is incorrect, since 2/5 is less than 1/2.

## Informal Geometry

Concepts of informal geometry are often absent from standardized assessment instruments. Not only has the *Curriculum and Evaluation Standards for School Mathematics* (National Council of Teachers of Mathematics 1989) emphasized the need for more spatial and geometric experiences in the mathematics curriculum, but the spatial and perceptual domain often offers students with special learning needs a broader base of topics in which their strengths can be revealed and celebrated. Moreover, strengths in spatial and perceptual skills may suggest alternative pathways through which arithmetic topics can be developed.

Geometric topics assessed should not be limited to the recognition and naming of two-dimensional geometric shapes. Instead, students should be

Fig. 5.18. Alternative approach to traditional arithmetic approaches

asked to recognize, copy, and maneuver both three-dimensional and two-dimensional geometric figures. Students' performance in geometric activities provides further opportunities in which aspects of their mathematical profiles can be revealed. When asked to produce geometric designs from memory, some students build carefully in a piece-by-piece fashion until the overall design emerges (Mathematics Learning Style I), whereas other students first approach the overall designs, with successive refinements until the exact design emerges (Mathematics Learning Style II). See figure 5.19.

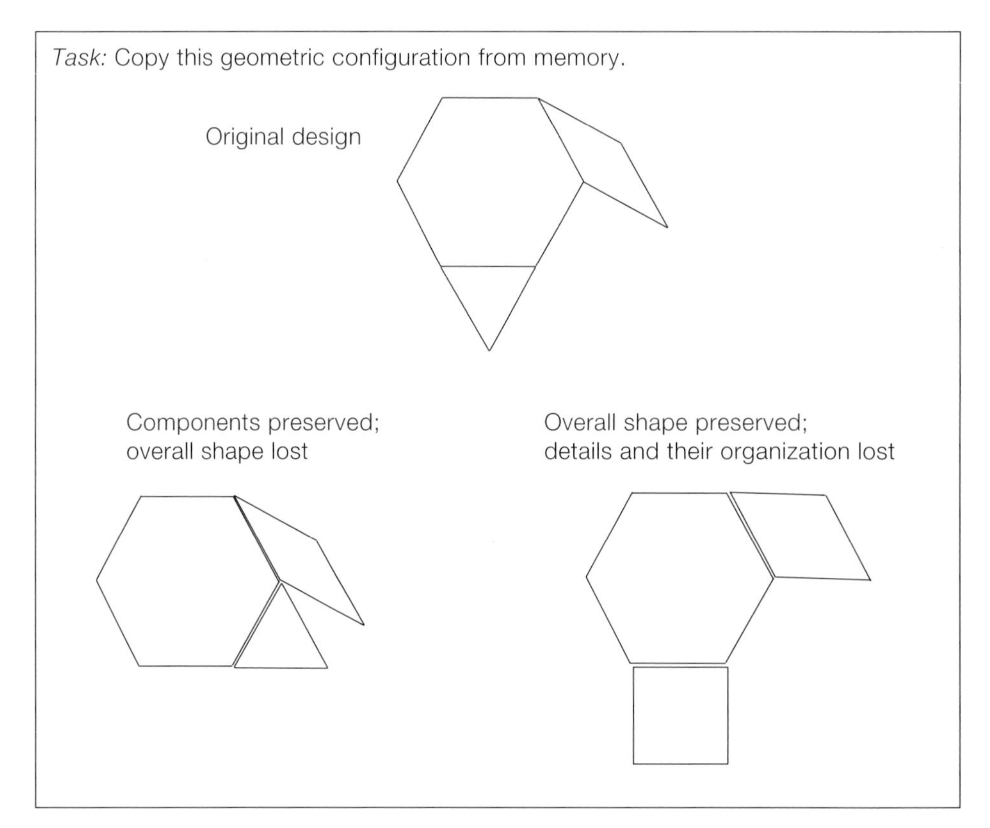


Fig. 5.19. Alternative approaches to geometric configurations

Activities with two-dimensional and three-dimensional geometric configurations also furnish opportunities to examine available organizational and integration skills. Though these approaches are manifest in the geometric domain, they also suggest the effectiveness of approaches that students would bring to arithmetic circumstances.

## Informal Graphing and Data Analysis

The inclusion of informal graphing and data analysis in assessment activities parallels the inclusion of such topics in the regular curriculum and the increasing demand for these skills in real-life situations. Moreover, these topics serve as a context in which basic problem-solving skills can be elicited. Students should be given graphs to read and interpret so that they can demonstrate how well they can use symbolic representation for real-life situations and logical reasoning to extrapolate conclusions from data. Students should be given bar graphs, line graphs, charts, and maps from which they can discern relationships and form opinions. Assessment of these topics should focus on the quality of the interpretations and the sophistication of the conclusions generated. Success in these activities would suggest that students have strengths in graphing and data analysis skills that will be available to them in other subject areas as well as in life and workplace situations. Moreover, success in graphing activities would suggest that students' strengths in problem solving are mobilized when data are interpreted pictorially and are explicitly structured.

## Problem Solving and Logical Reasoning

The *Curriculum and Evaluation Standards in School Mathematics* (National Council of Teachers of Mathematics 1989) has emphasized the importance of problem solving across all domains of mathematics. Students' effectiveness in drawing conclusions, appreciating relationships, and posing new questions in a variety of problem-solving situations should be assessed. Real-life applications should be included, but they should represent relevant situations from which students must identify relevant information, discard superfluous data, choose an approach, and follow through to a conclusion. Students should also be asked to discern patterns and relationships governing analogies, sequences, sets of elements, and intuitive functions. Teachers should observe whether students can continue a pattern or relationship presented to them and describe it in more general terms. Patterns and relationships can be represented with concrete models, such as attribute blocks (Marolda 1993), by pictorial representations, or in symbolic form. In both the teaching and the assessment process, analyzing patterns and relationships enhances logical thinking skills and is, therefore, a most worthwhile activity.

After the student has been assessed in the broad range of mathematical content, it is important to evaluate the domain of mathematical activity that is most comfortable for the student. Too often students with special learning needs report that "I can't do math" when in fact they mean they have difficulty with arithmetic. They do not realize that mathematics encompasses much more than arithmetic. It is especially important for all students to recognize that they can participate in mathematics but that there may be a particular type of mathematics that is more suited to their learning predispositions. Despite challenges in arithmetic activities, topics in geometry or logic may offer opportunities for successful mathematical experiences.

It is especially important for all students to recognize that they can participate in mathematics but that there may be a particular type of mathematics that is more suited to their learning predispositions.

Assessment activities can be conducted in one-to-one situations or with small groups. They can be pursued in single or multiple sessions. The one-to-one setting, with multiple sessions and flexible time allowances, often offers the optimal opportunities to observe students' strengths and vulnerabilities. Assessment need not be restricted to separate settings, however. Opportunities for assessment can emerge in the classroom where the interface of *the child* and *the world of mathematics* is dynamically portrayed.

## INTERFACE OF THE CHILD/WORLD SYSTEM

Once the *child* and the *world* aspects of the child/world system have been delineated, a final consideration must be directed to their interactions, with a view to occasions of "match" and "mismatch." The challenge to teachers and evaluators of students is to achieve a comprehensive understanding of these interactions with a view to maximizing the matches and minimizing the mismatches.

When assessment is conducted in a separate setting, the evaluator primarily focuses on the definitions of *match* and *mismatch* that characterize the mathematical profile of a specific student. Once the mathematical profile is defined, the teacher must respond by designing appropriate instructional experiences that are sensitive to the learning differences and the special learning needs that are defined.

Assessment can also occur within the context of classroom work. Although the previous sections of this chapter have described specific mathematical situations in which the *child* and *world* are revealed, it is important to recognize that those occasions are not limited to specific testing sessions but are often found in the regular curriculum of the classroom. The classroom teacher must recognize those opportunities where the teaching experience can be used as part of the assessment process as well. In the classroom, the teacher must present—and then capitalize on the variety of—ongoing mathematical experiences in both a diagnostic and a teaching fashion. The particular presentations of topics should be designed so that they reflect the various contexts of mathematics—specifically, arithmetic, geometric, and logical activities—and are also sensitive to the varied learning styles the students bring to mathematics. Lessons should be presented and observed with a special view to both the preferred modalities with which concepts are interpreted and the preferred processing styles in which topics are pursued. In order to convert classroom activities to assessment opportunities, the teacher must find ample opportunities to observe students, record the qualitative as well as the skill aspects of their performance, and compare the students' effectiveness over a variety of mathematical experiences. The key to assessment in any setting, however, is the opportunity to observe carefully, record those observations fully, and then interpret the observations within the context of a relevant and meaningful cognitive model.

In order to convert classroom activities to assessment opportunities, the teacher must find ample opportunities to observe students, record the qualitative as well as the skill aspects of their performance, and compare the students' effectiveness over a variety of mathematical experiences.

Once mathematical profiles have been generated and the particular learning approaches of students with special learning needs are defined, approaches and strategies can be designed to make use of the areas of relative strengths and minimize the instances of relative vulnerability. Two contrasting student protocols are shown in figures 5.20 and 5.21 to illustrate the dynamics of the child/world system and demonstrate how to translate findings from assessment experiences into more effective classroom participation for the students with special learning needs. The characteristics of students with Mathematics Learning Style I, as well as those with Mathematics Learning Style II, are described in both general and mathematical terms. The protocols, which were generated by teachers from their classroom observations, describe the nature of the *child*, the quality of the child's interaction with the *world* of mathematics, and some specific teaching and learning implications of the matches and mismatches that result.

| Cognitive and Behavioral Correlates | Mathematical Behaviors | Teaching and Learning Implications and Strategies |
|---|---|---|
| Highly reliant on verbal skills | Concepts suffer in the course of calculations | Emphasize the meaning of each concept or procedure in verbal terms |
| Sequencing skills available | Approaches situations as recipes | Encourage students to use subvocalization to direct procedures |
| Not facile at appreciating relationships | Problems arise when there are too many steps or recipes | Use discrete (set) models for quantity |
| Translates perceptual situations into verbal terms | No awareness if recipe chosen is inappropriate | Have students keep a notebook of important mathematical "recipes" using their own words |
| Uncomfortable with demands for quick assessment of situations | Difficulties selecting appropriate procedures from repertoire of learned skills | Encourage students to interpret and restate problems verbally |
| Difficulty with versatility | Estimation skills elusive unless strong format is presented | Model complex problems with similar problems in simpler terms |
| Intuitions or insights not readily available | Fraction procedures available, despite instability in fraction concepts | Give explanations before or after procedure, not while student is working |
| Difficulty with generalizations | "Talks through" all procedures | |
| Prefers "How?" to "Why?" | Vulnerable under perceptual demands | |
| Demands THE approach | | |
| Doesn't "daydream" enough | | |
| Deliberate, cautious style | | |

Fig. 5.20. Protocol of child/world system for Mathematics Learning Style I (from Davidson and Marolda [1993])

Assessment experiences designed for students with special learning needs should serve to define and then strengthen the interface of the child/world system. Effective assessment in mathematics must closely examine students' achievement and learning approaches and lead to personalized recommendations for specially tailored instructional materials and strategies. With mathematical profiles so defined, topics can be selected with respect to those in which students have particular strengths and adjusted for those in which students have particular vulnerabilities. Comprehensive and carefully designed assessment efforts are mandatory in order to insure that findings emanating from them lead to the design of mathematical approaches that help achieve the mathematical potential of students with special learning needs.

| Cognitive and Behavioral Correlates | Mathematical Behaviors | Teaching and Learning Implications and Strategies |
|---|---|---|
| Perceptive | Arithmetic solutions imprecise but always in the ballpark | Introduce perceptual models for calculation |
| Intuitive | Uses estimation strategies spontaneously | Encourage estimation |
| Prefers perceptual stimuli and often reinterprets abstract situations visually or pictorially | Generates arithmetic facts from related facts (e.g., doubles) | Encourage a variety of models |
| Solutions have good "flavor" but are not precise | Can appreciate a choice of approach in word problems but has difficulty in following through | Require and reward general solutions as well as precise ones |
| Appears to be able to think, but performance makes you wonder | Likes informal geometry | Encourage students to draw pictures to interpret word problems |
| Rather low achievement but described as "can do anything" | Resists paper and pencil; "careless errors" | Help students to create their own troubleshooting notebooks of personal demons and typical examples |
| Impatient with one topic; needs constant change | Impulsive in responses; second solutions better | Introduce geometric topics as a vital part of the curriculum |
| Easily draws conclusions, forms opinions, likes discussions | Recognizes incorrect answers but can't correct errors | Maintain a good sense of humor in response to apparently casual style |
| Does not proofread or rewrite | Excels with manipulatives | |
| Likes to set up experiments but not record results | Reports, "I like math, but I'm not good at it." | |
| Tends to work backward on multiple-choice tests | | |

Fig. 5.21. Protocol of child/world system for Mathematics Learning Style II (from Davidson and Marolda [1993])

## REFERENCES


Bernstein, Jane Holmes, and Deborah P. Waber. "Developmental Neuropsychological Assessment: The Systemic Approach." In *Neuromethods*, edited by Alan A. Boulton, Glen B. Baker, and Merrill Hiscock, pp. 311–71. Clifton, N.J.: Humana Press, 1990.

Davidson, Patricia S. *Mathematics Learning Viewed from a Neurobiological Model for Intellectual Functioning.* 2 vols. Washington, D.C.: National Institute of Education, 1983. NIE Grant no. NIE-G-79-0089. ERIC Document Reproduction Service nos. ED 239914 (vol. 1) and 239915 (vol. 2).

Davidson, Patricia S., and Maria R. Marolda. "Mathematics Diagnostic/Prescriptive Inventory." Unpublished manuscript, 1993.

Gregory, Richard. *The Oxford Companion to the Mind.* New York: Oxford University Press, 1987.

Holmes, Jane M. "Testing." In *Assessment of Developmental Disorders: A Neuropsychological Approach*, by Rita Rudel, Jane M. Holmes, and Joan Rudel Pardes, pp. 116–201. New York: Basic Books Publishers, 1988.

Loviglio, Lorraine. "Mathematics: A Tale of Two Hemispheres." *Massachusetts Teacher* (January/February 1981): 8–12.

Marolda, Maria. *Activities with Attributes.* Palo Alto, Calif.: Dale Seymour Publications, 1993.

National Council of Teachers of Mathematics. *Curriculum and Evaluation Standards for School Mathematics.* Reston, Va.: The Council, 1989.

———. *Mathematics Assessment: Myths, Models, Good Questions, and Practical Suggestions.* Edited by Jean Kerr Stenmark. Reston, Va.: The Council, 1991.

Piaget, Jean. *Child's Conception of the World.* Totowa, N.J.: Littlefield Adams, 1965.

Stenmark, Jean Kerr. *Assessment Alternatives in Mathematics: An Overview of Assessment Techniques That Promote Learning.* Berkeley: EQUALS, Lawrence Hall of Science, University of California, 1989.

**Maria Marolda** and **Patricia Davidson** have collaborated on curriculum development and staff development in mathematics since 1968 and on assessment research and diagnostic testing since 1977. Marolda is senior mathematics specialist in the Learning Disabilities Program of the Department of Neurology at Children's Hospital in Boston. Davidson is professor of mathematics and director of the Critical and Creative Thinking Graduate Program at the University of Massachusetts Boston, and has been a chief examiner in mathematics and chair of the Examining Board for the International Baccalaureate diploma program for the past five years.

MRC
M–
M+
×
7
9
+
6
2
3
=

6

# Planning for Instruction

## The Individualized Education Plan and the Mathematics Individualized Learning Plan

*Carole Greenes*
*Frank Garfunkel*
*Melissa DeBussey*

FOR years, planning for instruction for students in mathematics (as well as in other content areas) was prescribed by curriculum developers in the form of invariant sequences of topics. This procedure was based on the belief that there was a single correct way to learn a subject. For example, the study of subtraction always followed the study of addition; division always followed multiplication; algebra always followed arithmetic.

Teachers did their best to present concepts and skills in ways that would make sense to the greatest number of students but rarely strayed from the scope and sequence of the textbook. All lesson and unit plans were subject-centered and assumed that students were motivated to learn, were attentive, had appropriate levels of skill and knowledge, and—above all—possessed a clear understanding about the role of schooling in their lives. These were the givens in regular education.

Students without motivation, appropriate levels of skills and knowledge for their age and grade, and normal means of communication were either excluded from regular education or relegated to its back alleys—special classes or special schools—out of sight, out of mind. There was never any question about who or what was at fault. It was not the schools, the teachers, the teaching, the materials, the textbooks; it was the students who did not fit. It was the students who needed to be changed.

Specially designed instruction meets the unique needs of special education students.

This lack of "fit," resulting in the exclusion of school-aged children from public schools and from most schooling, was the major impetus for the development of child-centered learning plans—for Individualized Education Plans (IEPs). In the first section of this chapter, the rationale for the IEP is present-

IEPs do not address the needs of the gifted.

ed along with descriptions of its content and its development and impact on education, particularly mathematics education. IEPs are intended for special education students, are mandated by law, and are funded by specific programs.

However, IEPs do not address the needs of other students who require tailored educational experiences—students such as the gifted, those at the lower range of "normal" achievement, and those with specific gaps in need of remediation. For these students, as well as those qualifying for IEPs, we introduce and describe the Mathematics Individualized Learning Plan (MILP) in the second section of the chapter. The MILP is a new model currently being developed for use with students in the Chelsea Public Schools in Chelsea, Massachusetts, by Carole Greenes and Melissa DeBussey, with funding from the Polaroid Foundation.

## THE INDIVIDUALIZED EDUCATION PLAN (IEP)

The IEP was designed to remedy what the courts (Public Law 94-142, The Education of the Handicapped Act of 1975) had concluded was a denial of equal protection without due process. This fundamental principle of American

law, which was inscribed in the Fourteenth Amendment to the Constitution, is the basis for the development of the IEP construct. The interpretation of this principle was that schools had to change so that they could and would provide *all* students with appropriate education. The vehicle for this change was to be the IEP, a written document approved by the school and the parents, that would specify how the schools would change to meet the needs of the students.

The Individualized Education Plan (IEP) construct, as it has been described in judicial opinions and in statutes since the mid-1970s, is much more than what the words denote. It is "individual," but it includes other students, teachers, related services, and facilities; it is "education," but it includes related services that are necessary in order for education to take place; and it is a "plan," but it includes material about the student's past, present, and future.

## The Content of the IEP

The IEP connects children with necessary services.

The IEP has become the linchpin that connects children with special needs, curriculum, services, parents, teachers, and administrators. In practice the IEP is used in a great variety of ways, but it has enormous potential for transforming the nature of education for special students because of the ways in which it ties the assessment of performance to instruction and establishes a basis for school accountability for student learning. The Individuals with Disabilities Education Act of 1991, or IDEA, specifies that the IEP include the following:

- A description of current levels of educational performance
- Annual goals and short-term objectives
- Educational services to be provided

Many states have elaborated on the IDEA specifications and augmented the list. For example, in Massachusetts (Massachusetts Department of Education 1986) the IEP must include descriptions of—

- measurable physical constraints on performance;
- learning style;
- methodology and teaching approach;
- specialized materials and equipment;
- physical education services;
- parent-child instruction;
- the location of services to be provided;
- daily and yearly schedules;
- provision for participation in regular education;
- transportation requirements;
- the schedule and nature of periodic formal reviews.

An example of an IEP for a student with difficulties in mathematics is included in appendix C. The student, Eddie, was retained twice at the grade 5 level because of weak skills in language arts and mathematics. The IEP proposes placing Eddie in a grade 7 classroom in a K–8 elementary school and having him participate in a pull-out program that focuses on reading, written language, mathematics, and study skills. Specifically, Eddie will receive special education instruction in mathematics and reading/language arts for two periods a day (80 minutes) in the school's learning center for students with special needs. As can be seen in the IEP objectives and as is characteristic of most such plans currently, the content of the mathematics is *primarily computation*—though in Eddie's situation some attention to the development of a "growing ease with word problems" is also prescribed. The sections that follow furnish insights into the IEP process and provide ideas for broadening the scope and quality of individualized plans for mathematics.

## The Development and Impact of the IEP

An IEP is developed by a team of professionals and parents after a comprehensive multidisciplinary evaluation of the child has been completed. The IEP educational goals and objectives are either designed by the IEP team or selected from templates of predeveloped IEPs that most closely conform to the student's needs. Computer programs exist that facilitate this process. In this event, canned goals, objectives, performance levels, and assessments are selected by number and transformed into personalized prose statements.

Some programs may be commercial packages with preset compendia of goals. Other programs may be customized or developed by schools or school districts to match their particular goals. Regardless of how the IEP is developed (manually, from a standard package, or from custom software), the importance of accurately defining individualized goals, methods for helping the student achieve those goals, and assessment strategies for documenting progress toward meeting the goals cannot be overemphasized. If IEPs can be criticized, it is generally when these tasks have not been adequately addressed.

Crucial to an understanding of the IEP construct is how it fits into the due process entitlement of children with disabilities: it is the fulcrum of that entitlement. Parents (or in some instances, adolescents) can reject the development of the IEP, or they can reject the IEP at any time. Schools cannot politely reply to the objections of parents by acknowledging and then ignoring them. Schools are required to respond, and if parents are not satisfied with the response, school administrators must argue their case before an impartial hearing officer (HO) in an administrative hearing. If the resolution of the hearing is not satisfactory to either the parents or the schools, the decision reached at the hearing can be appealed in state or federal courts.

Parents must be well informed and educated about the goals, objectives, and approaches of the IEP process.

Because of the mandated role for parental participation, it is crucial that parents be well informed and educated about the goals, objectives, and approaches of the IEP process in order to enlist their support. Although the IEP is not a binding legal contract between the school district and parents, it is a legal agreement that is made by the school district with the parents to provide services for a child. To date, school districts have not been fined or punished in other ways for not providing services called for in IEPs, although they often have had to pay parents' attorney's fees. However, it is not uncommon for school districts to be brought before hearing officers in compliance hearings, where the HO orders the school district to comply with the HO's rulings.

Thus, it should be clear that the IEP is not just an "educational plan" that well-meaning and skilled teachers develop and put into place in order to facilitate their teaching and the learning of their students. The IEP must specify objectives, goals, and time schedules for achieving them. But the added dimensions—the formal parent involvement, advocacy, periodic review, due process, and service entitlements—bring two additional factors into the educational equation: accountability for the success of the individual child, and the burden of developing programs and services that will enable *all* children to benefit from education.

There is no guarantee that parents will use the process or, if they use it, that it will be used effectively. But both state and federal regulations make it clear that parents "have the right to be present at and participate in *all* meetings of the team in which the IEP is being developed and written" (Massachusetts Department of Education 1986, p. 35). The process is there to be used, and consequently it has the potential for playing an integral role in the development of the curriculum. Because this process calls for changes in schools to meet the special needs of certain students, it also has the potential for having major impact on the total curriculum of the school.

The teaching of mathematics becomes different with IEPs, as does all teaching. Parental participation in developing the IEP may require that parents

understand the content, the pedagogy, and the rationale for both. If they do not understand, there is a mechanism they can use to help them—the rejection of the IEP. When this occurs, a hearing officer may require, in writing, a description of the content or teaching methods, or the rationale for either or both. The added dimension—structures and procedures for accountability—transforms education. The power of professionals to decide things unilaterally is significantly reduced; the power of parents to question teachers and schools is enhanced.

Structures and procedures for accountability transform education.

This raises critical educational and pedagogical issues. IEPs are often reviewed by local special education administrators, by state educational agencies, by advocacy organizations, and by parents in order to assess the level of the school district's compliance with these IEPs and with state and federal special education statutes and regulations. There are two ways to review the quality of IEPs. The first is procedural: Has the process been honored? Do parents know about their rights? Have the teachers (classroom *and* content specialists), parents, psychologists, and other relevant professionals participated? Are there connections among the initial referral, the evaluation, the curriculum, and the instruction?

The second is substantive: Does the IEP make sense for the student? Is it a bridge that connects where the student came from, where the student is, and the goals for the student? Are goals that are commensurate with objectives considered worthy by education experts today? With respect to mathematics, does the IEP reflect goals and objectives highlighted in the NCTM *Curriculum and Evaluation Standards for School Mathematics* (1989)?

Are mathematics teachers who understand these Standards involved in planning the IEP, which will further insure that goals linked to rich mathematical experiences beyond computation are included? Are mechanisms identified for assessing student progress toward these goals? Does the IEP relate mathematics education to the student's reading ability? To science understanding? To other curricular areas? Is the IEP coherent? Is it intelligible?

## IEPs and Mathematics Subject Matter Plans

Traditionally, what mathematics students are taught in grades K–12 has been largely determined by the textbook (Mathematical Sciences Education Board 1990). Even in classrooms with well-prepared, creative teachers who are willing to depart from the text, the departures have generally been in the form of activities, investigations, and the use of models and materials to enhance the acquisition of the textbook's skills or concepts—*not* departures from the content or the order of the presentation of the content.

Textbook lessons are usually synonymous with daily lesson plans. The scope and sequence in mathematics textbooks and workbooks has been determined by education specialists—educators and mathematicians—in accordance with assumptions about how *most* students can learn the subject matter. The organization of textbooks promotes the notion that learning mathematics is linear; that in order to be successful, all students must proceed through the text, page by page, and master each skill or concept before progressing to the next. Underlying the use of textbooks is the belief that all or most students will be able to learn the subject matter in the same order and in the same way.

The IEP construct takes a different tack. The IEP assumes that different students learn in different ways, many of which may be at variance with the way subject matter is traditionally taught, organized, and presented in textbooks. Therefore, based on a multidimensional evaluation of the student, alternative sequences, instructional strategies, and schedules must be developed. The IEP is an alternative to the norm, to the textbook way. It becomes important when educational policy dictates that we deal with students who cannot "make it" within the normative or standard way of teaching.

The IEP assumes students learn in different ways.

When students do not learn the material being presented, a common assumption is that they are not bright enough, are lazy, are not doing the required work, do not pay attention, or are not motivated. Rarely is it assumed that the material is not being taught appropriately—that the fault lies with the teacher and the teaching. The IEP can be viewed as a mechanism that realigns the teaching-learning process by placing the burden on the teacher and the school rather than on the student.

It is not surprising that the process of designing alternative systems for teaching and learning requires a different list of players from that used for textbooks. The alternative, represented by the IEP, is to be based on evaluations from professionals, who are not ordinarily involved in educational planning, and from parents, who are not usually involved in curriculum planning.

## Pull-Out and In-Class Models of Instruction

Over the past fifteen to twenty years, the controversy about "pull out" programs for special-needs students has been increasing. When schools first

began to assume responsibility for some children with disabilities, the general practice was to remove them from regular classrooms and place them in separate classrooms, usually in separate buildings. The ostensible rationale for this practice was that separate classes and schools could provide specialized teaching and learning environments that would benefit students who could not learn in regular classrooms. The benefits of such removal and placement, however, have never been demonstrated.

More recently, the practice of placing students with special needs in separate classrooms with special education teachers—resource rooms or learning centers—for only part of each day has become popular. It was, and still is, argued that the protected environment of a resource room or a learning center will maximize the abilities of students who cannot benefit from regular classrooms. Again, as with separate classrooms and schools, the benefits of partial pull-out programs have not been demonstrated.

Given the failure of proponents of separation to demonstrate that removal for all or part of each school day is beneficial, opponents argue that being pulled out of the regular classroom, in the absence of strong evidence favoring such separation, is unjustifiably discriminatory. Further, they claim that the removal and separation of children with disabilities is not materially different from removing and separating children on the basis of race or ethnicity.

From an educational perspective, it has also been argued that pulling children out of regular classroom activities is disruptive both to the children being pulled out and to other students in the class; that education in separate environments with different teachers leads to the discontinuity of instruction; and that leaving the classroom is stigmatizing to children who are removed (Routman 1991).

An alternative model for providing individualized attention to children with special needs is to offer that assistance in the regular classroom by having regular and special education teachers coteach, by involving students in peer teaching, and by engaging students in cooperative learning activities. This model has a number of possible advantages. From the perspective of the special education student, the special education teacher may be seen as part of the regular learning environment, someone to go to for help when needed.

Individualized attention can be provided through coteaching, peer teaching, and cooperative learning.

This ability to get help on the spot places much of the responsibility for when and how to get help on the student instead of its being dictated by an arbitrary schedule. Working together with the same students means that the teachers, regular and special, would be able to group students on solid pedagogical grounds rather than on the suspect distinction between students with and without disabilities (Robinson 1991).

Working together, the regular and special education teachers would be able to goup students on solid pedagogical grounds.

There are other compelling advantages to an alternative model that views teaching and learning as cooperative and collaborative activities. Teachers will have opportunities to view each other working with a wide range of students, increasing the likelihood that difficult-to-teach students will be taught in ways that are conducive to their learning and that teachers will not be trapped by overly narrow ways of viewing and working with their students.

## THE MATHEMATICS INDIVIDUALIZED LEARNING PLAN (MILP)

Lack of motivation, membership in a peer culture characterized by anti-intellectual values, personality factors that interfere with the relationship between pupil and teacher or between pupil and pupil, and inadequate programming of instructional sequences—all can contribute to unsatisfactory school learning in mathematics. Needless to say, the causes for students not learning mathematics are often extremely complex and must be assumed to be a function of both the environment (teacher, teaching, school, culture) and the student. Regardless of the cause(s), the development of Mathematics Individualized Learning Plans (MILPs), which identify students' strengths and weaknesses and their educational needs and recommend appropriate instructional sequences, experiences, and materials, would be beneficial.

Whereas the IEP is not intended for gifted and talented students, the MILP is appropriate for such students. There are students who effectively "drop out" of mathematics education because the curriculum does not provide sufficient challenge to maintain and enhance their interests or expand their repertoire of knowledge, skills, and processes. High achieving, talented students, as well as those at the other end of the achievement spectrum, would profit from alternative programming.

## The Development of the MILP

As with IEPs, the creation of an MILP can be done by hand or with the aid of computer programs. For special-needs students, the MILP may serve as a part of an IEP. For students at risk of being placed in special education, the MILP may be treated as a stand-alone educational document. For special-needs students and low achievers in mathematics, the MILP should be developed jointly by the teacher, who is the content expert; the special educator, who knows strategies of assessment and pedagogy; the parents, who will support the learning process; and if the problems are complex, other professionals, such as, for example, neurologists.

For high achieving students, the MILP should be developed by the teacher with subject matter expertise. At the elementary school level, the content expert may not be the regular classroom teacher but rather the school system's mathematics specialist. With high functioning students, the MILP planning team might consist of four members: the classroom teacher, the mathematics specialist, the program director for gifted and talented, and the parent.

## Components of the MILP

The Mathematics Individualized Learning Plan has five components: (1) curriculum objectives; (2) strategies for assessing achievement of the objectives; (3) instructional materials for introducing, maintaining, and enriching an understanding of the mathematical concepts, skills, and strategies identified in the objectives; (4) pedagogical approaches that foster the acquisition of concepts, skills, and strategies; and (5) links with other subject areas.

### *Curriculum objectives*

Fundamental to a learning plan in any subject area are the criteria against which learning can be measured. These criteria should be in the form of clearly specified curriculum objectives or goals (i.e., what we want students to know and to be able to do). In 1980 in *An Agenda for Action* and again in 1989 in its *Curriculum and Evaluation Standards for School Mathematics*, the National Council of Teachers of Mathematics stated that the focus of school mathematics, at all levels, should be on reasoning and problem solving and not on arithmetic computation and the manipulation of algebraic symbols. Students should learn to reason mathematically to be able to bring to bear, at the appropriate time, those mathematical skills, concepts, and strategies necessary for the resolution of problems. This new focus on problem solving and reasoning demands new and varying types of curriculum objectives. Some objectives will be process oriented—for example:

- Organize information in a table.
- Draw inferences from measures of central tendency (mean, median, and mode) and range.

Other objectives will focus on the application of concepts, skills, or factual information:

- Identify varying collections of bills and coins having the same value.
- Formulate problems from data obtained from an experiment.

Still other objectives will focus on the understanding of mathematical concepts or terminology:

- Compare and order three- and four-digit numbers.
- Use the formula to find the circumference of a circle.

### *Strategies for assessing student achievement*

The importance of having strategies and materials for ascertaining what students know and the degree of their "knowing" (i.e., how well they know it) cannot be overemphasized. Often students proceed through school without any sense of how well they have grasped concepts and skills. Often too, teachers and parents do not have a clear idea of what their children have achieved to date or what they are capable of doing. Without such knowledge, teachers cannot plan appropriate instruction, teachers cannot individualize learning, and parents cannot assist with the learning process.

Objectives may require alternative types of assessment.

***Assessment strategies determined by objectives.*** Because of the varying nature of learning objectives, they may require different types of assessment. The achievement of some objectives may be evaluated by using paper-and-pencil techniques, such as multiple-choice and short-answer questions, essays, and reports. The multiple-choice test format may be used, for example, to assess students' abilities to "identify sets of equivalent fractions." The short-answer format may be used to determine if students are able to "write large numbers in scientific notation form." The essay format may be used to ascertain students' abilities to "list outcomes from experiments and use these data to make predictions" or to justify their solutions to complex problems. The report format may be used to gain insight into students' understanding of the "characteristics of misleading graphs."

Other objectives may lend themselves to assessment through some form of artistic expression, such as acting out the solution to a problem, drawing a representation of a neighborhood, using computer graphics to construct representations of three-dimensional figures, or carrying out a simple experiment to see whether, for example, heads are more likely to appear than tails in a trial of fifty tosses of a coin.

Some assessments should be made of students as they work individually; others should examine how students work in groups and contribute to the efforts of the groups. Learning mathematics is basically a social activity. Students interact with teachers and other experts as they explore topics and gather and analyze information. Students interact with one another as they

Learning mathematics is a social activity.

discuss, brainstorm, wrestle with ideas, and conduct experiments, investigations, and surveys to solve problems. An analysis of behavioral patterns during group activities may indicate why certain children are not learning as well or as quickly as they should.

***Assessment strategies determined by student input and output: modality preferences.*** Current assessment processes, such as standardized achievement and end-of-chapter tests, often do not accurately portray what a student knows. For example, a student may be able to draw a diagram or a series of diagrams to explain how something functions but not be able to record the steps in prose. This may particularly occur with a child whose first language is not English or for whom writing is difficult because of poorly developed fine-motor coordination. For children with spatial and conceptual problems, the opposite may be true; those children may prefer writing to drawing or constructing. To be able to assess accurately what students know, assessment techniques must be sensitive to the output modality (writing, speaking, drawing, constructing) preferences of students.

Similarly, input modality preferences must be considered. For some students, solving a mathematics problem is overwhelming because of the enormous task demands of simultaneously having to decode and comprehend the text of the problem, decide on a solution plan, and carry out the plan. For students who have difficulty with decoding text, presenting the problem on audiotape often helps to reduce the task overload and thereby permit effort to be expended on the comprehension, planning, and solution steps. Other students may prefer the text because of the opportunity to read and reread (at their own pace) what is offered by the written word.

***Calculators and the assessment of process skills.*** Some students know when to add, subtract, multiply, divide, take a root, or square a number to solve problems but, because of the sequential nature of these algorithms, have difficulty carrying out the computational steps. Since the emphasis in mathematics education should be on problem solving and not on computation and since many curriculum objectives are process oriented, greater insight into students' problem-solving abilities may be acquired if students are permitted to use calculators while taking tests (as well as for classroom and homework assignments).

With MILPs, students cannot remain anonymous.

With students at both ends of the achievement spectrum mainstreamed into regular classes, anonymity is a major concern. When MILPs are used, students cannot remain anonymous in a classroom. The MILP demands that the teacher assess students' understanding at regular intervals. This multimodality, multidimensional type of assessment also gives the teacher a more comprehensive view of what the student knows and what he or she needs. As a result, instruction can be customized, and students will continually be challenged. They will no longer be passed from grade to grade just because they're "nice kids" or because they are "special."

Obviously, the varying types of assessment demand that teachers become experienced with these new methods. Teachers need to be able to evaluate students' productions (e.g., essays, drawings, simulations, models); observe and evaluate students' behaviors and processes as they engage in, for example, debates, investigations, and experiments; and interview students in order to gain insight into their problem-solving and reasoning abilities.

### *Instructional materials*

The third component of the MILP requires the identification of appropriate curriculum instructional materials. If a student is having difficulty with an objective or cluster of objectives, the recommendation may be to use manipulative materials, software tutorials, or homework activities to explore the concept in different ways. For students demonstrating or capable of exceptional performance, curricula may include more complex problem-solving activities or an interview of experts on the design of mathematical models to represent real-world phenomena (e.g., the eclipse of the sun). Curriculum materials, including textbooks and ancillary print materials, trade books, models, manipulatives, auditory tapes, videotapes, and software, must be selected on the bases of the student's achievement level, interests, and preferred input and output modalities.

***Selecting manipulative materials and models.*** For the past twenty years, there has been widespread use of manipulative (concrete) materials in the teaching of mathematics, particularly at the lower-grade levels, and unquestioning acceptance of the notion that the use of such materials helps to clarify mathematical relationships and processes for all children. This is true for some children but not for others. Some students may have particular difficulty with the concept of number, others with the concept of length or measurement, and still others with shape or other attributes.

Manipulative materials often combine these attributes in ways that may not address a particular student's deficit and may, perhaps, be confounding. For example, blocks are often of different sizes, shapes, and colors.

When faced with the task of counting a set of multiattribute blocks, students with attention deficits may find the different shapes, sizes, and colors distracting. The choice of manipulative materials should be driven by the mathematics they may clarify and the individual child's learning style, not by the availability of the materials.

***Selecting software.*** Typically, students with learning difficulties in mathematics have been directed to software that provides drill and practice on computational techniques, to the exclusion of programs that enhance the understanding of mathematical concepts and their applications. Underlying this practice is the belief that computational expertise is a necessary prerequisite to problem solving (i.e., "You can't learn to solve problems if you can't compute").

This belief is unfounded and not consistent with the current emphasis on problem solving in mathematics education for students of all abilities and at all grade levels. Regardless of achievement level, students will profit from, for example, exploring geometric relationships using Logo Turtle Graphics and the Geometric Supposer (Sunburst) programs. With Lego Logo and Microcomputer-Based Instrumentation Laboratory software, students can model mathematically and analyze phenomena from the physical and natural sciences. With other software from Technical Education Research Corporation, like, for example, The Factory and The Super Factory (Wings for Learning), students can develop their logical reasoning abilities. And with tool programs including spreadsheets, graphing, and equation-solving packages, students can explore and analyze mathematical functions.

Textbooks with limited prose are more abstract and more complex.

***Selecting text materials.*** Although the reading level is an important consideration in selecting a textbook and other printed materials, teachers must recognize that mathematics textbooks with limited prose are more abstract and more complex for students than texts with extensive prose. Sometime during the late 1960s and early 1970s, educators became convinced that if the amount of reading in mathematics textbooks was reduced, the mathematics would be easier for children (particularly low achievers) to understand and the problems would be easier to do.

This has not proved to be true. Without context, students have little or no understanding of the problems and no basis for judging the reasonableness of solutions. Although the reading and decoding may be burdensome for some children, they can be assisted with these tasks by a reader or by an audiotaped presentation of the problems. At no time, however, should mathematics be presented as computation or manipulation in the absence of application settings.

### *Pedagogical approaches*

Once a student's needs have been identified, it is important not only to identify appropriate curriculum materials but also to describe useful pedagogical techniques. For example, for a student who is having difficulty with a particular concept, the pedagogical recommendation may be to pair the student with a peer to construct a model that shows an application of the concept or to assign an adult tutor to review the material with the child. For others, the instructional recommendation may be to use a different input modality.

Using a calculator reduces the tedium of computation.

As mentioned earlier in the discussion of assessment strategies, some students have difficulty with mathematical problem solving because they are overburdened with the need simultaneously to read, decode, and comprehend a word problem as well as decide on its solution process. If the teacher or students tape-record the problem in advance, then the students can focus on its comprehension and solution as they listen. Likewise, using a calculator reduces the tedium of computation and places the emphasis of mathematical problem solving on the decision-making and step-by-step processes to be followed to solve the problems.

Pedagogical strategies need to be described not only for teachers but also for tutors and parents. Although parents are not teachers, they are expected to be participants or partners with their children in the learning process. The MILP should present ways in which parents can interact with their children to help them become successful learners.

### *Links with other subject areas*

A Mathematics Individualized Learning Plan should be developed in concert with plans for other subject areas. There are several reasons for this concurrent effort. First, many mathematical concepts, processes, and skills are central to the learning of other subjects.

For example, understanding the mathematical concept of proportionality is necessary to be able to solve percent problems, to interpret the scale of a map to find distances, to determine the dimensions of a scale model of a vehicle, to predict the number of bass in a stream from a sample, to increase the ingredients in a recipe, and to ascertain the size of one dose of a prescription drug. Experience with major mathematical concepts prior to their application in other subject areas would help to ensure success with those other subjects.

Many mathematical concepts and processes are central to other subjects.

Second, application settings for mathematics problems are often taken from the physical and natural sciences, the social sciences, and the arts. Limited knowledge of these other subjects may become a barrier to the successful solution of the mathematical problems. For example, simple machines (inclined plane, pulley, lever) often furnish the context for mathematics problems dealing with ratio and proportion. Using these machines in a mathematics setting, without prior exploration in a science class, may confuse rather than enlighten students. In the development of learning plans, effort should be made to orchestrate instruction so that application settings and mathematical concepts are presented concurrently or sequentially in order to support one another.

Third, the focus of mathematics education on problem solving is not unique. The abilities to understand and analyze a problem, to design a plan for solving the problem, to carry out the plan, and to evaluate the answer or solution are necessary for solving problems in all content areas. Thus, it is not surprising that many processes and behaviors are generic; their development should be fostered in all subjects. For example, the ability to "collect, organize, display, and analyze data" is an important goal in the study of biology, chemistry, physics, economics, and geography as well as mathematics. The ability to "identify an appropriate sample population for a given experiment" is a process useful in all the sciences. By scheduling the introduction, maintenance, enrichment, and assessment of these generic processes across subject areas, teachers will enable students to have more experience with the processes and to gain greater appreciation for their value and power.

Finally, the link between mathematics and reading must be underscored. Being able to read and understand mathematics texts and mathematics-application problems is central to the learning of mathematical concepts, skills, and processes. Unlike narrative prose, mathematics prose in instructional materials and problem situations is densely packed with important information and technical vocabulary and often lacks redundancy. This type of prose requires a slower reading rate, careful attention to detail, and several rereadings. Being aware of the characteristics of mathematics text,

along with techniques for reading the text, may enhance students' comprehension and achievement.

Appendix C contains an MILP constructed for Lisa (6 years, 7 months), a second-grade student with special talents in mathematics, reading, and written expression. The MILP focuses on further development of Lisa's problem-solving skills, with particular attention to strategies for solving nonroutine problems. Emphasis is to be placed on connecting Lisa's artistic and creative story-telling talents and her interest in dinosaurs to her study of mathematics. Lisa will receive specialized instruction primarily in her regular grade 2 classroom. She will join students in grades 3 and 4 for some investigations and projects.

## A COMPARISON OF METHODS FOR PLANNING INSTRUCTION FOR STUDENTS WITH SPECIAL NEEDS IN MATHEMATICS

Table 6.1 summarizes the differences among three approaches to planning instruction in mathematics for meeting the varying needs of students: (1) the classroom-based approach, (2) the IEP, and (3) the new model introduced in this chapter, the MILP.

Table 6.1
*Three approaches to planning in mathematics*

| Questions | Classroom-Based | IEP | MILP |
|---|---|---|---|
| Who is it for? | All students, including students at risk, special education students, and talented students | Special education students | Special education students, talented students, and students with gaps |
| What must occur before the plan is discussed? | Subject-matter assessment | Multidisciplinary evaluation | Multidisciplinary evaluation |
| Who participates? | Subject-matter specialists | Administrator, teacher(s), psychologist, social worker, nurse, parent(s) | Classroom teacher, special educator, math specialist, parent* |
| What role do parents play? | They may or may not be informed | Participants, with veto power | Participants |
| What kind of document is the plan? | Academic plan | Agreement | Academic program |
| What is included in the plan? | Mathematics content and methods | Subject matter, methods, related services | Objectives, assessment strategies, instructional materials, pedagogical approach, and links to other subjects |
| What is the duration of the plan? | Teacher's decision | One year or until the parents reject it | One year |
| Can plan be rejected by parents? | No | Yes | No** |
| How are rejection or questions resolved? | Informally | Impartial hearing | Informally** |

*Plus other if part of IEP
**Unless part of IEP

## THE IMPLEMENTATION OF THE MILP

The movement toward the inclusion of more and more diverse students into regular classes means that the distinction between special and regular education will soon become obsolete. IEPs are increasingly being carried out within the context of regular education. Constructions like the proposed MILP will facilitate this process by broadening the target population to include a more diverse range of students with special needs. It is now becoming more and more apparent that separating children into homogeneous enclaves—schools and classes—is counterproductive. The separation adds little or nothing educationally and subtracts much intellectually, socially, and politically (Oakes 1986, 1990; Slavin 1988).

The distinction between special and regular education will soon become obsolete.

If MILPs are to be effective, they will need to include certain IEP elements:

Parents must be involved in evaluation and planning; the plan should be multidisciplinary; and the MILP must be an agreement among teachers, parents, and students that is supported by the school administration. Periodic reviews, at intervals of perhaps nine weeks, should be carried out to assess the success of the plan and to revise or rewrite it when necessary.

IEPs came about because courts ordered them and legislatures wrote statutes requiring them. We hope MILPs will come about because mathematics educators realize that the franchise of schooling—the right of all students to receive an education equal to their potential—cannot be achieved in any other way. Schools need to be reshaped so that all students can receive "equal protection." This cannot happen without some form of individualization that has, as its hallmarks, accountability and competence.

The question of how to achieve accountability still remains. The formal grievance process that is part of the IEP is one way to ensure accountability. Should it be extended to cover all individualized learning plans, such as the MILP, or can accountability be accomplished by less formal means? To the extent that planning involves increasing numbers of participants with varied backgrounds, more attention needs to be paid to the dynamics of the group decision-making process and to ways of giving participants sufficient background information.

## BIBLIOGRAPHY


Department of Health, Education, and Welfare Office of Education. "Education of Handicapped Children, Implementation of Part B of the Education of the Handicapped Act." *Federal Register* 42, no. 163, 23 August 1977.

Individuals with Disabilities Education Act. U.S. Code. Vol. 10, chapter 33, as amended by PL 101-476, sec. 1401(a)(20) (1991).

Massachusetts Department of Education. *Chapter 766 Regulations.* Boston: The Department, 1986.

Mathematical Sciences Education Board, National Research Council. *Reshaping School Mathematics: A Philosophy and Framework for Curriculum.* Washington, D.C.: National Academy Press, 1990.

National Council of Teachers of Mathematics. *An Agenda for Action.* Reston, Va.: The Council, 1980.

———. *Curriculum and Evaluation Standards for School Mathematics.* Reston, Va.: The Council, 1989.

Oakes, Jeannie. *Keeping Track: How Schools Structure Inequality.* New Haven, Conn.: Yale University Press, 1986.

———. *Multiplying Inequalities: The Effect of Race, Social Class and Tracking on Opportunities to Learn Mathematics and Science.* Santa Monica, Calif.: Rand Corp., 1990.

Robinson, Karen. "Pull Out or Put In?" In *With Promise: Redefining Reading and Writing for "Special" Students*, edited by Susan Stires. Portsmouth, N.H.: Heinemann Educational Books, 1991.

Routman, Regie. *Invitations: Changing as Teachers and Learners, K–12.* Portsmouth, N.H.: Heinemann Educational Books, 1991.

Slavin, Robert E., ed. *School and Classroom Organization.* Hillsdale, N.J.: Lawrence Erlbaum Associates, 1988.

**Carole Greenes** is a professor of mathematics education and associate dean of the School of Education, Boston University. She is principally interested in how children think about and solve mathematical problems. She heads the curriculum development and staff-training components of the Boston University–Chelsea Public Schools Project.

**Frank Garfunkel** is a professor of special education in the School of Education, Boston University. He is a nationally known expert in special education law and public policy.

**Melissa DeBussey** is a graduate of the College of Liberal Arts (mathematics) and the School of Education (mathematics education and special education) of Boston University. She is currently Space Operations Officer for the U.S. Air Force.

7

# Accommodating Special Needs

*Nancy S. Bley*

*in collaboration with*

*Carol A. Thornton*

THE setting might be a self-contained special education classroom. It might be a mainstreamed classroom. Imagine walking into the room to find students working in groups of twos or threes to solve a mathematics problem that has been posed. Any number of those in the class have "special learning needs."

The students discuss the problem. Some prefer initially to think through the problem independently. Others interact immediately with a partner. In the end, they listen to one another and explain their thinking, often using some manipulative, a drawing, or a calculator result to derive, confirm, or illustrate a point. The students, individually or with a partner, write up their conclusions in some way. Different solution approaches emerge and are shared within groups and eventually with the class.

The scenario above suggests one teacher's approach for accommodating and challenging students with different ways of learning. Because individual teaching and learning styles evolve in various ways, other scenarios might well have been presented with the theme of implementing mathematics instruction based on NCTM's *Curriculum and Evaluation Standards* (1989) in settings involving students with special learning needs. Two common threads that affect daily planning for, and instructional decision making in behalf of, special-needs students in the group underlie these scenarios:

- A repertoire of assessment and instructional strategies, including many discussed in other chapters of this book, that enable the teacher to coordinate daily planning to further the best interests and learning potential of students with different ways of learning
- An understanding of specific ways to accommodate different learning needs or disabilities of students in the class

In reexamining efforts to support appropriate mathematics instruction for students with special learning needs, this chapter will focus on ways of accommodating different learning needs and disabilities and will supplement the discussions found in other chapters regarding appropriate assessment and instructional strategies.

## REACHING STUDENTS WITH DIFFERENT LEARNING NEEDS AND DISABILITIES

Broadly defined, students with special needs are individuals who, for one or more reasons, require certain adjustments to their instructional program. The adjustment for special needs might be temporary, such as for the student with a broken arm, or long term, as represented by a visually impaired student who has trouble seeing the chalkboard or reading standard-sized print in a textbook. Teachers can make these adjustments on a day-by-day basis only when they are aware of what is required and of how and when to implement changes. (See chapter 4 and appendix B for a more specific description of major categories of special needs. Some students, of course, have multiple disabilities, which must be taken into account.)

First and foremost, it is important to remember that students with special learning needs are individuals with many of the same needs as other students. Whatever the disability, the student involved is basically more *like* than *different from* other peers. It is also important to be sensitive to the fact that the experience of a disability may affect an individual's self-image and can also influence a student's social or emotional development, relationships with peers and adults, and functioning in group situations. Hence, one important maxim in dealing with special-needs students is that rather than "doing for" them, we should help them to help themselves so they become as independent as possible.

Teachers should know the range of behaviors and the variation to be expected from each disability.

Classroom teachers are becoming increasingly aware of the need to interact closely with special education staff to learn as much as possible about the extent of any disability and how it might affect individual students. In general, teachers should know the range of behaviors and the variation to be expected from each disability. Several questions can guide a teacher in assessing a disability's impact on learning:

- What learning strengths are intact?
- What learning abilities are suppressed by each disability?
- What learning or teaching styles are more compatible with specific learning needs?

Answers to these questions and information in the student's file should provide further input and perspective for day-by-day planning and interaction. The child's educational history should furnish information concerning effective instructional approaches, gaps in content background, and levels of achievement and aptitude. (Chapter 5 presents assessment ideas that should also prove useful in this regard.) Interactions with the family should offer valuable information about the child's self-image, the degree to which the child may have been protected or challenged, and the attitude of the family toward mathematics. Finally, on a daily basis the teacher might observe the quality and quantity of the interactions between the individual and his or her peer group. This type of information reveals the student's approach to life, attitudes about school work, and feelings about mathematics.

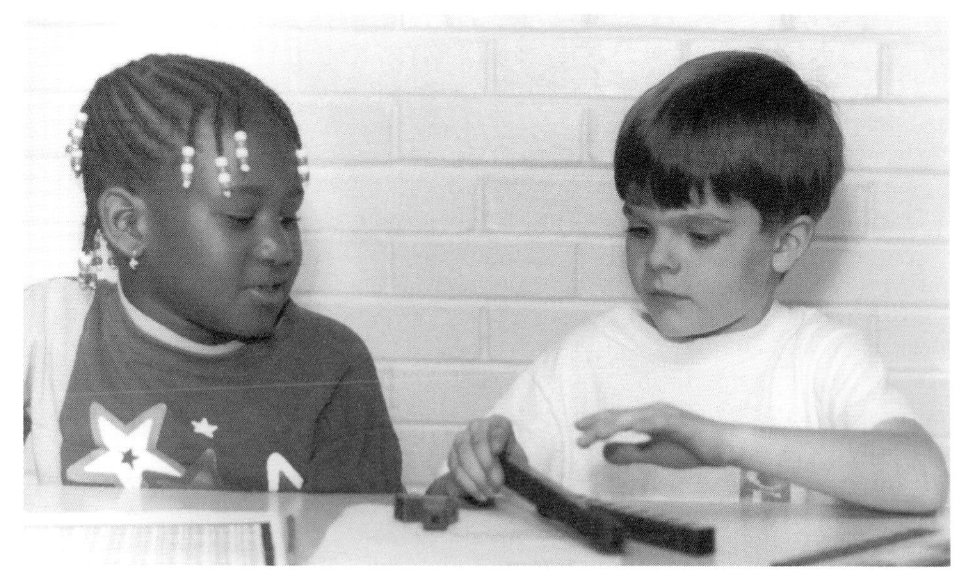

## Specific Disabilities

This section proposes suggestions for accommodating different disabilities within the daily mathematics lesson. In particular, ideas for accommodating students with attention difficulties, hearing deficits, learning disabilities, mental disabilities, physical disabilities, social or emotional disorders, and visual disabilities are presented, as well as general guidelines for working with students who are gifted or talented in mathematics. These suggestions are presented to provide some insight for dealing effectively with special needs in the mathematics classroom. Special education teachers who are familiar with the ideas will be in a position to clarify or offer additional suggestions to teachers who are not.

### *Students with attention deficit disorders*

For students with Attention Deficit Disorder (ADD) or Attention Deficit Hyperactivity Disorder (ADHD), attending to relevant aspects of tasks is not automatic. Most ADD and ADHD students want to learn, and particularly as they gain an acceptance and understanding of their disability, they are also very interested in learning what will help them become successful students. Such intervention strategies as the following have proved helpful:

- Establish routines and discuss them so students will know what is to happen and when.
- Use active-learning approaches and provide other legitimate outlets for the release of excess energy.
- Clearly communicate expectations and consequences for desirable and undesirable behaviors.
- Communicate to the child your acceptance and understanding of the disability.
- Emphasize and reward the positive as much as possible. When appropriate, send home a note to parents so they can further praise and encourage.

- Plan ahead. Create an area with a minimum of visual, auditory, or changing stimuli, where students who are highly distractible can sometimes work. Encourage students to recognize when they can benefit from working in these locations and to request doing so when they think it is necessary.
- Keep oral (and written) directions short and simple but clear.
- Establish eye contact before making important verbal statements.
- Use color highlighting at the chalkboard, on transparencies, and on activity sheets to help attract and hold the student's attention on key points.

Routine is an important part of structure, and it helps hyperactive and distractible students to be less impulsive and better organized: "Each morning, math always comes after spelling. We usually start with a question or a problem." For most of these students, structure affords a degree of security. The amount of structure required to promote socially acceptable behavior will vary from individual to individual, depending on the task and on the level of distractibility. The goal is to find the minimum degree of structure necessary to allow learning to occur. Then, as the student begins to establish some inner controls over behavior, the structure can be gradually reduced or modified. Class rules (a few!), with relevant consequences clearly understood and consistency in follow-through, are important.

Routine is an important part of structure, and it helps hyperactive students to be less impulsive and better organized.

Hyperactive students may need experiences requiring more active learning and gross-motor movement. These students often benefit from cooperative learning groups, since this approach serves as a legitimate setting for them to sit on their knees, walk around the room, and interrupt conversations in the normal give-and-take of students' talking to one another. By clarifying at the start of a session the group's expectations for behavior and individual responsibilities within the group, as well as directions, the teacher might well avoid disruptions due to unpleasant behavior during group work.

Since a great deal of energy is required for ADD and ADHD students to focus their attention, they may have trouble responding quickly and logically to oral questions. This situation can be alleviated by reminding a student to think first, then share the response with a partner before volunteering to share the information with the class (McTighe and Lyman 1988).

Although the ability to recall basic facts allows students with attention problems to succeed in estimation, mental math, and other aspects of computation, teachers need to understand that students' distractibility and difficulty with attention may, in fact, impede their ability to recall known facts. A curriculum that emphasizes problem solving, number sense, and estimation over speed of response and allows calculators to be used as needed alleviates much stress and enables these students to progress conceptually despite their disability.

Many students with ADD or ADHD are on some form of medication. In order for them to participate successfully in an instructional session, it is extremely important that the classroom teacher is—

- aware that the student is taking medication, even if the medication is taken at home before coming to school;
- knowledgeable about the side effects of such medication;
- aware of signs indicating that the medication is wearing down;
- conscientious about seeing to it that the student takes the medication during the day if that is part of the program.

Too often students begin to appear to be disruptive or are accused of not working, being lazy, or forgetting when in fact a problem has developed with the medication.

Conversely, just as it is important for the classroom teacher to be knowledgeable about a student's medication, it is equally important that neither the student nor the teacher use the medication as an excuse or as a reason for failing to do a task. Many students are neither on medication nor need to be on it. Others may still exhibit some of the signs of ADHD or ADD even when on the medication. To become actively in control of one's behaviors is a learning process. The teacher should encourage students to monitor their own performance and relay the information to the doctor, who can then prescribe the appropriate dosage.

### *Students with hearing disabilities*

A small percentage of school-aged children are deaf or hard of hearing. Of primary concern is the extent to which the loss of hearing affects the student's ability to communicate. Whereas some students who require special education because of stuttering, articulation, and voice or communication difficulties have good hearing abilities, students who have had a hearing loss from birth or from an early age typically have speech or language problems. The later the hearing loss, the less disabled the individual is in language development and general communication skills.

Students with a hearing loss may appear not to understand mathematical concepts when in fact they are not understanding the *language* that is being used.

Students with a hearing loss may appear not to understand mathematical concepts when in fact they are not understanding the *language* that is being used. When ways are found to reach these students, often through visual-tactile approaches and modeling, they often show good understanding and good ability to apply this understanding. The following suggestions have proved helpful in enhancing mathematics instruction for students with a hearing loss:

- Establish eye contact before speaking to a student with a hearing loss and before making important statements.
- Speak in a natural tone of voice. Do not exaggerate tone, pitch, or lip movements to facilitate speech reading.
- Use an overhead projector when possible so the student will be able to see your face. When appropriate, give the student with the hearing loss a photocopy of your transparencies. When it is necessary to write on the chalkboard, write and talk at separate times.
- Teach from a position where your face is well lighted. Do not stand with your back to a glaring light or a window, since the glare makes it difficult to see your face and lip movements.
- Be sensitive to seating arrangements, and permit the student with a hearing loss to change seats whenever necessary.
- Write out important words and ideas whenever possible.
- Relate learning to life situations with which the student is familiar.
- Emphasize visual-tactile approaches. Use and encourage students to use manipulative or visual aids as often as possible and appropriate.
- If a child wears an amplification device, keep down the level of noise so that the child can sort out the important sounds.
- Avoid interrupting students when they are speaking. Allow them to finish their thought before responding. They need to learn to finish their own sentences.
- Unless the hearing loss is recent and the student knows how to take good notes or is proficient on a computer, have a hearing peer use carbon paper when taking notes. The carbon copy can then be given to the student with the hearing loss.

Students who have not acquired adequate speech-reading skills will require an interpreter. Students who use manual or total communication will greatly appreciate the teacher and classmates who learn and use some signs. Such an effort will demonstrate that people important to the student are really trying to communicate.

### *Students with learning disabilities*

Students with learning disabilities (LD) are *specifically* affected in the development of their abilities to speak, perceive, think, read, write, spell, or calculate. Although a student's disabilities may be hidden, they can be extremely debilitating.

Because these students do not learn in the same way other students do, it clearly is insufficient just to know that a student "has learning disabilities." It is necessary that all teachers and paraprofessionals working with the individual understand the specific nature and extent of the disability—whether it involves (in any combination of ways) auditory, visual, or motor reception or response; whether memory, perception, language, writing, or other factors are affected; whether the difficulty surfaces in an inability to reason abstractly or in a consistent tendency for reversals or includes any spatial or temporal disorientation.

Students with learning disabilities have specifically different learning needs. If the disability extensively affects learning, highly structured, carefully incremented programs in self-contained classrooms or private schools may be necessary. The more typical scenario is that students with learning disabilities are supported through regular pull-out sessions or by team teaching with special education support, where the regular classroom teacher assumes an important responsibility for orchestrating independent and small- and large-group learning experiences to accommodate the student's special needs in a mainstreamed setting.

The initial approach in any of these programs typically is to understand and build on the student's learning *strengths* and gradually, together with the student, to discover appropriate, effective ways to compensate for any learning *weakness*. If the child learns best by looking, the teacher includes a visual component. Initially, to eliminate repeated failures, no demand is made that requires students to rely solely on their deficit area. By working closely with the school psychologist and special education staff, teachers can gain perspectives about different disabilities, how they affect individuals in the class, and how they might best be taken into account during mathematics instruction.

The initial approach in LD programs is to understand and build on the student's learning *strengths*.

Because of the many different problems that students with learning disabilities may have, each requires treatment on an individual basis—thus the Individualized Education Program (IEP) or Mathematics Individualized Learning Plan (MILP) described in chapter 6 and appendix C. Teachers must be sensitively attuned to the specifics of each child's learning differences. To do so requires that time be spent examining the child's file and consulting with special education specialists and with the individual student and parents.

The students themselves often offer the most perceptive insights on how they learn best. Many students with spatial or abstract-reasoning difficulties realize, for example, their need to use manipulatives well beyond the time that their peers require these aids. This need underscores the importance of making manipulatives readily accessible to help these students understand, apply, and retain concepts.

With a few exceptions (e.g., students with mental disabilities who also have a specific learning disability), most students with learning disabilities are quite capable and *can* learn when their particular way of learning is identified and respected. Given the importance of mental math and estimation, for example, approaching these topics visually, tactilely, verbally, and in an unhurried fashion is important for nurturing the success of students with memory or abstract-reasoning disabilities. In order to estimate or compute mentally, one must be able to retain and process a great deal of information simultaneously, which these students often find quite difficult. Most, however, can understand, learn, and effectively use estimation and mental-computation strategies when hundred charts (see suggestions presented in chapter 10) or counting lines (see fig. 7.1) are used in conjunction with appropriate visual, auditory, and kinesthetic-tactile approaches.

Most students with learning disabilities are quite capable and *can* learn when their particular way of learning is identified and respected.

The illustration in figure 7.1, for example, might help these students make such simple estimations as that involving grandparents of a classmate who will soon celebrate their forty-third wedding anniversary. "Should I say they have been married about *forty* years or about *fifty* years?" Referring to the counting line, a student might finger-trace the distance from 43 to 40 and from 43 to 50 to help determine tactilely that 43 is closer to 40 than to 50 and to conclude that "the Browns have been married a little more than forty years." Encouraging the students to use visual charts and other aids to assist understanding during early experiences with topics like these and throughout the learning process until they have internally established a good visual frame of reference paves the path to their success.

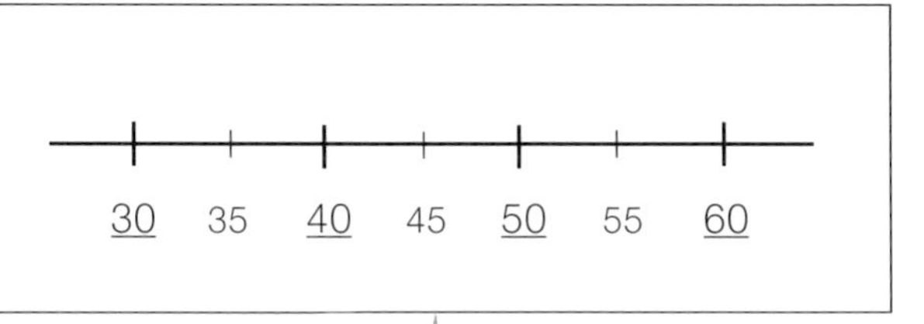


Fig. 7.1. Counting line

Similar approaches will be helpful to students with auditory memory or auditory perception difficulties. Additionally, these students should frequently be requested to paraphrase what they hear. This approach not only assures that they understand what is being discussed but also helps them retain important information. Like many other students, they must "verbalize in order to internalize."

Another example of finding and using a particular way of learning involves calculator use by students with specific visual difficulties. If required to use scientific or graphing calculators, for example, students with visual-perception disabilities may have particular difficulty reading the small-print labels on multifunction keys. If a large-key calculator is not available, supply an enlarged paper sample of the calculator and work with individual students to color highlight the key labels to which they can refer to facilitate their locating needed keys.

A different need will be indicated by students who have visual-motor or visual-memory difficulties in connection with calculator use. Allow these students to use printing calculators so they can refer to the printed tape to check whether they have been accurate in their entries.

The tape is also a useful reference for students who frequently lose their place or forget whether a particular number has been entered.

Some students with learning disabilities may have problems with laterality. They shift hand preference by sometimes using the right hand, sometimes the left. A choice of handedness should be made and consistently reinforced by the teacher.

Other students may have a perseveration problem that causes the child to dwell conceptually or physically on one task or problem; such children cannot shift without help. Perseveration needs to be stopped whenever noted and the student's attention shifted to something else. In written work, color coding, verbal reminders, and spatially organized pages on which individual problems or response spaces are boxed often help students maintain awareness of the need to "change sets."

It may be more appropriate to allow some students with visual or motor disabilities to respond into a tape recorder than to require them to write down their results. Students with expressive-language difficulties might be given the option of writing reports (perhaps on a computer) or drawing pictures rather than presenting explanations orally. Gradually, as the students gain confidence, the teacher can encourage them to share an explanation orally with a classmate with whom they are comfortable. When summarizing material for the class, it may be helpful to use the overhead projector and then give a photocopy of the transparency to students who have short- or long-term-memory, figure-ground, or other perceptual disabilities.

Students with nonverbal learning disabilities (see, e.g., Foss [1991]), who often encounter abstract-reasoning difficulties, may exhibit competence in low-level skills like computation but have considerable trouble understanding concepts and applying problem-solving strategies. These students can begin to learn and reason, however, when the teacher employs a whole-language approach to mathematics; drawing on familiar experiences, this method uses, and encourages students to use, their own language, invented but accurate algorithms, and familiar physical objects or drawings to meaningfully interpret and express mathematical concepts or solution approaches.

> The students need to verbalize what they do and frequently explain or justify the statements they make.

These students need to verbalize what they do and frequently explain or justify the statements they make, for only through self-monitoring attempts like these will they become aware of what makes mathematical sense and what does not. This metacognitive approach helps them understand and retain material and helps them be less impulsive in their responses to problems. Another aspect of this approach, well aligned with recent recommendations for improving school mathematics instruction, emphasizes encouraging students to identify, sort out, organize, and verbalize the "big ideas" or relationships in a lesson or unit and to relate the broad steps of a solution strategy. This need for verbalization can often be met within the larger group by asking students at strategic times to tell a partner the big idea of an activity or problem.

Plan ahead. If students have reading difficulties, place them with a partner or in a compatible cooperative setting where others can provide help or do the reading. Assign other more accommodating tasks to the student in need. When activity or text pages are needed, bookmarks, color cues, highlighting,

or boxes (see Bley and Thornton [forthcoming]) may help structure the page for students with visual-perception problems.

Like many other students with disabilities, those with learning disabilities may be easily frustrated. Establishing clearly defined, reasonable, short-term goals—while making the student aware of both the goals *and* their achievement—will foster progress and build self-confidence. Focusing on the positive by pointing out what was achieved yesterday, what has been achieved today, and what can be achieved tomorrow gives a sense of success. Recognizing personal progress, not comparison to others, is essential.

Students with learning disabilities represent the largest proportion of students with special needs in the school population. Because of their disabilities, these students (like others with disabilities) typically require more time to engage successfully in problem solving and routine mathematical tasks. Additionally, they have to exert more effort and energy than their nondisabled peers to succeed and hence tire more readily. This situation suggests the need for a "less is more" approach, guided by the careful selection of problems and tasks. (For detailed suggestions in this regard, please refer to the subsequent chapters of this book, to Bley and Thornton [forthcoming], and to Davidson and Marolda [1993] for more detailed mathematics teaching and learning suggestions for students with learning disabilities.)

### *Students with mental disabilities*

Students exhibit varying degrees of mental disabilities, ranging from moderate and severe disabilities that render them unable to learn mathematics (see the ideas presented in chapter 18) to those that cause students to learn much more slowly and reach a plateau in reasoning ability at an earlier chronological age than their peers. Although sometimes able to be educated in the regular classroom, the latter students have lower memory, attention, and language abilities and acquire concepts and new understandings at a *much* slower pace than their nondisabled peers. Moreover, because of their lower ability and frustration levels, they often exhibit inappropriate social behaviors.

Focusing on the functional applications of mathematics that will enhance future employability and living status and developing socially appropriate behaviors are important thrusts of the curriculum for students with mild mental disabilities. In particular, emphases on practical problem solving; number and operation sense; and measurement, time, and money concepts and skills are important. When mildly disabled students are mainstreamed for some aspects of the mathematics curriculum, the support of the special educator in designing and identifying ways in which these students can be involved and successful is important. With all special-needs students, this approach is necessary to any successful behavior plan.

All students learn best when they have strong understandings on which to build. Fortunately, today's emphases on experience-based mathematics learning, on open-ended problem solving and applications, on cooperative learning, on calculator use, and on performance-based assessment provide a better forum for active participation by students with mild mental disabilities than the step-by-step, sequential computational programs that were prominent in the past.

In mainstreamed settings, students with mild mental disabilities, who tend to have relatively good rote recall, can benefit from teamwork in which they

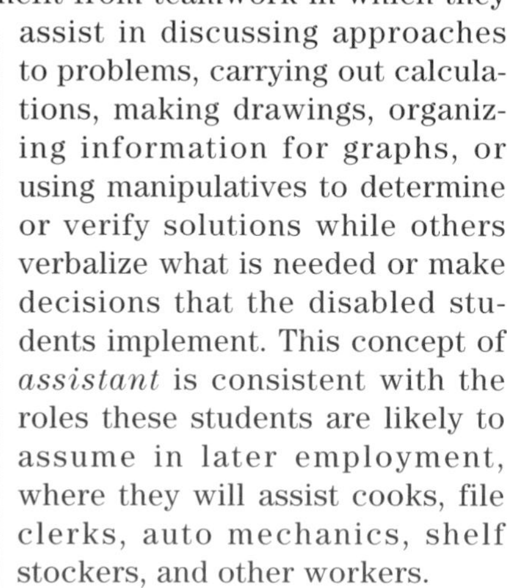

assist in discussing approaches to problems, carrying out calculations, making drawings, organizing information for graphs, or using manipulatives to determine or verify solutions while others verbalize what is needed or make decisions that the disabled students implement. This concept of *assistant* is consistent with the roles these students are likely to assume in later employment, where they will assist cooks, file clerks, auto mechanics, shelf stockers, and other workers.

In order for students with mild mental disabilities to be successful in partner or team situations, a risk-free, accepting, patient attitude must prevail. Sometimes this entails the teacher's taking other students aside to define the expectations (and reasons) for interacting with the disabled student.

Students with mild mental disabilities do best with consistent expectations for behavior and, generally, with clear, simple directions. To demonstrate their understanding, these students should be encouraged to verbalize basic expectations and directions in their own words, and they should be encouraged to ask relevant questions. In order for them to grow in their responsibility to their partner or group and in their understanding of successive problem-solving experiences, participation should begin at a very simple level.

Students with mild mental disabilities do best with clear, simple directions.

At first, for example, the student might use a printing calculator (one with a tape printout so there is a written record of what is keyed for easy checking by the individual or by others in the group) to enter numbers and compute as directed. Over time, the student may learn to read the results and might announce or record them for the group.

Initially engaging these students in the same or a similar role with different partners or groups allows for needed "overlearning" and more readily frees their energies so they can go beyond the participation task and begin attending to the mathematics of the experience. Interacting with peers in this way increases "normalcy" in social behavior and language patterns and establishes a forum for higher-level thinking and understanding. These students, like others, will have breakthrough insights that will contribute substantially to solving a problem or to a group discussion.

### *Students with physical disabilities*

There are four groupings of physical disabilities: (1) musculoskeletal conditions; (2) congenital defects of the lungs, heart, and extremities; (3) neurological impairments, such as cerebral palsy, caused by lack of oxygen to the fetal brain (which accounts for about half of all physical disabilities); and (4) accidents and other physical impairments. Most students with physical disabilities are quite capable of learning and achieving at the same level as their peers, even though they may work slower or require communication or other adaptive devices, such as arm rails, computers with pointers, cut-out desks adjusted to the appropriate height, or lap trays that fit onto their wheelchairs.

Most students with physical disabilities are capable of learning and achieving at the same level as their peers, even though they may work slower or require adaptive devices.

If a functional impairment interferes with the education of a student, some type of adaptive device, if currently unavailable, can be designed. Contact personnel at your state board of education for assistance. Often an occupational or physical therapist may have an inexpensive suggestion. For example, if students find it difficult to steady a book for reading or paper and pencil for writing, arm weights can be used in conjunction with a rubberized sheet of Dycem, available from most orthopedic supply stores. Classroom teachers, special educators, nurses or other medical specialists, speech therapists, psychologists, and occupational and physical therapists compose the basic team responsible for tailoring the individual programs and adaptive devices needed by the student.

If coordination is impaired and successful calculator use is impeded, students can learn to work independently by using an adapted or large-key calculator or a calculator on a computer with such adaptive devices as pointers. Other students may need to work with a partner and dictate or observe while another enters data into the calculator. Students should be made aware that actively attending to what is keyed into the calculator is essential to the decision making associated with a situation. This strategy deserves careful monitoring by teachers.

A major instructional goal should be to help students with physical disabilities to develop self-sufficiency. It is important to know how and when to help physically disabled students. For example, even though a buddy system may be necessary at times, picking up a student who keeps falling is not recommended (unless the medical staff advises differently), for doing so can promote a sense of dependence in that child. In a similar vein and whenever possible, students with physical disabilities should be encouraged to do their own work—even if it takes much longer—so that self-esteem and independence can be fostered. In order to sustain this independence, it may be necessary to allow these students to complete a shorter assignment or to be given the assignment in advance so they can plan and use their time in the best way for them.

Because of the intense preoccupation with self-help skills during the early years or the early period of disability, many students may not have been able personally to experience much of the world. Thus they may not have traveled, gone camping, ridden public transportation, or ridden an escalator. Classroom and problem-solving experiences must take these facts into consideration. These students should be offered as many field trips and enriched experiences as possible so as to broaden their perspectives for mathematical applications and uses.

### *Students with social or emotional disorders*

There is no simple way to reach students who have social (behavioral) or emotional disorders in the mathematics classroom. Careful collaboration with special education or medical staff is frequently necessary for dealing with students who are anxious or withdrawn, grossly immature, or socially aggressive or who exhibit severe conduct disorders.

These students typically require direct instruction in basic social skills and in appropriate communication strategies. All teachers and school personnel dealing with these students must respond as a team so as to maintain consistency in dealing with the disability. The focus should be on rewarding students' best efforts. Set realistic expectations and, whenever possible, offer immediate, public praise for appropriate effort and behaviors while ignoring inappropriate behaviors or correcting them privately. Consequences for inappropriate behaviors should be determined up front and be discussed and clearly understood by the students involved.

All teachers dealing with the students must respond as a team so as to maintain consistency in dealing with the disability.

Students with social or emotional disorders typically benefit from highly structured approaches and an environment that promotes predictable outcomes for behaviors. Consistency is important and includes regular routines during instruction. Activities with a high success factor are very important for these individuals.

Group work and cooperation, as well as the sharing of abilities on group projects, can foster better social interaction in these students. When they are placed in cooperative learning groups, careful thought must be given to the selection of the other team members, to the proximity of the group to other groups and to the teacher, and particularly to these students' strengths.

Often, simply affording students with social or emotional disorders the opportunity to succeed in front of peers is enough to keep undesirable behavior under control, at least temporarily, or to redirect the student should a problem arise. Most students enjoy learning and want to show what they know. They may need help in doing so in appropriate ways, but when this help is given in a consistent, firm, realistic, and structured setting, students who otherwise give in to undesirable behavior or emotional outbursts often can participate appropriately during instruction.

### *Students with visual disabilities*

One child out of every ten entering school has some visual impairment. One out of every thousand children (the lowest incidence of all the disabling conditions) has a visual impairment that cannot be corrected and needs special education support and special adaptations for functioning in the classroom (Kirk and Gallagher 1986). Some of these students are partially sighted; others have no vision. It is helpful to check the medical files for a summary of each student's vision diagnosis and prognosis and to request clarification from the special education staff, the school nurse, parents, or a medical specialist regarding the implications for the mathematics program.

In general, tactile perception is extremely important to these students and must be the basis of all mathematics instruction. In conjunction with auditory input, it is the major way in which students with visual disabilities can experience mathematics in relation to their world. Teachers should be sensitive to the fact that exploring an object tactilely necessitates much more effort than merely looking at something. Sight is a unifying factor in experiencing objects and occurs simultaneously, not successively, as when students with visual disabilities experience an object through touch.

Tactile perception is extremely important to students with visual disabilities and must be the basis of all mathematics instruction.

Active learning is essential for students with visual disabilities. Activities requiring motor skills—and learning in general—take more time for these students than for their sighted peers. During whole-group sessions, students who are partially sighted generally prefer to be seated near the center of the discussion and near the overhead projector or the chalkboard if either is used. Especially at the upper-grade levels, peer teaching and notetaking by sighted classmates can help. Sessions might also be tape recorded for playback by individual students at a later time.

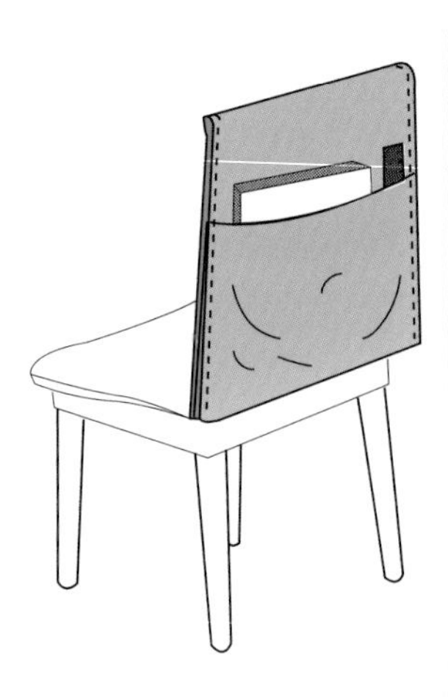

Fig. 7.2. A "math chairpack"

A curricular emphasis on estimation, problem solving, and decision making is important. Students with visual disabilities do best when they are allowed to use their *reasoning and thinking strategies* and are not pressured by time constraints and computational accuracy. These students should rely more on estimation and be allowed to use a calculator with headphones and a voice synthesizer for more complicated computations. To foster independence and self-directed work, this and other mathematics materials should be available for ready use. One option is to supply a "math chairpack" for each student's desk (see fig. 7.2) for storing the more commonly used manipulatives.

Mobility is a major issue for students who are visually impaired because the ability to get around is essential to any independence in living. It is useful for children with visual disabilities to determine and map out mobility patterns in the classroom and major parts of the school building so they can be somewhat independent during the school day. Be patient if they are slow; do not do for them what they can accomplish themselves. A rotating buddy system can also be established to provide the necessary help in mobility, for instance, in fire drills and during certain aspects of mathematics instruction (e.g., notetaking). The teacher and sighted students alike should strive to be specific when speaking to students with visual disabilities. "Your paper is over there" will have little meaning. "Your paper is on the second desk from the front—the row by the windows" is more informative.

The following general suggestions may help create a supportive environment in the mathematics classroom for partially sighted students:

- Use the student's name to invite attention.
- Encourage students to use whatever residual vision they have. Holding visual materials close to their eyes may be necessary, and it is not harmful.
- Provide preferential yet flexible seating (near the center of the discussion and near the overhead projector or chalkboard if either is used). Going to other parts of the classroom for small-group work and changing seats from time to time offer good orientation experiences.
- Maintain a well-lighted work space. (Some exceptions, for example, albinos, are very light sensitive and need a more shaded work space.)
- Provide students with a slant-top desk and a book holder (e.g., from the American Printing House for the Blind) to keep the book at a comfortable angle for reading.
- Allow the option of writing on the chalkboard, even as part of cooperative group work, instead of bending low over a paper (as doing so is very tiring!).
- Use cream-colored paper and a black felt-tippd pen or other appropriate materials to avoid glare but furnish a good contrast for writing.
- Plan ahead. Request large-print versions of your mathematics textbook through your state office of education or from the American Printing House for the Blind. These versions or a cassette tape of the text will help ease reading strain.

- Talk while writing on the chalkboard. Use thick yellow or orange chalk whenever possible.
- Use pressure-sensitive or textured materials for drawings so numbers and shapes can be illustrated for or by students who have visual disabilities. Some of these special materials are also available from the American Printing House for the Blind.

When dealing with students having severe visual impairments, additional modifications are sometimes necessary:

- When coming into close proximity with a student who is blind, identify yourself. Doing so allows the student to become oriented both to your presence and to your position (and possibly avoid an awkward collision!).
- Establish obstacle-free paths for the student's travel within the classroom and determine fixed places for the storage of mathematics materials.
- Emphasize active-learning and visual-tactile approaches for mathematics instruction. Use a label-tape printer and a brailler to adapt mathematical games and activities. For example, one could use dice with braille or raised dots on the various faces of the cube.
- Give students a commercial tray or a cookie sheet with raised edges to define a systematic and bounded working area for activities.
- Let the student take your arm when being led. Be sure to walk slightly ahead of the student.

More modifications of regular classroom procedures are needed to accommodate visually impaired students than those with other deficits. A creative teacher willing to devote time and thought to the matter can almost always modify a classroom environment to accommodate these special needs.

### *Students who are gifted or talented in mathematics*

We have been challenged by the authors of chapters 2 and 4 to rethink and redefine our ideas about being gifted in mathematics. Both sets of authors appreciate the fact that giftedness can emerge in many different ways. The authors of chapter 2 point out that students who can compute easily are often mistakenly assumed to be mathematically talented, even though some of these students may lack important understandings, intuitions, and insights. These authors highlight the value of a rich, hands-on, problem-centered mathematics curriculum that serves as a natural forum for identifying and nurturing the mathematical or spatial strengths of creative problem solvers. The authors of chapter 4 support this recognition of "many talents." They review a model of multiple intelligences in which mathematical-logical and visual-spatial talents are but two of the domains that might be considered and that may influence the way in which students interact with mathematics.

These facts broaden our perspectives and approaches for seeking out and dealing with students who are mathematically gifted or talented. The keys to doing so, particularly at the lower levels, are a strong commitment to a broad, balanced, problem-solving approach for developing, extending, and applying mathematical concepts and skills—and tuning in on the way different students respond. This approach requires great flexibility, creativity, and sensitivity on the part of the teacher and the willingness to observe and listen closely to students as they explain and justify their thinking.

As the mathematics curriculum becomes richer and more challenging in this way; as different solution approaches are expected and respected; as students explore, predict, verify, and share their thinking; and as teachers build on students' thinking, the special talents of individual students can be discovered and accommodated.

Gifted and talented students tend to be curious and natural learners.

Gifted and talented students tend to be curious and natural learners. Once their potential is tapped, they tend to be capable, creative, committed individuals. Often their thinking is quite divergent and nontraditional. Continuing to give rich mathematical experiences to these students is a challenging, exciting, and serious responsibility.

Particularly as students become older, their programming needs change. Although school size and staff often dictate what is possible, it clearly is desirable to allow students who are gifted or talented in mathematics to interact with like-minded students on a regular basis. This interaction can provide the forum for students to grapple with significant mathematical ideas and problems and fully challenge their talent, initiative, and creativity.

Some authors (e.g., House [1987], Maker [1986], Maker and Schiever [1989], Sorenson [1988]) have argued that, at the upper-grade levels, these interactions should be encouraged through programs or classes that are *qualitatively different*. Accelerated versions of the regular program or a series of isolated, unconnected activities or projects are undesirable. Some larger districts offer special pull-out programs in mathematics, special classes, or learning centers for upper-elementary-level students to address this need. Most high schools offer different mathematics tracks to accommodate the needs of different students.

The essential components of programs for mathematically talented students have been identified by House (1987). Such programs—

- involve good mathematics, "not just a collection of games, tricks, puzzles, or isolated topics" (p. 47);
- emphasize applications and problem solving;
- are based on sound pedagogy;
- are designed and facilitated by competent teachers;

- invite and nurture higher-order thinking skills;
- emphasize communication by expecting students to "read, write, listen, speak, and think mathematically" (p. 48);
- nurture good study skills and work habits;
- accommodate individual differences;
- encourage and nurture creativity;
- employ a broad variety of learning resources from textbooks, calculators, and computers to television, concrete materials, and outside resource persons;
- relate mathematics to other content areas of the school program;
- are well planned and coordinated;
- include ongoing evaluation of students' progress and of the program's effectiveness;
- are responsive to individual students' concerns and needs;
- are flexible enough to allow students "to move without prejudice in or out of the program as their needs change" (p. 49);
- involve a certain status and prestige for both students and teachers.

An increasing number of school districts and states have established special classes; regional centers; and magnet, mentor, or summer programs to accommodate middle or high school students who are gifted or talented in mathematics and science (House 1987). Chapter 15, for example, describes goals, flexible options, and promising practices for working with high school students who are gifted and talented in mathematics. Many of the ideas in that chapter can be adapted to work with gifted and talented middle school students enrolled in special "high ability level" mathematics classes.

Middle school teachers in regular classrooms can extend the efforts of lower-level programs and focus on providing a rich, broad, problem-centered curriculum that recognizes, accommodates, and nurtures the creative talent of their gifted students. This approach, often necessary for teachers in smaller school districts that do not have the staff or sufficient numbers of students to create special programs, can be enhanced by offering on a regular basis *carefully planned* enrichment activities and problem-solving extensions for the most capable students.

Ideally, any enrichment experiences should not be in addition to the regular assignment. Rather, they should *replace* certain aspects of the assignments given to the other students and, over the year, have an identifiable content focus. This focus might be exploring a standard topic in greater depth, exploring nonstandard content, or informally studying advanced content. Whichever focus is selected, the development of important thinking processes and the understanding of important mathematical concepts should be emphasized.

Despite the fact that the reality of the classroom challenges and sometimes taxes the time and energies of elementary and middle school teachers to respond to, and nurture the potential of, gifted and talented students, a growing number of teachers are being more successful by—

- basing instruction on problems and activities that invite different solution approaches and many levels of solution so that less talented students can participate in the task with more talented individuals and all can experience individual success;

- making differentiated assignments on the basis of students' ability and interest;
- asking open-ended questions that invite different levels of response;
- asking students to *predict* outcomes before initiating a solution;
- expecting and encouraging different solution approaches;
- using flexible grouping schemes that—
  1. sometimes involve more-talented students with their peers in opportunities to grow socially or exercise appropriate leadership roles;
  2. sometimes involve more homogeneous settings in which, after solving the problem at hand, the more-talented students can interact with other capable students on solving or creating problem extensions.
- seeking additional resources to supplement the basic program.
- asking open-ended end-of-lesson questions or posing problems for which the teacher will not accept an answer so as to encourage the students to explore various solution strategies. (Many gifted students will be stimulated to reflect on and answer the question or solve the problem!)
- planning specific times (in addition to unexpected opportunities) to discuss projects, special assignments, and ideas with these special students.

A problem-centered approach to teaching and learning has many advantages. While nurturing and challenging creative talent, this approach, often grounded in collaborative problem solving, becomes a forum for the exchange of ideas and accommodates heterogeneity in ways that are not possible in other learning environments. The difficult task is creating good problems that allow for multiple solution strategies and, ideally, multiple correct solutions. The administrators' challenge is to give teachers time to collaborate in creating prototypical problems for this purpose.

Most teachers try to accommodate the needs of their gifted and talented students by making differentiated assignments. This strategy works only when teachers establish time for interaction and regular feedback. Gifted students otherwise may gain minimal comprehension or begin to consider the assignment merely as busy work to be done so the teacher can work with others. Although these students can and often do teach themselves, they benefit from periodic interaction and guidance. The goal of these sessions should be to ask questions that lead the students to make predictions, to validate or justify their thinking, to consider alternatives, or to extend their thinking.

Teachers of gifted students need to establish time for interaction and regular feedback.

Regular interactions also afford an opportunity to get to know the students as individuals and to really listen to their ideas. These are the times to encourage them to ask unusual or even unanswerable questions. It is important to accept, encourage, and respect creative approaches and efforts even if, in a particular situation, they do not lead (or seem to lead) to a problem solution. It is also advisable to encourage reflection so that important connections and big ideas are recognized.

These discussions might also serve as opportunities to help students be organized, articulate, and precise about exploring and reporting on mathe-

matical ideas and problems. Or they may open the door for counsel and advice related to peer acceptance or social skills.

During the early school years, most gifted students can process information rapidly without pencil and paper. As they progress in school and the information and details in problems become more complex, many of these individuals find themselves feeling less confident because they cannot remember, use, and simultaneously process all the information. They may feel awkward and perhaps "dumb" if they have to write things down, they may become very angry, or they may lose their previous interest in mathematics. Particularly susceptible are gifted students who have a short-term-memory disability or students with ADD or ADHD.

One approach for helping these students avoid the frustration they feel when they must write down information in order to avoid error is to ask them to be notetakers for other students, perhaps for those who cannot process so rapidly but who are able to understand and apply written information. Alternatively, encourage students with memory or attention difficulties to demonstrate their understanding through physical models or drawing or to verbalize what they do or think so others can record.

Like their peers, students who are gifted or talented have special learning styles and needs in the mathematics classroom. It is important to understand their particular way of learning and to both challenge and stretch their thinking and creative problem-solving abilities in order to nurture their potential for, and growth in, thinking mathematically.

## WINDOWS OF OPPORTUNITY: REACHING ALL STUDENTS WITH SPECIAL LEARNING NEEDS

The teacher's role in day-to-day decision making and planning for mathematics instruction is complex. The effort to reach all students involves these parameters among others:

- An understanding of the thinking and levels of mathematical understanding demonstrated by different students
- A knowledge of the mathematics to be learned and the big ideas that should be emphasized for individual students
- A sensitivity to different learning strengths and disabilities and an understanding of how best to design and accommodate learning tasks to meet various needs
- The ability to coordinate all these factors—to plan and orchestrate problem-based learning activities in ways that take into account differences in children's thinking, abilities, and potentials

This chapter has focused on ways to accommodate students with different ways of learning during mathematics instruction—one important component of instructional decision making. It has stressed understanding students' learning strengths, as well as any disabilities, and the need to help students assess their potentials and limitations realistically while learning to compensate in deficit areas. Several major suggestions emerge from the discussions of different categories of special needs that affect daily planning and instructional activities in a constructivist mathematics classroom with a thoughtful teacher:

- Create a climate of trust, acceptance, and respect in the classroom.
- Emphasize a variety of learning approaches (visual, auditory, kinesthetic-tactile) to meet different learning needs.
- Accommodate different learning needs and styles by accepting and encouraging different and personally meaningful solution approaches and strategies and ways of recording or presenting solutions and ideas. As facilitators of learning, teachers must be willing to accept responses that students may have arrived at through variant but accurate paths.
- Provide adequate blocks of time for students to work through important mathematical ideas. Sometimes units of work must be spread over several days for "learning breakthroughs" to surface.
- Accept the fact that more often than not, an accumulation of rich mathematical experiences that build on (or "connect to") students' understandings, experiences, and interests is more important and appropriate than day-by-day mastery of isolated concepts or skills.
- Allow special-needs students to use buddy systems and participate in cooperative-learning-group activities in accountable ways, since this approach can provide important support to learning.
- Frequently ask students to paraphrase directions and their understanding of

important mathematical ideas and procedures. Many special-needs students internalize concepts and retain them only when they "hear themselves speak."

- Realize that most special-needs students tire easily because extra energy must be expended in coping with a disability. A "less is more" stance must be taken by carefully selecting problem tasks and by trimming routine assignments.
- Whenever possible, allow students to use their learning strengths to approach new learnings. Eventually students should be encouraged to use compensatory strategies in deficit areas so they can deal holistically with topics.
- Emphasize the positive in progress made toward reasonable goals.

Overriding all these considerations is the fact that students with special needs are, first and foremost, students with the same basic needs and concerns as their peers. If their teachers furnish rich mathematical experiences in such constructivist settings as those modeled in the remaining chapters of this book, these students, like their peers, will be significantly empowered to think mathematically in a way that is consistent with their learning abilities and potential.

## BIBLIOGRAPHY


Bley, Nancy S., and Carol A. Thornton. *Teaching Mathematics to Students with Learning Disabilities.* 3rd ed. Austin, Tex.: Pro-Ed, in press.

Davidson, Patricia S., and Maria R. Marolda. "Mathematics Diagnostic/Prescriptive Inventory." Unpublished manuscript, 1993.

Foss, Jean M. "Nonverbal Learning Disabilities and Remedial Interventions." *Annals of Dyslexia* 41 (1991): 120–40.

House, Peggy A., ed. *Providing Opportunities for the Mathematically Gifted, K–12.* Reston, Va.: National Council of Teachers of Mathematics, 1987.

Kirk, Samuel A., and James J. Gallagher. *Educating Exceptional Children.* 5th ed. Boston, Mass.: Houghton Mifflin Co., 1986.

McTighe, Jay, and Frank T. Lyman, Jr. "Cueing Thinking in the Classroom: The Promise of Theory-Embedded Tools." *Educational Leadership* 45 (April 1988): 18–24.

Maker, C. June. *Critical Issues in Gifted Education: Defensible Programs for the Gifted.* Austin, Tex.: Pro-Ed, 1986.

Maker, C. June, and Shirley W. Schiever. *Critical Issues in Gifted Education: Defensible Programs for Cultural and Ethnic Minorities.* Austin, Tex.: Pro-Ed, 1989.

National Council of Teachers of Mathematics. *Curriculum and Evaluation Standards for School Mathematics.* Reston, Va.: The Council, 1989.

Sorenson, Juanita S., ed. *The Gifted Program Handbook: With a Special Model for Small School Districts.* Palo Alto, Calif.: Dale Seymour Publications, 1988.

**Nancy Bley** is the academic coordinator at Park Century School, a school in Los Angeles for students with learning disabilties. She has made a number of presentations at annual meetings of the Learning Disabilities Association of America and of the National Council of Teachers of Mathematics. In addition to writing a number of articles in journals on computers and mathematics teaching, she has coauthored a book on teaching mathematics to the learning disabled.

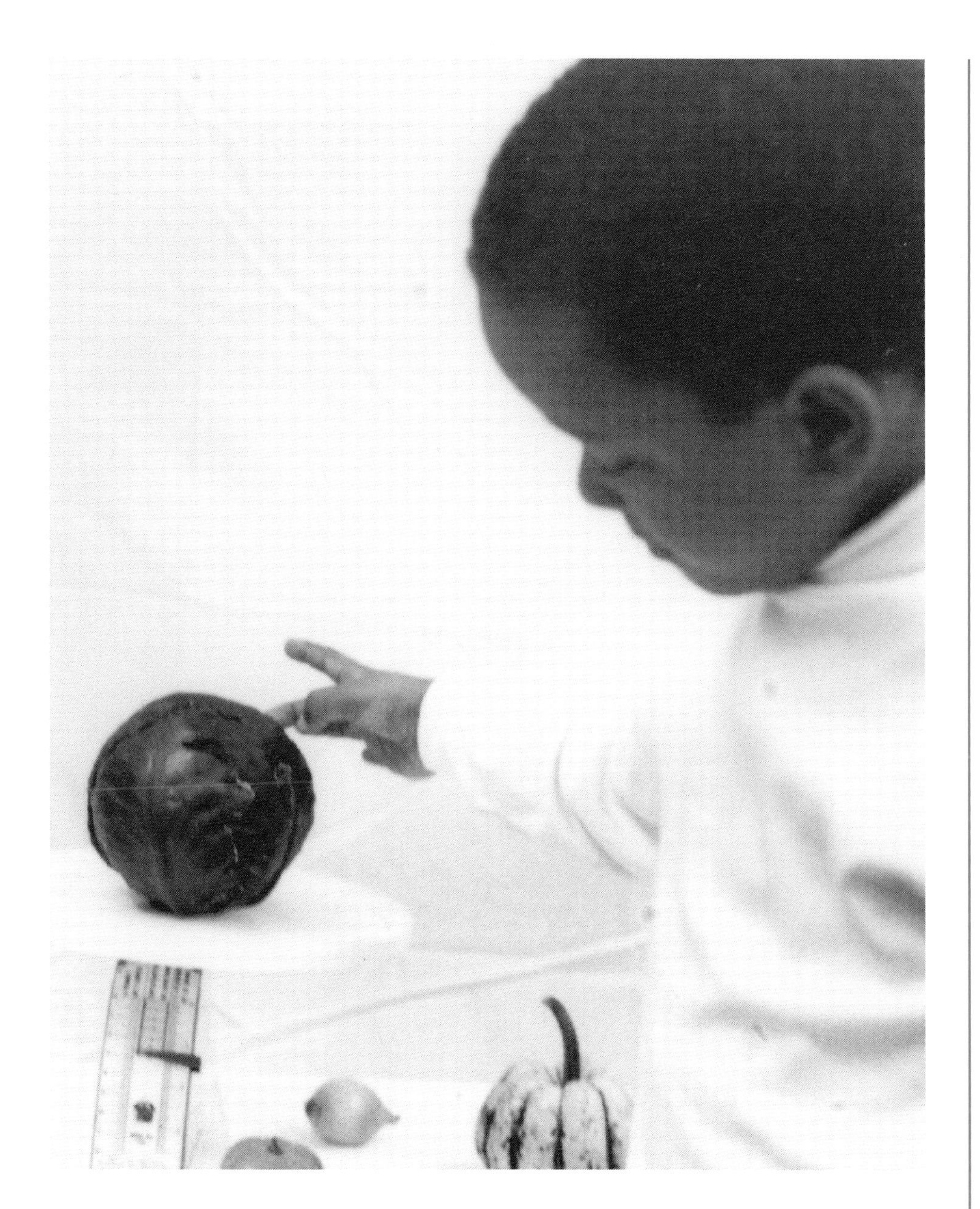

# Part 2

## Classroom Episodes

8

# Mathematical Thinking in Early Childhood

*Judith K. Wells*
*S. Kay Dunlap*

*in collaboration with*

*Beverly Hopcraft Sullivan*
*Mary Leo Bourisseau*
*Sharron Williams*

For young children, mathematics does not exist. It is only created by children from the language of their experience. For most children, the language of mathematics has its beginnings in oral language and the development of critical concepts through problem-solving experiences involving the following activities:

- Sorting and classifying
- Comparing and estimating
- Patterning and logical thinking
- Estimating and measuring
- Exploring shapes
- Collecting, organizing, and displaying data
- Counting and exploring numbers

Many young children are identified for intervention programs in early childhood because of receptive- or expressive-language deficits. These children are disadvantaged linguistically. They are "at risk" conceptually in mathematics in comparison to other areas of their experience (Members of the Teaching Committee of the Mathematical Association 1987). By focusing on experiences that enrich quantitative and comparative language and that improve their ability to communicate using the language of mathematics, we contribute significantly to the development of their potential for thinking mathematically.

In order for children to become fluent in the language of mathematics, they must have opportunities to hear and speak the language (National Council of Teachers of Mathematics [NCTM] 1989). Language and basic mathematical understandings develop hand in hand when all children are given opportunities to—

- reflect and communicate mathematically during interactive play;
- extend their personal involvement with the language of mathematics derived from literature;
- interact with other children or adults in natural settings in mathematically interesting ways.

When such experiences are relevant to children's lives, they are likely to generate meanings for quantitative language and provide a significant basis for important mathematical concepts.

## ILLUSTRATIVE CLASSROOM EPISODES

A central theme of the *Curriculum and Evaluation Standards for School Mathematics* (NCTM 1989) is that knowledge should emerge from problem situations. In this spirit, the three classroom episodes that follow highlight interventions for differently abled children in classroom settings. Whenever possible, the mathematics is embedded in thematic units, presenting an opportunity for children to make connections that are personally meaningful. Thus children will see the usefulness of mathematics in everyday life (NCTM 1989). Ideas for extending both the intervention and the mathematics are then provided. Throughout, the examples emphasize reaching children and stretching their mathematical thinking through language.

### Close to One Pound

Mrs. Fitzpatrick's kindergarten pupils were working on a vegetable unit. The unit began with the story *Eating the Alphabet: Fruits and Vegetables from A to Z* (Ehlert 1989). As the children explored an assortment of vegetables, they identified each vegetable by name. They also observed many similarities and differences and began to sort the vegetables (e.g., by leaf and stem, roots and seeds, big and little, heavy and light).

Each child had an opportunity to hold different vegetables in each hand and compare them by estimating which was heavier or lighter. The teacher modeled and encouraged the language *heavy, heavier* and *light, lighter* as the children placed the vegetables on a balance scale to observe how a heavier vegetable caused the pan to drop. Mrs. Fitzpatrick took advantage of a teachable moment by posing the following problem:

The market was having a special. For one low price of two dollars, customers could mix or match any combination of vegetables up to one pound. What could be bought for two dollars?

As the children were working, Mrs. Fitzpatrick observed Terrell, a child with motor deficits, drop several of the radishes. Afterward he ended up selecting a large squash (an easier vegetable to handle). When he put the squash on the scale he said, "It's heavy! It's almost to the 'one.' I need more. Something little to get to the 'one.'" Mrs. Fitzpatrick said, "Yes, that's right, Terrell, it is close to one *pound*."

**Manipulative experiences like these nurtured Terrell's lagging motor development. Mrs. Fitzpatrick supplied vegetables of different sizes so Terrell could opt for larger vegetables when he had difficulty with the smaller ones.**

**Mrs. Fitzpatrick took advantage of a teachable moment to model appropriate vocabulary and hence nurture language development.**

"Can I have that carrot?" Terrell asked Lisa. "It's not very big. I think that will make it go to the 'one.'"

Mrs. Fitzpatrick smiled at Terrell's insight and asked, "Terrell, why do you want it to be close to the 'one'?"

Terrell responded, "Well, the special says for two dollars we can get one pound."

When the carrot was put onto the scale, Lisa said, "Oh, Terrell, it went past one pound."

"But not by much. I know what to do!" exclaimed Terrell. Mrs. Fitzpatrick watched as Terrell struggled in his effort to get to the green beans. She

**Because of Terrell's motor deficit, the gathering of information through the sense of touch was a valuable experience. The larger pans that the teacher had supplied for Terrell's scale allowed him to place the vegetables more easily.**

also saw him beam as, one by one, he managed to balance the beans on the scale. "There, it worked! Now I need to draw a picture of what made one." Terrell paused, looked at Mrs. Fitzpatrick, and then added, "pound." Mrs. Fitzpatrick noted with pleasure his use of the new vocabulary.

As Mrs. Fitzpatrick moved to another group of children, she saw Terrell take out paper and a crayon. Like his classmates, Terrell was intrigued by the task of finding and making a record of vegetables that weighed as close as possible to one pound.

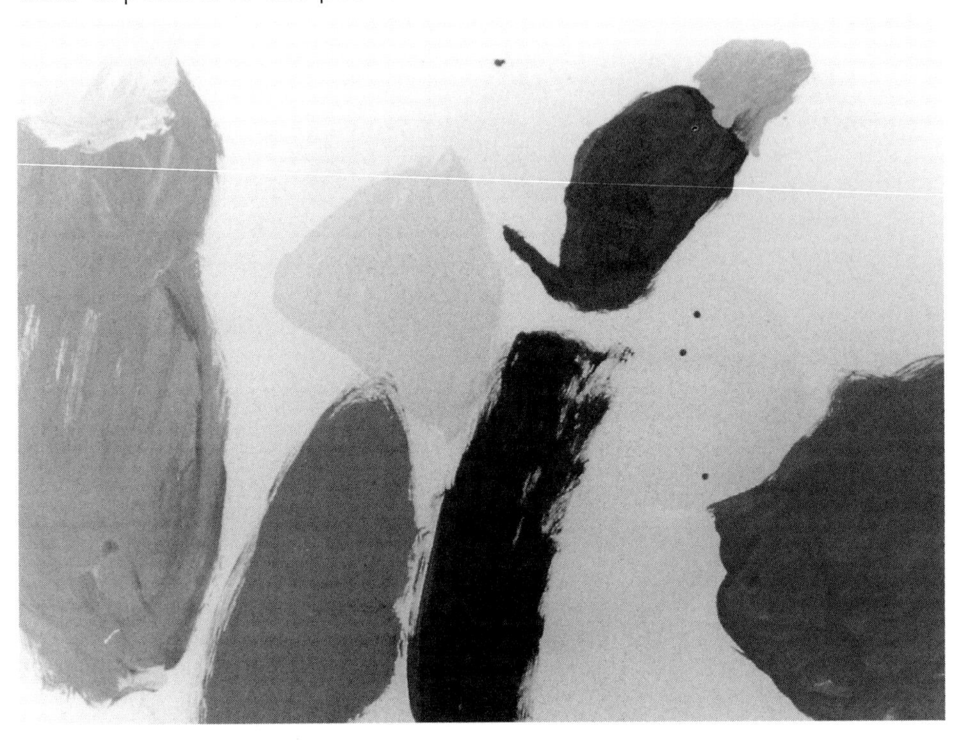

When he had finished, Terrell looked to see what the others had done. "I didn't use as many vegetables as Lisa," he announced. The children began comparing the numbers of vegetables they used. Mrs. Fitzpatrick encouraged some of the other children to line up the vegetables in a one-to-one matching to help them compare.

Mrs. Fitzpatrick followed up on Terrell's statement in the following way: "All of you were able to get 'close to one pound' with the vegetables that you chose. But everyone else used more vegetables than Terrell did. Why do you think Terrell was able to come close to a pound by using so few vegetables?" She was pleased with their thoughtful responses.

**The teacher pointed out that more than one way existed to solve the problem. She also stretched the children's thinking by asking why Terrell came close to a pound with so few vegetables.**

Later that day, the children took turns putting one vegetable in a box and describing the way it felt so that others could guess what was in the box. At Terrell's turn, he selected an ear of corn. "It's sort of smooth and sort of bumpy on the outside."

"Can you think of more clues, Terrell?" asked Mrs. Fitzpatrick.

"If I dig in a little bit, it feels hairy. If I dig a little more, it has lots of bumps. If I dig too hard, it's juicy."

Mrs. Fitzpatrick recorded the descriptors on paper as other children took turns describing the attributes of various vegetables. Later she planned to read the attributes and ask the children to match each with the vegetable. ❑

## Extending the Special Education Thrust

In some early childhood intervention programs, average and above average children with no apparent deficits are included as role models for children with special learning needs. In the area of communication, including mathematical language, it is often easier for children to model the language of a peer than that of an adult. The children often use their peers' ideas as building blocks to solve problems and to extend their own thinking. By constructively interacting with classmates, children construct knowledge, learn other ways to think about ideas, and clarify their own thinking (NCTM 1991). Incorporating these kinds of language experiences is especially important for children with special needs.

The following variations enable the modification of the previous lesson to meet the needs of language-deficient children in prekindergarten programs, including, for example,

- children with hearing deficits who have had limited experiences with language;
- children with physical disabilities who may have a limited experiential basis for articulating and understanding language because in their preschool years they focused on self-help skills in the home;
- children with mental disabilities who can profit from the conceptual emphasis associated with the language-rich settings.

### *Variation 1: Sorting*

In this extension, each child selects one vegetable to hold. After identifying two areas on the floor for the sorting process, the teacher asks a child to tell something about the vegetable (e.g., "My vegetable is green") and then directs the children holding a green vegetable to go to the designated area. The children holding a vegetable that is *not* green go to the other area. Then the group discusses why a particular vegetable belongs in the group. This activity could be repeated several times with other attributes. Alternatively, the teacher could show two groups of vegetables and ask the children to guess the sorting rule.

Next, as described above, the children hold different vegetables in each hand and compare them by estimating which is heavier or lighter, using terms like *heavy*, *heavier* and *light*, *lighter*. As in the activity, the children should note what causes the pan to go down (heavier vegetables) and be challenged to place additional vegetables in a pan to try to balance the two pans. As a follow-up activity, the children draw the vegetables to record their findings.

### *Variation 2: Verbalizing attributes*

In this variation, the children select five vegetables from those available. The teacher initiates a discussion of some of the attributes of the vegetables selected, emphasizing all the senses. The children close their eyes and turn

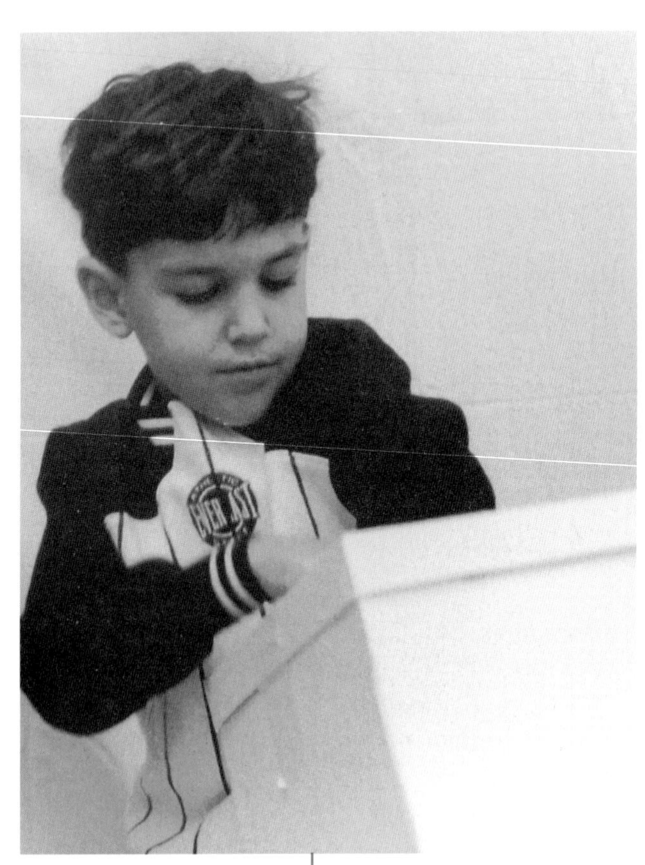

away while a duplicate of each vegetable is placed in a separate "feely box." Keep one of each vegetable in full view for the children's reference.

One child then feels the vegetable in the box and describes its attributes. The other children try to determine which kind of vegetable is in the box. Once a decision has been made, the box (with lid on) is placed next to the duplicate vegetable. After all the "feely boxes" have been sorted, they are opened and checked for correct placement. Alternatively, one child could work independently, feeling the vegetable, guessing it, placing it next to a matching vegetable, and opening the box to verify the match.

### Extending the Mathematics

The thematic approach to curricular planning and instruction allows the teacher to offer balanced integrated learning experiences for the children. Often the integration of mathematics with other areas of the curriculum and the development of a variety of mathematical strands are natural outgrowths of a selected theme (NCTM 1991). Children interact with each other and with the teacher as they bring enriched meanings to mathematical concepts.

In the first part of the classroom episode, the children noted that when they placed a vegetable on each pan of the balance scale, the one that was heavier lowered the pan. An extension of that phase of the activity might be to focus on using different nonstandard units (blocks, cubes, teddy bear counters, etc.) to weigh a particular vegetable. Ask, for example, How many Unifix cubes does it take to balance a carrot? How does that compare to the number of Unifix cubes needed to weigh an ear of corn?

The vegetable theme in the episode could also serve as a vehicle to introduce and explore three-dimensional shapes. The teacher presents a wooden geometric solid (e.g., a sphere) and asks the children to select the vegetable that has a shape most like the sphere's. (The children might select a radish, onion, or potato.) The introduction of other shapes, such as the cone (carrot or turnip) and cylinder (cucumber or carrot with ends cut off) could lead to a sorting activity and a discussion of the characteristics of the solids and the shapes of the vegetables.

The lesson with vegetables could be linked to an art lesson by having the children use cut vegetables dipped in paint to create vegetable prints. It would be quite natural to encourage the children to create or copy a pattern as they stamp with the vegetables. Furthermore, the children could create a pattern as they place cut vegetables on a skewer to make a shish kebob to be enjoyed during snack time.

## Move Over!

The first-grade children laughed as they listened to Mr. Henry read *Roll Over! A Counting Song* (Peek 1981), in which a little boy keeps rolling over and as he does so, ten imaginary animals fall out of bed, one after the other. "Shall we read it again so you can act it out?" In response to an enthusiastic yes, Mr. Henry assigned the roles. The group of "players" was mixed and included Amara, a child with significant developmental lags and memory deficits and several other children at risk, as well as youngsters who were average or above average in their functioning. As the children improvised by using a rug as the bed, the play went on, and "the little one said, 'Move over!'"

**Mr. Henry's practice was to pair children having lower-functioning skills with more capable peers. Amara was paired with a child with good verbal skills who was also a good, logical thinker.**

After the story, Mr. Henry instructed the children to pair up with their problem-solving partner for that day. He gave each pair a mat with a picture of a bed on it and a dozen cubes. "With your partner, use cubes for the characters in the story. Together show how many creatures were in the bed at the start. Then take turns. One should tell the story while the other acts it out with cubes."

Mr. Henry circulated among the children as they worked. He noticed that two groups had finished early and were creating their own variations of the story. He suggested to these children that they enter their story in the computer, which they did with his help.

And so the "Move Over!" story came about (see fig. 8.1). At the point in the story where two characters were in the bed, Amara suggested, "Let's have two fall out this time."

**Amara may never have initiated creating personal versions of the story, but she was afforded a very rich experience because of the problem-solving partner assigned to her that day.**

**This remark demonstrates that when children encounter a meaningful problem, they have an opportunity to reassess and justify their thinking as they resolve it.**

Juan argued, "But we can't do that because we'd have no one left."

Ally enthusiastically contributed, "That's okay. That would be zero." They all laughed and said, "Then *no one* would say, 'Move over!'" With Amara's help, the group was able to agree on the conclusion for the story.

> We wrote this funny story together:
>
> Ten in the bed and the little one said, "Move over!"
> And three fell out!
>
> Seven in the bed and the little one said, "Move over!"
> And two fell out!
>
> Five in the bed and the little one said, "Move over!"
> And three fell out!
>
> Two in the bed and the little one said, "Move over!"
> And two fell out!
>
> Zero in the bed and no one said, "Move over!"
> And no one got any sleep because they all hurt their heads!
>
> The End
>
> By Nikhol, Juan, Ally, Elizabeth, Amara

Fig. 8.1. A computer printout of the children's story

The next day the group shared "Move Over!" with the class. Everyone laughed at the ending. Afterward, Mr. Henry asked, "How could we show with numbers what happened in the story?"

He gave each pair of students a set of "–1," "–2," and "–3" cards. He then invited the children to listen as he retold their story and picked a card to match the "move over" action in the story. As the reading continued,

the children showed their "–1" card each time one character fell out of bed. When the story came to "and two fell out," he noticed that Amara hesitated in selecting the "–2" card and that she looked to others for reassurance. Before continuing, he called on Juan to tell why he chose the "–2" card. From time to time as the story progressed, the children told why a card was selected, and Mr. Henry noted that Amara grew more confident.

**Justifying responses is important. Because of Amara's lack of confidence in responding, however, Mr. Henry delayed calling on her to justify her choice of card. For the moment, he gives her further experiences to allow for overlearning.**

On another day, Mr. Henry read Gerstein's (1984) version of *Roll Over!* In this story the animals that are crowded into the child's bed fall out, one by one, until only the child is left. Afterward, Mr. Henry helped the children make a chart (see fig. 8.2) to show how this story, the computer-published "Move Over!" story, and *Roll Over! A Counting Song* were alike and how they were different. ☐

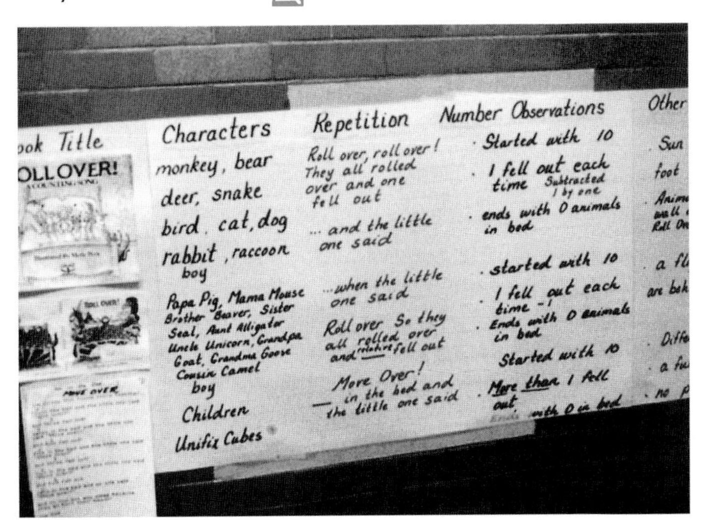


Fig. 8.2. The pupils' chart comparing the three stories

## Extending the Special Education Thrust

Most often, children who need additional challenges are readily identified by teachers. In making important connections in the learning process, these children make greater intellectual leaps than others. Like many children, the gifted and talented frequently draw from previous experiences and knowledge to embellish or access strategies to extend their own learning. Other characteristics of gifted children may include apparent boredom or dissatisfaction with the difficulty level of an activity. The slow rate at which an activity proceeds may also frustrate these children.

The previous lesson can easily be extended to provide opportunities to nurture higher-level thinking among very capable learners. Teachers may choose to pose related but more complex problems for some children, for example, "Suppose that ten characters were in the bed, and each time one more fell out than in the previous episode. How many times would we say, 'Roll over!' before no one was left in the bed?" Because using counters or cubes would still be appropriate to help solve the problem, the teacher would be able to meet the needs of children at various levels without noticeable differentiation.

Another extension involves the illustrations in *Roll Over! A Counting Song* and the concept of subtraction as part-part-whole. On each of the two facing pages of the book is a total of ten animals—some in the bed and some on the floor. The children use cubes to represent the animals as they place some animals in the bed and hide in their hand or under a bowl the ones that fell out of the bed. Then they ask their partner to tell how many fell out of the bed.

Children come to school with a wide variety of experiences and knowledge. It may therefore be necessary for the teacher to adapt the activity for more capable students by using larger numbers or by supplying a mixture of addition and subtraction cards. Alternatively, children who might benefit from additional challenges could work at the computer with an appropriate program.

### Extending the Mathematics

The theme of "in the bed" lends itself to other rich mathematical extensions. For example, the teacher or a student could record the corresponding number sentences as the story is told. This activity also affords an opportunity for the teacher to assess the children's level of understanding.

The teacher can continue the activity by first modeling alternative verses (e.g., if there were eight in the bed and the little one said, "Move over!" and three fell out, how many would still be in the bed?). Then the children could be paired and invited to create and record their own verses. Again, if appropriate, the children could be encouraged to write the corresponding number sentences, which could then be shared with the group.

Children who are tentative in their responses should be encouraged to use objects to help solve the problem. Likewise, other, more confident children should be encouraged to use objects only to verify their results. Later, the children can reverse the process by selecting a number sentence and identifying the appropriate verse.

Another variation of the story could include addition. Instead of having characters falling out of bed, the story could have the characters climbing back into bed. There were seven in the bed and two little ones said, "Move over! We're getting back into the bed!" (i.e., 7 + 2). The story could be told using only addition or both addition and subtraction. Further variations may come about as a result of reading *The Napping House* (Wood 1984), a story that uses the theme of many creatures climbing into a bed. The ending to this story includes a surprise that reverses the action. Additionally, the story could be dramatized or acted out with finger puppets.

From the standpoint of both instructional planning and individual assessment, it is imperative in selecting activities for children to keep in mind the unique and authentic interests and ideas of the children. Active listening helps teachers become aware of their pupils' interests and ideas.

## Hats

Ms. Weiner thought about yesterday and how much the children in her kindergarten group had enjoyed *Aunt Flossie's Hats (and Crab Cakes Later)* (Howard 1991). In the book, Aunt Flossie tells her niece stories about the hats in her collection, each of which remind her of a unique experience. Ms. Weiner looked forward to the follow-up session planned for today's lesson.

After the daily opening, Ms. Weiner sat with the children in a large circle. She made careful note of where Phoebe, a child with a degenerative hearing impairment, was seated in relation to her. "Do you remember

yesterday's story? Please tell your partner about one of Aunt Flossie's hats and why it was so special."

Ms. Weiner watched as Phoebe and Ann repositioned themselves so that they were directly facing each other. After the children had talked to their partners and had shared comments with the larger group, Ms. Weiner showed them an unusual hat from her childhood. She then told a story about why it was special to her.

"Did you remember to bring your special hat today?" she asked them. In turn, each child showed and told about a special hat.

As he was telling about his hat, Johnathan said, "Hey, I found a number in my hat! There's an eight right here on this tag." As several others found the hat sizes, Ms. Weiner recorded them on the chalkboard, and a discussion about the numbers ensued.

Taking advantage of the pupils' interest in hat sizes, Ms. Weiner asked, "How many of these learning links do you think it would take to go around *your* head?" Ms. Weiner recorded the responses on the chalkboard as the children announced their estimates. The children then continued to work with a partner to measure their heads with links. The children were surprised when they found very few differences in the sizes of their heads.

Ms. Weiner overheard Phoebe tell Ann, "I think Ms. Weiner's head will be a lot bigger than ours!" They then measured to see. They were again surprised to find that the teacher's head size was not much different from their own.

**Even though Ms. Weiner wore an auditory unit, she placed herself in the circle opposite Phoebe to make it easier for her pupil to speech read. Phoebe's hearing deficiency is degenerative, so Ms. Weiner was aware of the importance of helping Phoebe develop good speech-reading skills early.**

**Ms. Weiner often gauges her wait time by observing Phoebe.**

**The children have been cued to look in Phoebe's direction when speaking. The fact that the teacher records what is verbalized helps all students organize their thinking. This technique is especially valuable for Phoebe.**

**Games like this involve not only logical reasoning but also a great deal of language. While Phoebe still has partial hearing, she benefits from experiences rich in language, particularly if she is encouraged to speech read.**

**Because the background noise is greater during the partners' sharing, Phoebe is helped when the teacher invites the children to summarize their thinking in a more structured way. This structure helps all the children to organize and communicate their thinking.**

"Let's do more measuring and comparing," suggested Ms. Weiner. "How do you think your waist measure will compare to your head measure?" The children were even more surprised when they found very little difference in the two measures. They had expected their head to be much smaller than their waist.

Ms. Weiner then asked five children to put on their hats and line up so the others could see them. She helped the others make statements about the five hats: "The biggest hat is in the middle." "The first hat is red." "The dress-up hat is in between the winter hat and the baseball cap." "The tallest hat is last." As the statements were presented, Ms. Weiner restated each as she recorded it on the chalkboard.

With a group of four children, Ms. Weiner played the "I'm thinking of a hat" game. "It does not have three feathers. It has more than one color. It is not the last hat. Which is my hat?" she asked. "Think. Which hat could it be?" challenged Ms. Weiner. She then asked the children to tell a partner which hat they thought she had in mind and why. Then she pointed to each hat in turn and asked, "Could it be this hat? Why or why not?" As the children volunteered information, Ms. Weiner marked an "X" on the chalkboard behind each hat as it was eliminated. ❑

## Extending the Special Education Thrust

Typically, a child with behavior disabilities operates at least within the average ability range. To integrate the child successfully into a regular classroom, the teacher needs to establish consistent rules and procedures. Although the child may need structure, he or she should also be involved in risk-free decision making by carefully limiting the number of his or her choices.

The availability of too many choices acts as a stimulus, and the child gets lost in the process. When activities are too long or too complex, inappropriate behavior begins to escalate. Then the learning process is no longer the focal point, and the behavior-disabled child creates a disruption in the classroom. Teachers can minimize these disruptions by planning shorter lessons or activities at a more appropriate level. When the observant teacher learns to identify and recognize disruptive behavior patterns, many potential problems can be successfully averted.

If a child with a behavior disability had been in the room, the teacher might have elicited appropriate rules and guidelines for behavior from the group prior to telling the story about her hat. This approach is particularly important for the child with severe behavior problems. To help keep the child focused on the activity, it may be helpful to issue a passport or behavior checklist. During the activity, the behavior targeted on the passport or checklist should be positively reinforced.

As the children identify the hats with the sizes labeled, they could sort the hats by placing them in two designated areas of the room. Removing the hats from close proximity eliminates unnecessary stimuli and creates a point of order to begin the next segment of the lesson.

For ease of distribution, the teacher could give a chain of a predetermined number of links to each pair of pupils as they begin to measure. The teacher could also give the following privilege to those who finish the activity early: After the pupils record their estimates and actual measurements, then they may explore freely with the links until everyone is finished.

### Extending the Mathematics

Problem solving is a primary consideration in all mathematical instruction (NCTM 1989). Many problem-solving situations could be presented using not only *Jennie's Hat* (Keats 1966) but also *A Three Hat Day* (Geringer 1985). These problems will be both meaningful and relevant, since they evolve from the literature as well as from the children's authentic experiences.

In the first of these stories, Jennie is disappointed with her plain hat. Her friends the birds, recalling Jennie's past kindnesses, bring ornaments for her hat. After hearing the story, the children could create their own hats. Children might be given a specified amount of money to spend to transform their plain hats into fancy hats. Each ornament is assigned a price. The children make decisions on how to spend the money to decorate the hat. During the process of selecting and purchasing decorations for the hat from a classroom store, the children are engaged in an active problem-solving situation that also enhances the development of money concepts.

In the story *A Three Hat Day*, R. R. Pottle the Third collects hats. On sad days he wears two hats, but on very sad days he wears three. This story lends itself to several natural problem-solving applications. For example, the teacher could pose the following question: On a very sad day, in how many ways could R. R. Pottle wear his hats? The children could opt to solve this problem using one of various strategies:

- Act it out
- Make an organized list
- Draw a picture

The possible solutions could be recorded and shared with the group. For pupils who are more able, a problem extension might be to ask the children to solve the problem using four hats.

Another opportunity for problem solving could incorporate probability. For example, the teacher could pose the following: "R. R. Pottle the Third has three hats: a red hat, a blue hat, and a yellow hat. On a *very* good day, he wears just one hat. Let's suppose that he closes his eyes and picks *one* hat. Which hat do you think he might pick?" The experiment could be conducted using cubes and the results tallied. Discussion might center on questions like the following: Does one color of hat have a better chance than others of being picked? If R. R. Pottle the Third were to close his eyes and pick again, would he choose the same hat? Would the chances change if R. R. Pottle the Third had two red hats and one yellow hat? These questions invite further experimentation, comparison, and discussion.

## CREATING WINDOWS OF OPPORTUNITY

In the three classroom episodes in this chapter, the mathematical learning was based on teaching that reflects a philosophy of creating *windows of opportunity* for all young children, reaching in subtle yet appropriate ways to meet the needs of children who are differently abled. By framing mathematical learning with ideas consistent with the two *Standards* documents of the National Council of Teachers of Mathematics (1989, 1991), educators strengthen their resolve to ensure that these opportunities are authentic, child centered, meaningful, and culturally and developmentally appropriate. The key is to allow children the opportunity to design their own personal windows. In this way, mathematics comes to life through experiences that are seen, felt, heard, and created firsthand.

Through these three classroom episodes, we have seen many children experience the language of mathematics as they—

- experienced confidence through sharing and extending tentative ideas;
- interacted in socially constructed environments;
- internalized mathematics concepts by verbalizing their constructs;
- strived to become competent as autonomous learners.

The thematic approaches stimulated by literature contributed variety, excitement, and relevance to learning. If teachers view themselves as facilitators of learning, teaching and learning become transformative, opening endless new windows and opportunities for grasping and learning mathematics.

## REFERENCES

Ehlert, Lois. *Eating the Alphabet: Fruits and Vegetables from A to Z.* New York: Harcourt Brace Jovanovich, 1989.

Geringer, Laura. *A Three Hat Day.* New York: HarperCollins Children's Books, 1985.

Gerstein, Mordicai. *Roll Over!* New York: Crown Publishing Group, 1984.

Howard, Elizabeth Fitzgerald. *Aunt Flossie's Hats (and Crab Cakes Later).* New York: Clarion Books, 1991.

Keats, Ezra Jack. *Jennie's Hat.* New York: HarperCollins Children's Books, 1966.

Members of the Teaching Committee of the Mathematical Association (United Kingdom). *Math Talk.* Portsmouth, N.H.: Heinemann, 1987.

National Council of Teachers of Mathematics. *Curriculum and Evaluation Standards for School Mathematics.* Reston, Va.: The Council, 1989.

———. *Professional Standards for Teaching Mathematics.* Reston, Va.: The Council, 1991.

Peek, Merle. *Roll Over! A Counting Song.* New York: Clarion Books, 1981.

Wood, Audrey. *The Napping House.* New York: Harcourt Brace Jovanovich, 1984.

## SAMPLE RESOURCES FOR INTEGRATING EARLY CHILDHOOD MATHEMATICS WITH LITERATURE

Burns, Marilyn. *Math and Literature (K–3)*. Sausalito, Calif.: Math Solutions Publications, 1992.

Griffiths, Rachel, and Margaret Cline. *Books You Can Count On: Linking Mathematics and Literature*. Portsmouth, N.H.: Heinemann, 1991.

Members of the Teaching Committee of the Mathematical Association (United Kingdom). *Math Talk*. Portsmouth, N.H.: Heinemann, 1987.

Thiessen, Diane, and Margaret Matthias, eds. *The Wonderful World of Mathematics: A Critically Annotated List of Children's Books in Mathematics*. Reston, Va.: National Council of Teachers of Mathematics, 1992.

Thomas, Rebecca L. *Primaryplots: A Book Talk Guide for Use with Readers Ages 4–8*. New York: R. R. Bowker, 1989.

Welchman-Tischler, Rosamond. *How to Use Children's Literature to Teach Mathematics*. Reston, Va.: National Council of Teachers of Mathematics, 1992.

Whitin, David J., and Sondra Wilde. *Read Any Good Math Lately? Children's Books for Mathematics Learning, K–6*. Portsmouth, N.H.: Heinemann, 1992.

**Judith K. Wells** is an elementary mathematics staff support specialist with the Shaker Heights City School District in Shaker Heights, Ohio. She was formerly an instructor in the mathematics department at Illinois State University, where she taught mathematics methods and content courses for elementary, special, and early childhood education majors. She is interested in teacher education and staff development.

**S. Kay Dunlap** is a teacher with the Shaker Heights City School District. Her teaching experiences include first through third grade, reading improvement, and Chapter 1 classes. Additionally, she is a doctoral candidate in curriculum and instruction at Kent State University, Kent, Ohio. Her interests are in the areas of early literacy and teacher education.

**Beverly Hopcraft Sullivan** is an early childhood special education teacher with the Shaker Heights City School District. She received her master's degree at Kent State University, Kent, Ohio. She is certified to teach gifted students and those with learning disabilities and behavior disorders. She was formerly a second-grade teacher in Columbus, Ohio.

**Mary Leo Bourisseau** is currently a teacher of developmentally handicapped students in the Shaker Heights City School District. She has also worked with populations that have severe behavior handicaps.

**Sharron Williams** is a specialist in the education of the gifted with the Shaker Heights City School District and an educational consultant. She teaches mathematics to elementary school–aged students and conducts workshops on curriculum infusion. In addition, she writes curriculum for law-related education and is a group leader for multicultural awareness seminars.

Number
Wipe Out

9

# Number and Operation Sense

*Francis (Skip) Fennell*

*in collaboration with*

*Theodore E. Landis, Jr.*

***Is 16 closer to 10 or to 20?***

***About how many pennies can fit in your hand?***

***If the meal costs $49.57, how much of a tip should we leave?***

***When we multiply 6.2 × 3.5, is the product 21.7, 28.7, 17.3, or 217?***

***If a ten-year-old is five feet tall, about how tall will he or she be at age twenty?***

EACH of these questions deals with the elusive notion of number sense. What is this thing called number sense? It can be described as good intuition about numbers and their relationships (Howden 1989). Number sense includes an awareness and understanding about what numbers are, their relationships, their magnitude, and the relative effect of operating on numbers, including the use of mental mathematics and estimation. The National Research Council in its publication *Everybody Counts* has stated that "the major objective of elementary school mathematics should be to develop number sense" (1989, p. 46).

Although the expression *number sense* is a relatively new one to mathematics educators, teaching to encourage insight and a comfortable use of number and number concepts has been on the "agenda" of insightful teachers

---

Many of the instructional ideas presented in this manuscript come from lessons depicted in the National Eisenhower Project entitled Teaching for Number Sense Now! Reaching the NCTM Standards (see Fennell 1992). This project (#R168D00132-91) involved teachers and supervisors from Baltimore, Carroll, and Howard Counties in Maryland and the District of Columbia Public Schools. The project materials include three staff development videotapes and accompanying printed support materials.

for years. The *Curriculum and Evaluation Standards for School Mathematics* (National Council of Teachers of Mathematics 1989) has recommended that the K–8 curriculum include whole number concepts and skills and the development of number and number relationships so that *all* children can develop number sense for whole numbers, fractions, decimals, and integers. Clearly, number sense should be a major consideration for any contemporary mathematics curriculum.

This chapter focuses on the importance of number sense as it relates to number meaning, number relationships, number magnitude, and operation sense for all students and highlights how this emphasis has played an important role in programs for students with special needs. Number sense is not a special unit or educational "Band-Aid" of special lesson activities and games. More than anything else, number sense is the foundation from which all other mathematical concepts and ideas arise.

The current emphasis on number sense is all about the development of a sense of knowing what to do in situations involving numbers and their use. This emphasis characterizes a way of teaching. One of the major goals of a number sense approach to instruction is to encourage students to choose or create methods and techniques that take advantage of their own understandings as they strive to "make sense" of, or develop a "feel" for, numbers. Trafton (1989, p. 75) has indicated that number sense is something that unfolds. Thus, the goal of teaching for number sense is developmental, involving an ongoing emphasis on nurturing sensible ways of thinking in numeric situations.

Number sense also involves nurturing a special kind of "spatial" sense—that is, the ability to "see" numbers, note their size, and determine relationships among them. To some extent this ability depends on visual perception. Teachers and instructional materials often focus on the visual strengths of learners. Even manipulative materials rely on visual organization to enable students to "see" concepts or relationships being developed. For many young students, the acquisition of number sense is the enhancement of their visual perception.

Students need to be able to approach numbers with flexibility, knowing what to do and having a repertoire of techniques and strategies for such solutions. They should be observed creating different ways to model the same number, identifying the missing number in a pattern or sequence and being able to define the pattern, using benchmarks (reference points) to describe the magnitude of a number, and describing what happens when using a particular operation.

## ILLUSTRATIVE CLASSROOM EPISODES

The following classroom episodes illustrate examples of number sense. They highlight the importance of number meaning, number relationships, magnitude, and operation sense. The episodes also illustrate how special education teachers have been able to meet the needs of their students while reaching important number sense goals.

## Developing Number Meanings ...

### ... Using Counting-Board Activities

From past experience with her primary-level, educationally handicapped and learning disabled students, Ms. McPartland has come to value the use of counting boards for representing and visualizing numbers. She is convinced that such activities nurture number meanings in important ways.

Today, Ms. McPartland began the mathematics lesson by asking the children if they would like to use pasta pieces to make a "six board." Greeted with enthusiasm, she distributed pasta pieces, tagboard, and glue to each child, demonstrated what was intended, and challenged the children to "make *six* in many different ways." Then she stepped back to observe as the children worked.

Ms. McPartland knew that most of her students recognized "six" in standard arrangements (e.g., fig. 9.1) because of dice games they had played at home and at school. She was pleased by the way in which the children now were exploring with pasta on their counting boards and finding other visual patterns for the number (fig. 9.2).

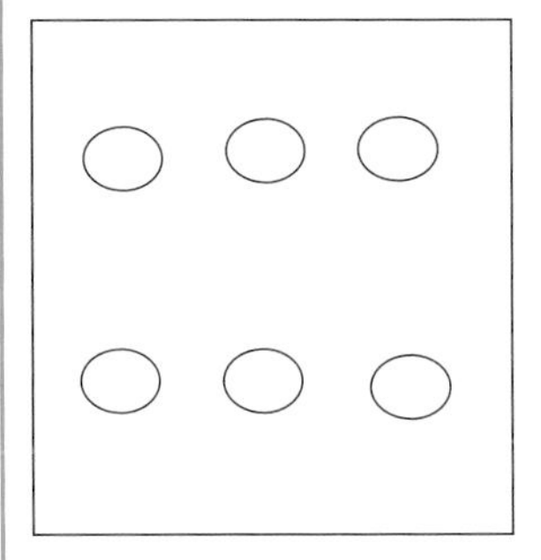

Fig. 9.1

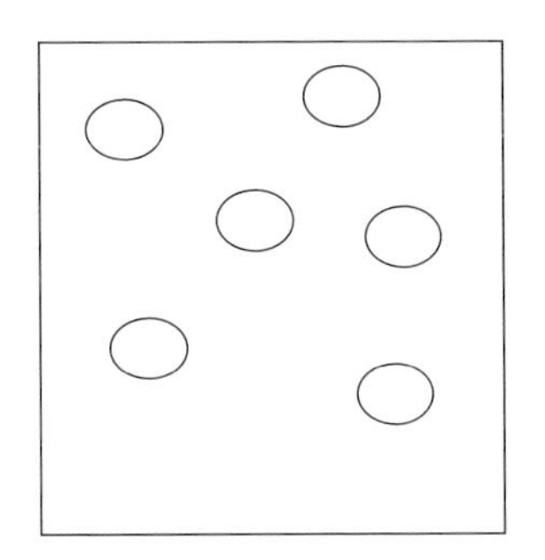

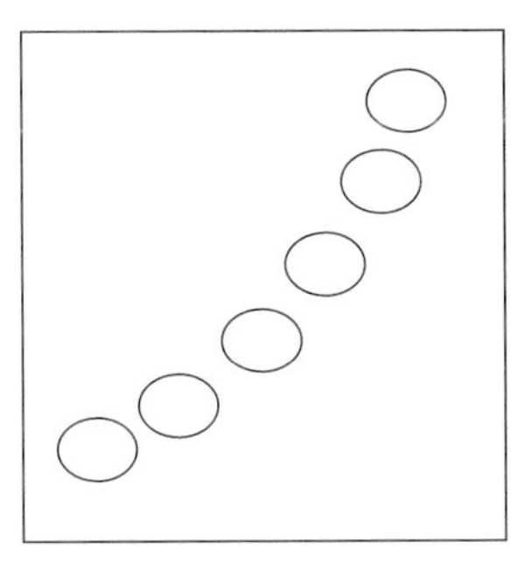

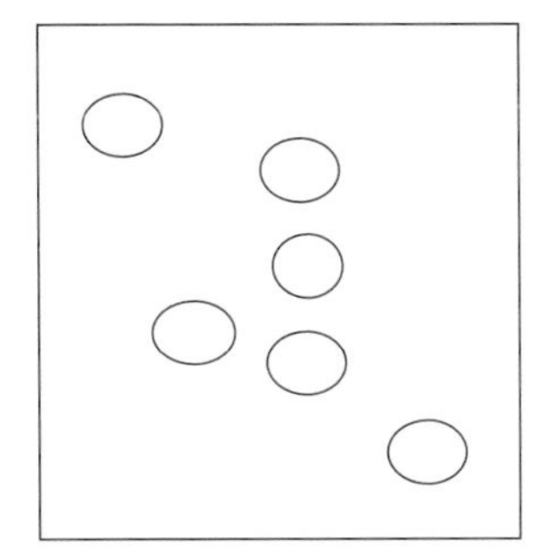

Fig. 9.2. Pasta cards for "six"

After the children had created their six-boards, Ms. McPartland involved all the children in a "sharing" session. Interaction centered on questions like the following:

How many pieces of pasta (objects) are on your counting board?

If you had one more than the number on your counting board, what number would you have?

Are there more than five pasta pieces on your counting board? More than ten?

How many more pasta pieces would you need to make eight? Ten?

**The importance of number meaning lies beyond the mere recognition of the number and "finger counting" to verify it. The importance of "seeing" numbers in different contexts helps to develop a flexibility for number use. Such activities enhance visual learning.**

The sharing time gave the children the opportunity to compare their thoughts with others and provided a forum for stretching their thinking about number. Ms. McPartland informally assessed the children as they responded. She noticed, for example, that Jamie did not hesitate when

**A step beyond counting boards would be to prepare tagboard or paper-plate dot cards. These may be used as flash cards. Again, it will be useful to prepare these in familiar and not-so-familiar arrays to help students visualize particular numbers in varied contexts.**

relating six to five or ten but that he had no strategy for determining how many more pasta pieces might be needed to make eight or ten. Ms. McPartland made the mental note to do more ten-frame work during the next weeks to help Jamie. "Maybe Ms. Gomez would have some good ideas."

### ... Using Ten-Frame Activities

Ms. Gomez valued ten-frame activities for both her primary- and her intermediate-level learning disabled students. They helped the children visually represent and compare numbers and explore sums of 10 and related differences, and they provided a visual foundation for strategies like adding and subtracting "beyond 10" for number facts like 6 + 8 (a ten frame and 4 more) or 13 − 5. (Further information for using a ten frame in the development of thinking strategies for mastering basic addition and subtraction facts can be found in Thornton [1990].)

Earlier in the year Ms. Gomez had just flashed dot cards to her students. They then used counters on their ten frames to represent the numbers seen. She varied this activity by calling out numbers, presenting numeral cards, or using counters on an overhead projector, which the children then illustrated with counters on their individual ten frames. Sometimes for a given number Ms. Gomez asked the students to use two-color counters to show the number in many different ways (fig. 9.3).

**The ten frame is a $2 \times 5$ grid that is a very valuable tool for establishing 5 and 10 as important benchmarks. The ten frame, which was developed by Wirtz (1980, pp. 5, 25), is also useful for collecting and organizing counters. Typically students fill their ten frames with counters going across each of the two rows.**

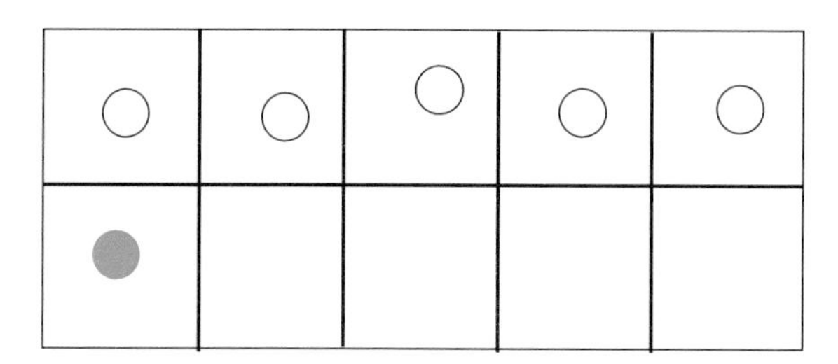

Fig. 9.3. "Six" could be 1 and 5; it also could be illustrated as 2 and 4, 3 and 3, 5 and 1, or 4 and 2.

Today Ms. Gomez had planned several extensions of these activities. As a warm-up, she used ten-frame flash cards and had the children whisper the number of stars seen on each card to a partner. She noted how quickly and confidently the children responded. "Our earlier ten-frame work seems to be paying off," she thought.

The card showing eight stars was next. After flashing it, she asked, "How did you figure that out so fast?"

Not hesitating, Shannon offered, "I just went 5—6, 7, 8 in my head."

Traci said, "I just knew it by looking!"

The children seemed proficient in number recognition, so Ms. Gomez introduced questions like the following as she continued the warm-up:

- How many spaces are filled? (6)
- How many "open" spaces are there? (4)
- How many more to make ten? (4)

Then Ms. Gomez asked the children to get out their own ten frames and some counters for a "show, tell, and share" activity. "I'll flash a ten-frame card and ask you to use counters to show and then tell your partner a number close to that on your card," she explained. "Ready?" At different times during the activity, Ms. Gomez asked for the following:

- "The number that is one more than the number shown." (2 more, 3 more)
- "The number that is one less than the number shown." (2 less, 3 less)

After several rounds of this activity, Shingo suggested they do "add or subtract"—"like we did last week." Ms. Gomez smiled and agreed. She had thought about doing just that and was pleased to see their interest. She started the activity by asking everyone to show "seven" on their frames. Then she wrote "9" on the chalkboard and challenged the students to tell their partners how many counters they would need to "add" or "subtract" to make nine.

Ms. Gomez waited a bit and then asked the group if they were ready to check. "OK, using counters, add or take away the number you said.... Did it turn out like you thought?"

"Sure," said Liz. "I just added two more to make nine."

Out of the corner of her eye, Ms. Gomez noted that Kimancha was one off but was being helped by her partner. She repeated this activity several times. Each time, after calling out a number that the children could show with counters on their ten frame, she wrote a number between 0 and 10 on the chalkboard. The children had to decide how many counters to add to, or subtract from, those on the frame to represent the number displayed.

As they shared solutions, Ms. Gomez wrote the number sentences on the chalkboard. She knew that later in the week she'd ask the students themselves

to work with their partner to record number sentences to match what they did with counters in the ten frame, and she believed that modeling now would help.

Glancing at the clock, Ms. Gomez decided to spend the remaining time focusing on story problems that could be solved using the ten frame. She invited the children to pair up and create their own problems for others to solve. One of the more unusual problems came from Jill and Brett:

> Jamie had 7 model cars and his brother had 6.
>
> How many more would they need to have a total of 20 cars?

Kimancha was quick to voice her initial reaction: "Twenty cars, Jill? That's more than the ten frame will hold!" Ms. Gomez didn't comment. She decided to let the students grapple with the problem. One solution came from Liz: "We could just put *two* ten frames together!" ❑

### Extending the Special Education Thrust

Ms. Gomez's lesson hinged on the students' use of ten frames. If children with orthopedic disabilities or visual or fine motor difficulties had been in the class, magnetic ten frames and magnetic counters might have been considered. These items can be used on magnetic boards, available from many school-supply companies, or placed directly onto some wheelchair desktops or work trays. The magnets help keep the counters in place, enabling these students to manage the manipulative materials more easily and hence better attend and participate more fully in the activities. Faculty and students in vocational shop programs are often willing to help construct such materials or make necessary modifications to laptop desks or trays so they can be used.

Similar materials might also be useful for students with attention deficit hyperactive disorders (ADHD). Alternatively, if the classroom is equipped with a magnetic chalkboard, such students with a partner might use magnetic counters on a large ten frame drawn on the chalkboard.

Some ADHD students, however, may be overly distracted by personally moving the counters on the ten frames, especially until they understand how these materials should be used. If on medication, they should not engage in these activities during parts of the day when their medication is wearing off; instead, writing numbers or number sentences to match what a partner does might be a better role for such students at times like these.

### Extending the Mathematics

One important extension of this lesson centers on developing the relationships between addition and subtraction. Working in pairs, for example, one child might use two-color counters on one (or two) ten frames to represent an addition number fact (part + part = whole). The other child might then illustrate the related subtraction fact (whole – part = part).

Alternatively, students might use counters on the ten frames to illustrate "families" of related addition-subtraction number facts (e.g., 6 + 7 = 13; 7 + 6 = 13; 13 – 7 = 6; 13 – 6 = 7). Working in pairs, as above, one child might use counters to illustrate a number fact while the other records the matching number sentence. Some students enjoy using triangle flash cards as in figure 9.4 in conjunction with this activity.

For "teen" number facts like 6 + 8 or 9 + 4, children might also be challenged to first predict how many more than ten the sum is, then use counters and two ten frames to verify their prediction. Or, for any "teen" number (e.g., 16), they might be challenged to use two ten frames to generate the different addends that make that sum (1 + 15, 2 + 14, 3 + 13, 4 + 12, 5 + 11, 6 + 10, 7 + 9, 8 + 8).

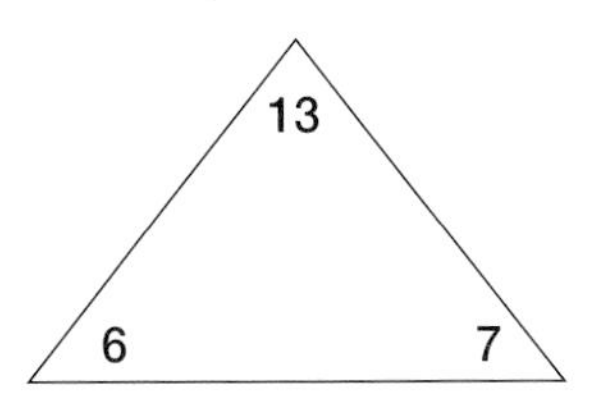

Fig. 9.4. Triangle flash card

Number meaning is the foundation for number sense. All too often special-needs students are expected to move far too quickly into symbolic computation activities with limited experiences involving numbers and their meanings and relationships. Using two ten frames in number-meaning activities like the one above gives visual and kinesthetic-tactile support for exploring sums to twenty.

## Number Relationships and Magnitude

Mr. Landis paused at the door of his intermediate-level LD (learning disabled) classroom. He noticed that several groups of students were taking advantage of the rainy-day "inside" recess by playing some of the math games they had just learned. Gail and Kwan had paired up against Amancha and Russ for "high-low"—each team trying to outdo the other by forming both the largest *and* the smallest three-digit number possible from the three number tiles each team had drawn. (See fig. 9.5.)

As he watched, Gail and Kwan recorded their numbers and started to "build" each of the numbers with base-ten blocks, for that was also part of the game. As they finished and compared their work to what Amancha and Russ had done, Mr. Landis noted that they were using the scoring

Fig. 9.5. The greatest number is 863; the smallest number is 368.

**Pairing students is a valuable way to help them develop confidence and group participation skills. In this activity, student pairs need time to create their greatest and smallest numbers and then to "build" their models. They also might be encouraged to verbalize *how* they determined the greatest and smallest numbers.**

agreed on in class: Each team that correctly created and modeled the largest and the smallest number possible from their tiles earned 1 point for each of the two amounts. The two teams then compared their numbers to determine who would earn the extra tally points for the "largest number" and the "smallest number" overall. Mr. Landis knew that if they were unsure or disagreed, they would refer to the page numbers of the large dictionary he had provided.

An enthusiastic outburst drew Mr. Landis's attention to the opposite side of the room. Calculator in hand, Stacey told Darrell, "There! I did it! I wiped out the 3." She reached for the laminated chart (fig. 9.6) to record the new number on the display.

| Number | Wipe Out | New Number |
|---|---|---|
| 4 576 | 5 | 4 076 |
| 789 | 7 | ? |
| 12 566 | 2 | ? |
| 754 | 4 | ? |
| 8 453 | 5 | ? |
| 33 870 | 8 | ? |
| 95 678 | 9 | ? |

Fig. 9.6. Laminated chart furnished by Mr. Landis for the "wipeout" game

**All students need the opportunity to experiment with calculators—for procedures as simple as entering numbers and as complex as storing amounts in the calculator's memory. "Wipeout" helps students understand the value of each digit within a large number.**

Mr. Landis smiled as Darrell's reply echoed one he himself had used frequently the day he taught this "wipeout" game to the class: "How did you do that, Stacey?"

When the game was played during class, Mr. Landis had required the students first to *predict* how a digit might be "wiped out" (replaced by a "0") on the display. He remembered how, that day, Stacey had quite readily (but incorrectly) suggested that "subtracting 50 from 6541 would wipe out the 5" in the number. When he overheard her "subtract 300" reply to Darrell today, he thought, "Good for you!" He walked over her way to voice his thoughts as the bell signaled the end of recess.

After the bustle of putting things away had subsided, Mr. Landis readied the class for a true-false warm-up activity he had planned to challenge the students to think about numbers close to 100—their size, what they mean, and how they can be used. "I want you to *think* about each statement I read, then tell your partner whether you believe it is true or false and *why* you think that way. When you reach an agreement, raise a hand to let me know. Ready?"

One by one, Mr. Landis read the statements in figure 9.7, and the students came to consensus on whether they were true or false. The last two questions invited many solutions from different perspectives, which he took time to hear. Then he asked the students to "think hard and work with a partner on these four questions" (fig. 9.8),which he presented on an overhead transparency.

There are about 100 desks in this classroom.
There are about 100 people in our class.
There are less than 100 people in our class.
There are more than 100 people in our grade.
There are more than 100 people in our school.
There are about 100 children in the school band.
There are about 100 people on your school bus.

Fig. 9.7. Mr. Landis's true-false statements

**100**

What one thing has the same value as 100 pennies?
About how many months equal 100 days?
About how many days are there in 100 hours?
Do you have any collection of about 100 objects? (If yes, describe.)

Fig. 9.8

**Display various physical models of 100 to serve as classroom benchmarks: 100 pennies, 100 counters, a 100 flat from base-ten materials, and a meterstick (100 cm). Special-needs students need materials to model and reinforce the mathematical ideas being presented. All activities should include "talk back" opportunities: "How did that work?" "How did you solve the problem?" "Show me how you did that one."**

After the warm-up, Mr. Landis asked the class to estimate the number of students in the school. The students worked in pairs. Each group had a calculator and paper and pencil. Henri and Jill thought that since there were 30 students in their class and 30 classes in the school, there must be about 900 students in the school. Once the students found a solution to the problem, they were then supposed to think whether their answer was reasonable or not. Mr. Landis wrote several answers from the student pairs on the chalkboard:

| | |
|---|---|
| Buff and Norma | 300 |
| Heather and Steffan | 3000 |
| Brett and Mia | 800 |
| Henri and Jill | 900 |

Each pair discussed how they determined their responses. The class agreed that Brett and Mia and Henri and Jill had reasonable answers. Brett

**Reasonableness is elusive. Students who understand number and quantity develop benchmarks that help them determine reasonableness. It is important for teachers to question students on the reasonableness of their responses. All students need numerous opportunities to establish and reinforce their benchmarks. The more students have an opportunity to develop such benchmarks, the greater the likelihood that their benchmarks will be used as the basis for other estimates.**

and Mia said that they used 800 for their estimate because they knew that their class was the largest in the school and that some classes had only about twenty students in them. So they thought, like Henri and Jill, 30 times 30 and then subtracted 100. After the explanation, the class decided that the estimates Buff and Norma and Heather and Steffan made were unreasonable.

Mr. Landis then presented the following problem:

> The population of the state of Maryland is about 5 million. Using Maryland's population as a benchmark, what is a good estimate for the population of the United States?

Using a calculator, Rey and Rico multiplied 5 times 50 (for the fifty states), got 250, and announced that an estimate of the United States population could be 250 million. At first some students thought this idea would work, but Heshmat convinced them that using a small state like Maryland as the "average" population state for the fifty-state estimate was not a good idea. He suggested they pick a larger state for the population estimate. Noting that it was just about time for music, Mr. Landis suggested that either they each write a paragraph supporting Heshmat's suggestion and research which state would be better or they present a different solution. □

## Extending the Special Education Thrust

The classroom episode above focused on nurturing several aspects of number sense for multidigit numbers. A modification involving color highlights might be considered for learning disabled students with abstract reasoning, memory, or reversal difficulties, since the use of color can capitalize on visual perception strengths to fix and retain concepts. After these students "build" a number with different-sized blocks (flats, longs, units), they might be encouraged to decide on a particular color for each size—hundreds, tens, and ones—and use colored pencils to record each number they represent. To consolidate learning, the students should also be encouraged to verbally interpret what they have written—specifically, what the different digits mean in relation to the blocks and (focusing on larger-valued digits) *how one can determine* that one number is greater than the other.

For students who are partially sighted, tracing the tile numbers with glue makes a raised surface that may help them better read the numbers on the tiles. Alternatively, some partially sighted students may require larger numerals and high contrast, such as that furnished by using a thick, bold, black felt-tipped pen on white or yellow index cards.

For the "wipeout" activity, a printing calculator may help some students because it provides a printed record of each key pressed, to which they can later refer. The printing option has proved useful for students with memory and visual perception deficits, as well as for those with abstract reasoning and memory difficulties.

As an outgrowth of the lesson itself, student teams might work together to create pages for a "Book of Estimates." *Estimated* data like state population, school population, class size, and other data of interest to the students might be presented by student teams on separate pages. This information might then be collated into a book and stored in a three-ring or theme binder, which will allow additional pages to be added as the year progresses. This project would have special value for students with language, organizational, memory, or abstract reasoning difficulties. Such students could be teamed with accommodating peers for gathering and presenting the data.

### Extending the Mathematics

Natural extensions of the "high-low" activity can challenge the thinking of mathematically talented or gifted students. For example, after creating the largest number possible from three tiles, these students might be asked to present their solutions to related questions like the following:

- What tile(s) would you need to make your number 10 times larger? 100 times larger?
- What number is one-half the largest number you can make?
- What number is one-half of the smallest number you can make?

These students might be further challenged to create and provide answers for similar questions for others in the class.

#### *Benchmarks*

*All* students benefit from an emphasis on using benchmarks—familiar numbers that form a ready frame of reference—for decision making. For example, "If your hand can hold twenty candies, about how many candies can fit in a box this size?" Or, "If Tommy takes 95 paces to walk from his desk to the cafeteria, about how many paces would he take to walk from his desk to the classroom across the hall? Would the number of paces be more or less for a very tall adult?" Encouraging students to use benchmarks as a regular basis for decision making can have valuable payoffs in the development of number sense.

## Operation Sense

Ms. Turnipseed's intermediate-level learning disabled students were just beginning to explore ideas of multiplication and division. She started math class today by asking how many paperback books she would have if she unpacked four boxes of books, and each box had three books.

Ruby said that a way to determine the total number of books was just to add 3 + 3 + 3 + 3. "Did anyone think of a different way of calculating the total number of books?" Ms. Turnipseed queried.

Bobby suggested that "you could use a calculator." Several other students organized counters into four groups of three, matching the action in the problem presented, and counted twelve books. Yumi suggested 4 × 3 = 12 as a number sentence for this situation.

Miss Turnipseed wrote Yumi's suggestion on the chalkboard for all to see, emphasizing that if "4 *times* I unpack a box of 3 books," then 12 books

**An important consideration should be the relative importance of a concept or skill for the child in daily life. Reasonableness is real-life math.**

**Multiplication is more meaningful to students when it is related to something they know—addition. Use counters, the number line, or arrays to model multiplication. Students should come to realize that multiplication is far more efficient than repeated addition or counting.**

**Operation sense is related to understanding the effects operations have on numbers. Students who understand operations and how they work have operation sense. We might ask students what happens when we multiply? (*Things get bigger*. Always? What about $1/3 \times 1/4$?) Operation sense also relates to the interpretation of an operation's answer, as in the division problems just completed.**

total would be unpacked. Together they compared $3 + 3 + 3 + 3 = 12$ to the multiplication sentence $4 \times 3 = 12$ and decided they had found two ways to express the same idea.

"When would it be a good time to multiply?" Ms. Turnipseed asked.

Beth checked with her partner and quickly offered their opinion: "A good time to multiply would be when all the numbers are the same, like $3 + 3 + 3 + 3$."

"And when would it be a good time to add?" she asked.

Bobby chimed in, "When the numbers are different."

Galen thought a moment and said, "It's like at the grocery store where my mom works. She showed me. When somebody buys several things at the same price, she doesn't have to punch in each one. If they buy six things, she can push a button that says, six at the price—whatever it is—and the machine tells the total cost. That's multiplying, right?"

"Right, Galen," said Ms. Turnipseed. "When several things cost the same, you can multiply the cost by the number of items to get the total." The class agreed that multiplication, more than anything else, was an efficient, fast way to add numbers that are the same.

Interested in challenging her students to compare and contrast the four operations (an important aspect of "operation sense"), Ms. Turnipseed wrote 32 and 17 on the chalkboard and suggested that they add, subtract, multiply, and divide by entering 32, the operation sign, then 17 on their calculators. "Take turns with your partner. One may use the calculator; the other should write the number sentence."

After the students had finished the task, Ms. Turnipseed wrote their results on the chalkboard as the basis for discussion. In an effort to nurture operation sense among her students, she had carefully planned the questions she wanted to surface in this discussion:

- What can you say about the sum when you added these two numbers? (It is larger than both addends.)
- What can you say about the difference between 32 and 17? (It is smaller than the number you subtracted from.)
- What can you say about the product of these two numbers? (Either way you multiply, the product is 544; the product is a lot bigger than either factor.)
- What happens when 32 is divided by 17? (The quotient is a bit less than 2; the answer includes a remainder; the quotient is much smaller than the dividend or divisor.)
- What can you say about operations on whole numbers generally as a result of this "experiment"? (When you add or multiply, the answer is usually larger than the number at the start; the order in which you add or multiply terms doesn't matter—you get the same result. When you subtract or divide, the answer is usually smaller than the number at the start.) What about fractions? Do the results of our experiment hold? (They do for addition and subtraction, but not for multiplication and division.) □

## Extending the Special Education Thrust

Recognizing which operation to use in a given problem situation is one characteristic of operation sense. To be successful in this task, students must recognize that a given situation fits the "pattern" of addition (or multiplication, or whatever the operation might be). Typically this recognition develops over time and is nurtured by frequent, varied experiences that invite children to solve a wide range of problem situations, discuss different solution approaches, and reflect on what happens in these different situations.

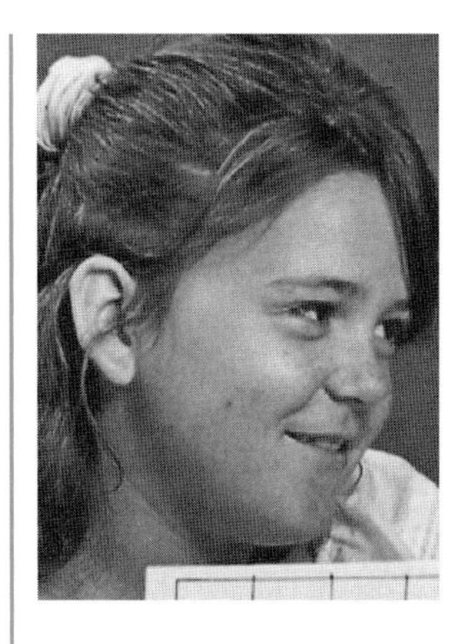

Ms. Turnipseed's lesson was one of a series that was designed to nurture an ongoing development of operation sense. The plan for the day was to focus on reviewing the meaning of multiplication and then discussing how different operations are "like" or "different" from each other.

Students with long- and short-term memory deficits will benefit from using counters to help verify their responses to questions like those asked by Ms. Turnipseed above. To emphasize overlearning, students should also be encouraged to justify their responses orally or in writing and, when appropriate, by actually computing.

Verbal interaction is particularly critical for students deficient in abstract reasoning or language development. In particular, these students should be challenged to answer "why" and "how" questions in relation to problem situations. These kinds of questions nurture conceptual understandings and also provide formative assessment opportunities relative to the student's level of understanding. All students, but particularly those who need extra time to process their responses, need time and encouragement, so wait them out.

Ms. Turnipseed's students used calculators for the last activity. Children who are slow in processing information should be provided with a printing calculator or be encouraged to keep a written record of their calculator keying. All too often these students "lose" the display on their calculator or forget the sequence used to obtain a particular response. Opportunity to review this sequence using a visual cue such as a printout tape or written record will help these students recall the process(es) and thinking used in obtaining the solution.

Another use of a visual cue is to use picture problems displaying different operations and how they work. Learning disabled students may benefit from a step-by-step poster of how certain algorithms work. This visual cue will allow the teacher more time to work with individual students while others begin to assume responsibility for using algorithms on their own.

## Extending the Mathematics

Operation sense is much more than just understanding how an operation works. An extension of the previous lesson, actually carried out by Ms. Turnipseed, focused on an important aspect of operation sense for division: appropriately interpreting the remainder in practical situations. In the lesson she presented these two problems to her students:

1. The Owls basketball team went to a summer camp. Twenty-nine players went. The players stayed in cabins with five players staying in each cabin. How many cabins had Owls players staying there?

2. Lee and Brett wanted to rent a paddle boat. Paddle boats cost $8 an hour. The boys had $20. How many hours could the boys use the boat?

The students worked in pairs to solve each of the problems. Each pair had a calculator and paper and pencil. For problem 1, Janie and Stan answered 5 r 4. Lisa and Linda answered 6.

Ms. Turnipseed asked the student pairs to explain their answers. Janie said that there were 5 fives in 29, with 4 left over, but Greg said that that didn't solve the problem. He said that Lisa and Linda's answer was correct because all the Owls are staying in six cabins. Five of the cabins are completely full of Owl players, and the sixth cabin has four Owl players. Lisa and Linda showed how they modeled the solution by grouping twenty-nine counters into groups of five (fig. 9.9). They showed the class that there were five complete groups of five. These represented the cabins full of Owl players. They also showed that although the sixth cabin had only four players, six cabins were needed for all the Owls to stay at camp. Computationally, the remainder of 4 was rounded up to make the answer 6.

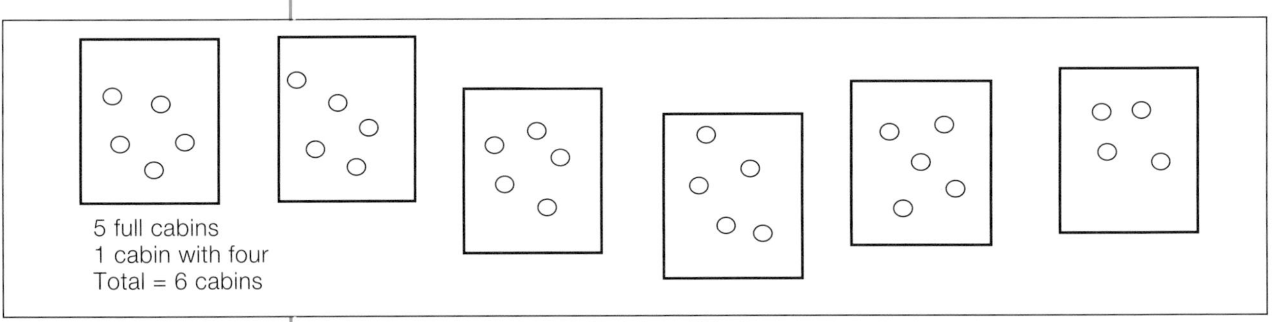


Fig. 9.9. Modeling of $29 \div 5$

In problem 2, Jessica and Liz said that the boys could rent a boat for 2 hours. Greg and Angela said the boys could rent the boats for 2 r 4 hours. Ms. Turnipseed used actual money and had students act out a solution to this problem. A student acted as the cashier and collected a $20 bill from two students, one acting as Lee and one as Brett. He changed this to show two payments of $8 each, with $4 left over as change. The boys could rent the paddle boat for 2 hours, and they had $4 left (a remainder). Computationally, the remainder of 4 was dropped. It was not used in the answer. (See fig. 9.10.)

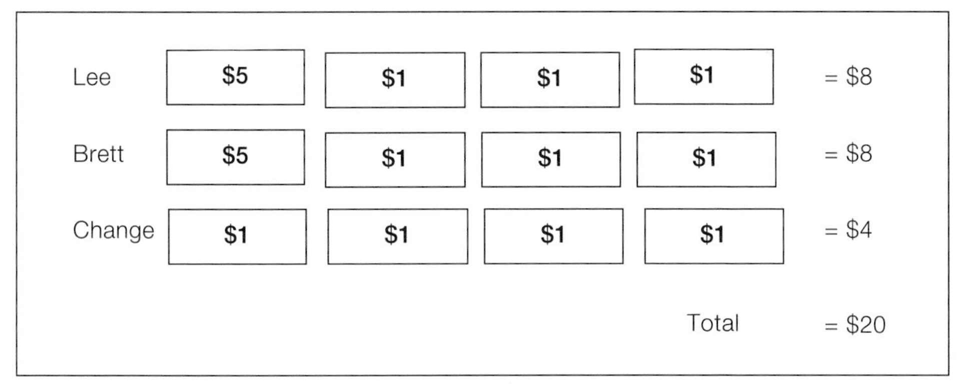


Fig. 9.10. Modeling of 20 ÷ 8 with money

Ms. Turnipseed then pointed out a similarity. In both problems, there was a remainder of 4. In the first problem the quotient was rounded up because all the campers would need a place to stay regardless of full cabins (five in a group). In the second problem the quotient stayed the same. The remainder did not affect the quotient, since the rental cost was for a full hour. (See fig. 9.11.) Both of these problems present real-life ways to use number sense as applied to a particular operation—division.

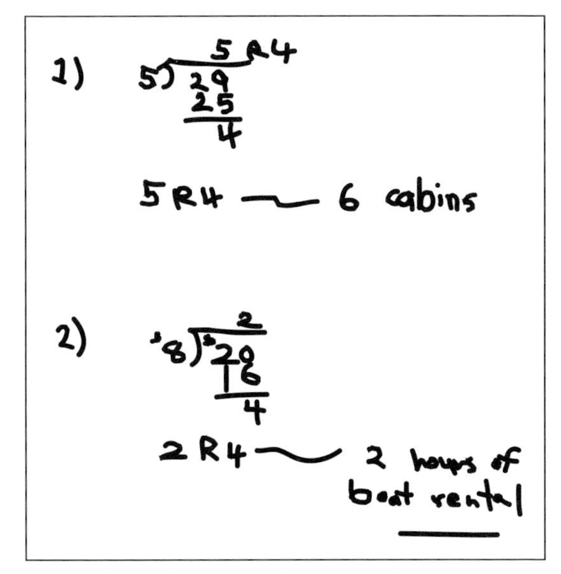


Fig. 9.11. Computational examples

But operation sense involves all the types of numbers that students encounter, rational as well as whole numbers (Reys 1991). As students become more comfortable with numbers, they should be able to respond to situations like the following, which will cause them to reflect on number meaning, relationships, and magnitude relative to an operation.

The school is collecting soup-can labels. To date, they have collected more than 50 000 labels. They are trying to collect a million labels. Answer the following:

1. Is the school close to collecting a million labels? How do you know?
2. Do you need to use an operation (+, –, ×, ÷) to solve #1 above?
3. If you wanted to know exactly how many more labels the school needed to collect to reach a million, what operation would you use? How would you solve the problem?

I have given each of you 100 counters. Using the counters, answer the following:

1. How much is 1/2 of your counters? How do you know?
2. Is 1/2 of 100 smaller or larger than 100?
3. Would 50% of 100 be smaller or larger than 100?

Show a group of 12 of your counters. Think about the following, then answer.

1. How much is 1/2 of the 12 counters? 1/3? 1/4?
2. If the counters represented 12 pieces of candy, and I wanted to give 1/2 of the candies to my sister and 1/3 to my brother, how many candies would I have left? What operations could you use to solve this problem?
3. Is 50% of 12 counters more or less than 12? How do you know?

## CREATING WINDOWS OF OPPORTUNITY

Everyone needs number sense. We live in a world where estimating time, money, amounts, distances, and even how much flour to put in a recipe is taken for granted. But number sense is more than just estimation. Far too many students in all types of classes, some called special education, do not experience a number-sense approach to understanding and using mathematics. Such an approach focuses, as this chapter did, on number meaning, number relationships, magnitude, and operation sense.

Students need to understand what numbers are and how they work. They need to know how and when to estimate, and how operations actually work. A number-sense approach to teaching fills this need and opens windows to student communication, involvement, and "reasonable" thinking about numbers.

Teachers who use manipulatives to develop number concepts and who ask the type of questions that challenge students to think about number and number relationships recognize the important role number sense plays in nurturing mathematical understanding. In such a rich program students are more likely to develop important intuitions about numbers and their uses and will be empowered to better deal with quantitative problems and opportunities in day-to-day situations.

## REFERENCES


Fennell, Francis (Skip). *Teaching for Number Sense Now!* Washington, D.C.: United States Department of Education, 1992.

Howden, Hilde. "Teaching Number Sense." *Arithmetic Teacher* 36 (February 1989): 6–11.

National Council of Teachers of Mathematics. *Curriculum and Evaluation Standards for School Mathematics.* Reston, Va.: The Council, 1989.

National Research Council. *Everybody Counts: A Report to the Nation on the Future of Mathematics Education.* Washington, D.C.: National Academy Press, 1989.

Reys, Barbara J. *Developing Number Sense in the Middle Grades.* Reston, Va.: National Council of Teachers of Mathematics, 1991.

Thornton, Carol A. "Strategies for the Basic Facts." In *Mathematics for the Young Child*, edited by Joseph N. Payne, pp. 133–51. Reston, Va.: National Council of Teachers of Mathematics, 1990.

Trafton, Paul R. "Reflections on the Number Sense Conference." In *Establishing Foundations for Research on Number Sense and Related Topics*, Report of a Conference, edited by Judith T. Sowder and Bonnie P. Schappelle, pp. 74–77. San Diego: San Diego State University, Center for Research in Mathematics and Science Education, 1989.

Wirtz, Robert. *New Beginnings—The Guide to Think, Talk, Read.* Monterey, Calif.: Curriculum Development Associates, 1980.



**Francis (Skip) Fennell,** currently a member of the NCTM Board of Directors, is a professor of education at Western Maryland College. For more than twenty years his teaching, writing, and work on federal projects have focused on elementary school mathematics, with particular emphasis on number sense, diagnosis, assessment, and problem solving.

**Ned Landis** is a special education resource teacher in the Carroll County (Maryland) Public Schools. He has worked in the field of special education for more than twenty-four years. He also coordinates the Curriculum Materials Collection and is an adjunct professor of education at Western Maryland College.

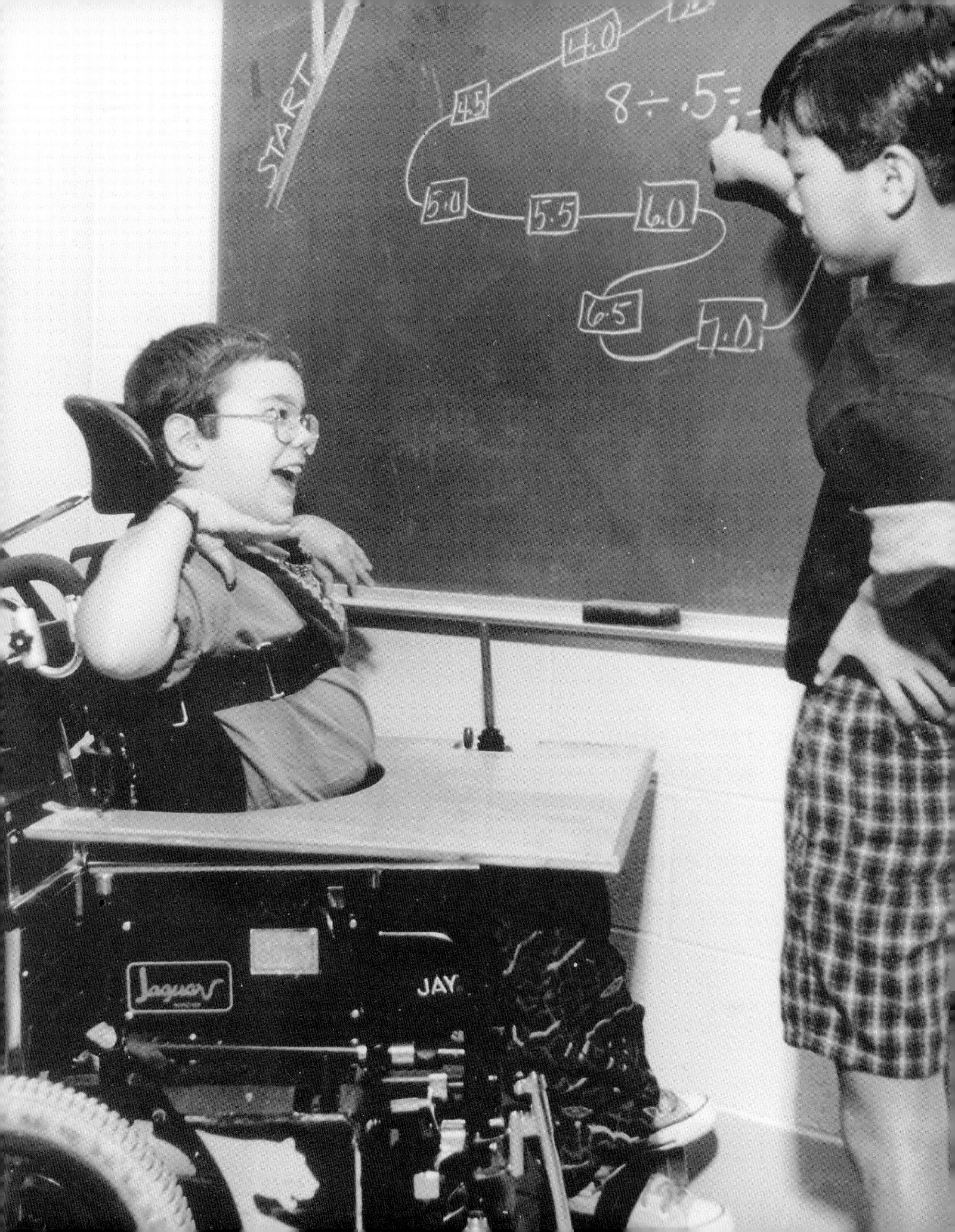
START
4.0
4.5
8 ÷ .5 =
5.0
5.5
6.0
6.5
7.0
Jaguar
JAY

# 10

# Computation Sense

*Carol A. Thornton*
*Graham A. Jones*
*in collaboration with*
*E. Paula Crowley*
*Paula J. Smith*

***Imagine a student immersed in a problem situation requiring a calculation. Before starting to compute, the child wonders:***

- ***Is addition appropriate here, or ... ?***
- ***Will an estimate do?***
- ***Can I do this mentally? Should I work this out on paper or use a calculator?***

***After calculating, the child asks another question:***

- ***Is that answer reasonable?***

STUDENTS who ask questions like these clearly have a broad concept of number and operations and demonstrate a willingness to use number sense in computational work. In other words, these students exhibit "computation sense."

Although computation sense has been an important part of the mathematics curriculum for a long while, it is only since the emergence of the *Curriculum and Evaluation Standards for School Mathematics* (National Council of Teachers of Mathematics [NCTM] 1989) that the learning of computation sense for *all* students has come into sharp focus. The thrust of this chapter is to examine issues and ideas related to the development of computation sense, especially for students having special needs.

Although often unexcited about mastering written computational techniques, gifted and talented students typically have strong intuitions and insights about what operation(s) to use in a situation and realize the general effect of adding, subtracting, multiplying, or dividing two numbers. These

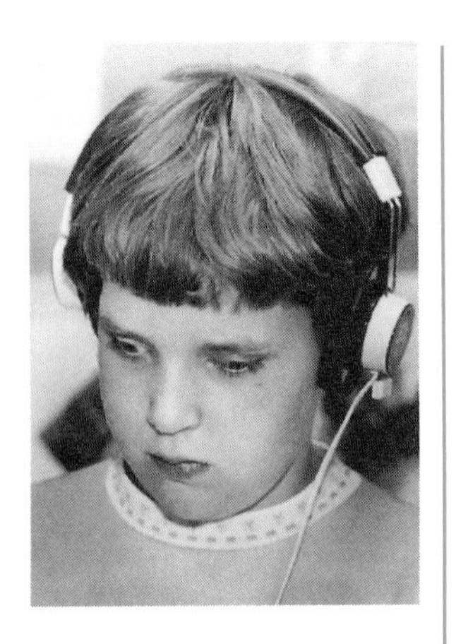

students also are likely to recognize the significance of relations between operations, such as those between addition and subtraction or multiplication and division. More important, they value and regularly make use of estimation, mental math, and technology in computational situations.

Yet, in spite of the fact that the typical curriculum for students with learning deficits has been computation-driven, many such students experience difficulty in some or all aspects of computation sense. Regrettably, the teaching-learning focus for students with learning problems has too often been on skills rather than on conceptual thinking reflected in processes such as estimation, mental computation, and the selection of the most appropriate computational procedure for solving a problem. *How do we nurture the development of good computation sense for these learners?*

## DEVELOPING MEANING FOR THE OPERATIONS

Central to the nurturing of computation sense is the development of meaning for the operations. This development focuses on fostering conceptual understandings for addition, subtraction, multiplication, and division, which, of necessity, are prerequisite to any discussion of computation sense.

The *Curriculum and Evaluation Standards for School Mathematics* (NCTM 1989) identifies four components essential to this understanding:

- Recognizing conditions in real-world situations that reflect a particular operation
- Gaining awareness of, and the ability to use, the properties of an operation
- Seeing relationships between operations, particularly the inverse relations
- Developing insights into the effects of an operation on a pair of numbers (e.g., recognizing that adding 7 to 24 produces a far smaller change than multiplying 24 by 7)

These four components are characteristic of operation sense. Students who exhibit these characteristics manifest operation sense and hence an ability to apply operations meaningfully and fluently.

Specific teaching-learning examples encapsulating these components and other aspects of concept development in relation to the operations are outlined in the *Curriculum and Evaluation Standards for School Mathematics* and are discussed in greater detail by Carpenter, Carey, and Kouba (1990), Irons and Irons (1989), and Rathmell and Huinker (1989). In nurturing *meaning for operations*, which is prerequisite to the development of computation sense, teachers need to monitor and assess students' abilities to—

- physically model or mentally solve a wide variety of problems;
- use appropriate mathematical language to describe each problem situation;
- write number sentences or otherwise meaningfully record and discuss problem solutions;

- recognize that some problems might be correctly solved (e.g., by classmates) in a variety of ways;
- recognize that various problem structures can be represented by a single operation;
- demonstrate operation sense.

Teachers can build on children's informal knowledge about an operation by providing repeated opportunities for them to listen to oral problem situations; use counters to solve them or mentally solve them; and then record, discuss, and compare solutions. Such problems might be based on literary themes (e.g., Burns 1992; Stumpf and Thornton 1994a, 1994b) or use other contexts meaningful to the children. Over time, teachers can nurture growth in children's understandings by offering carefully constructed problems that gradually increase in difficulty (Carpenter, Carey, and Kouba 1990; Stumpf and Thornton 1994a, 1994b). As an extension and vital part of this process, students should be encouraged to formulate and solve their own (and others') problems.

Initially only small numbers might be used in the problems. As children work, teachers can expect and encourage children to use different solution approaches. As teachers watch and listen, they can analyze children's thinking to answer two basic questions (Carpenter, Carey, and Kouba 1990) as a basis for future planning:

- What types of problems can a child successfully solve?
- What solution processes does a child use?

The natural extension of this work is to incorporate multidigit numbers, fractions, and decimals, to nurture and reinforce the development of place-value understandings (e.g., Thornton and Bohn 1993; Thornton, Jones, and Hill 1994), and to provide a context for both mental and written computation.

## DEVELOPING COMPUTING STRATEGIES

The second aspect of nurturing computation sense involves developing students' abilities to select and use a variety of computing strategies. This development occurs when both number and operation sense are applied to computing with whole numbers, fractions, and decimals.

For all students, but especially for those with learning deficits, a repertoire of suggested experiences is emerging. A principal element, typically neglected, is to provide readiness experiences for computation set in problem-solving contexts, as suggested above, in which children are presented with a wide range of problem types and are encouraged to solve, record, and discuss their solutions in personally meaningful ways. Many teachers also are beginning to realize that in order to foster computation sense, it may be necessary to place a purposeful emphasis on mental mathematics, estimation, and calculator work prior to or instead of formal attention to standard written algorithms.

Many resources are available to assist teachers in this effort (e.g., Coburn 1989a, 1989b; Hope, Leutzinger, Reys, and Reys 1988; Hope, Reys, and Reys

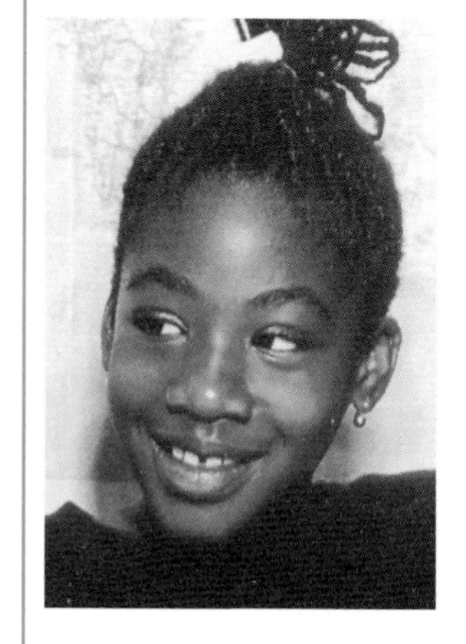

1987, 1988; NCTM 1986; Thornton and Behnke 1990; Thornton and Warner 1990; Thornton and Brown 1990; Thornton, Burns, Barnard, Peterson, and Lockett 1990). Students whose only experience is with standard paper-and-pencil calculation procedures too often become dependent on written algorithms. They lack the confidence to "let go" and develop that fluency of thinking necessary for efficiency and success with mental estimates or calculations in day-to-day situations requiring computation.

Teachers are finding that for many students engaged in programs with objectives like those outlined above, explicit teaching of standard algorithms (particularly for the addition and subtraction of whole numbers) is often unnecessary. The children themselves create acceptable alternatives, adopt "standard" approaches independently, or prefer to carry out the calculations mentally.

*If* standard algorithms are taught, then written procedures should be approached in ways that nurture intuitions. Instead of *telling* children the steps or "rules" for completing written computations, children can be engaged in "hands on" modeling of a solution and discussion of "big ideas." For example, when using concrete materials, children might be encouraged to ask one another questions like the following: "Are like units being combined?" "Is there enough to trade for a larger unit?"

Detailed teaching suggestions consistent with the approach summarized above are given in other sources (e.g., Bley and Thornton 1994; Rathmell and Trafton 1990; Payne, Towsley, and Huinker 1990; Cobb and Merkel 1989; Tucker 1989). For those students who experience difficulty with mental or written computation because the basic facts have not been mastered, suggestions are provided by Thornton (1994a, 1994b, 1990) and Bley and Thornton (1994).

For example, students may be individually assessed to determine exactly which facts are troublesome. *Known* facts can be crossed off on an individual "Cross Out" sheet, as in figure 10.1. Students could then use the chart as, little by little, they master additional facts. The goal is for the chart to become "blacker and blacker" as facts are gradually learned. Careful use of facts in problem settings, the sharing of "thinking strategies" to help students learn unknown facts (Thornton 1994a, 1994b, 1990), and, in extreme cases, compensatory strategies like color coding or finger tracing (Bley and Thornton 1994) may help students reach this goal, a prerequisite to success with both mental and written computation.

In fostering computation sense, teachers need to monitor and assess students' abilities to—

- select and use appropriate methods for computing from mental, paper-and-pencil, calculator, or computer techniques;
- use a variety of estimation, mental, and written computation techniques to solve problems involving whole numbers, fractions, and decimals;
- determine whether results are reasonable.

Investigating how an answer is obtained, whether it is a sensible one, and whether better, alternative solutions exist are appropriate points of focus for

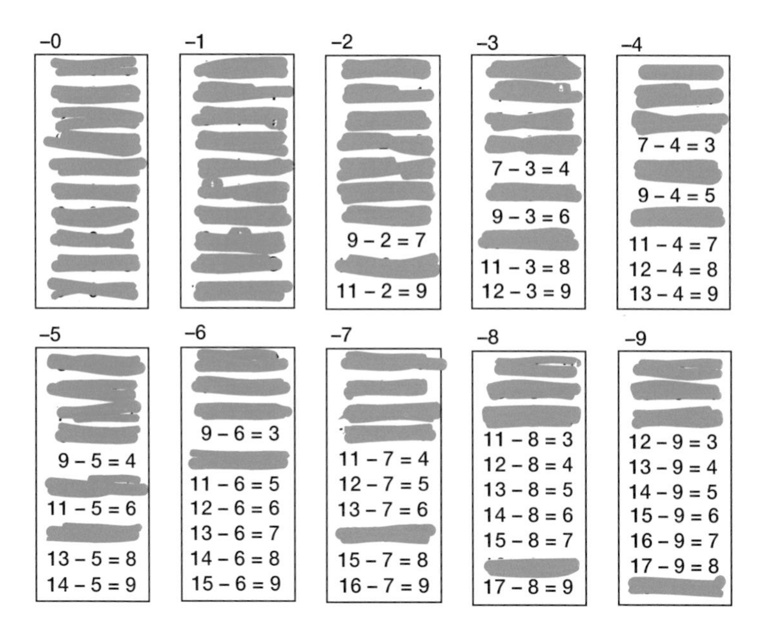


Fig. 10.1. Sample "Cross Out" sheet

just the solution itself. A classroom environment that encourages students to question, explore alternative solution strategies, verify, and generally "make sense" of computational results stimulates the development of computation sense. Allowing students to become active participants by sharing their solution approaches, reasoning, and conclusions through class discussions and through writing is an integral part of this process.

## ILLUSTRATIVE CLASSROOM EPISODES

The following three teaching episodes illustrate how different teachers have nurtured computation sense in classroom situations while accommodating the special needs of individual learners. The first episode highlights the power of the hundreds chart in developing intuitions and a visual reference for the mental addition and subtraction of two-digit numbers. The second demonstrates how a teacher develops intuitions and meaning for division with fractions. The third focuses on selecting the appropriate computation strategy for a given situation. Each episode is followed by a section that suggests how a lesson might have been adapted for students with differing special needs and a section that provides mathematical extensions related to the lesson.

### "100 Wins It!"

Mrs. Jasper thought back over the past few weeks and realized that her second-grade students had come a long way this spring! Initially they had just counted on and counted back by tens (and then by tens and ones) from different numbers on a hundreds chart. Using

the chart enabled all her students, including Janie, to move readily to mental additions and subtractions involving two-digit numbers with 9 and 11 and two-digit numbers with multiples of 10.

**The visual and tactile experiences on the hundreds chart, coupled with illustrating or writing matching number sentences, paved the way to Janie's success.**

Janie was a child with learning disabilities and short-term memory deficits. Mrs. Jasper had previously noticed that Janie benefited from visual and tactile experiences. She was not surprised that Janie especially enjoyed using a peep-hole card on the hundreds chart (see fig. 10.2).

When children worked with partners, Mrs. Jasper made sure that at first Janie manipulated the peep-hole card on the hundreds chart while her partner recorded the matching addition or subtraction sentence. This link to the written number sentence was an important connection, and after Janie had seen her partner record several number sentences, she began to get the idea and told her partner what to write. Eventually the two of them traded roles.

**Janie requires more time to internalize important connections. Working with a partner, she gradually links what is being said and done to the written number sentence.**

Mrs. Jasper's past experiences with other groups has led her to value the hundreds chart as a powerful visual tool for ushering many students to using mental strategies for adding and subtracting multidigit numbers. Even now, many of her second-grade students feel quite comfortable using the hundreds chart to add *any* pair of two-digit numbers. When the children occasionally use a calculator or cube trains of ten and extra ones to check, they are always pleased that they can add on the hundreds chart faster than using either a calculator or cubes. A few of her more capable students have never depended heavily on the chart and are now doing these calculations mentally.

Wanting to build on the successes of all her students, Mrs. Jasper has decided to base today's lesson on the game "100 Wins It!" (see fig. 10.3). She has used this game or variations of it with other groups.

| 1 | 2 | 3 | 4 | 5 | 6 | 7 | 8 | 9 | 10 |
|---|---|---|---|---|---|---|---|---|---|
| 11 | 12 | 13 | 14 | 15 | 16 | 17 | 18 | 19 | 20 |
| 21 | 22 | 23 | 24 | 25 | 26 | 27 | 28 | 29 | 30 |
| 31 | 32 | 33 | 34 | 35 | 36 | 37 | 38 | 39 | 40 |
| 41 | | | | 45 | 46 | 47 | 48 | 49 | 50 |
| 51 | | 53 | | 55 | 56 | 57 | 58 | 59 | 60 |
| 61 | | | | 65 | 66 | 67 | 68 | 69 | 70 |
| 71 | 72 | 73 | 74 | 75 | 76 | 77 | 78 | 79 | 80 |
| 81 | 82 | 83 | 84 | 85 | 86 | 87 | 88 | 89 | 90 |
| 91 | 92 | 93 | 94 | 95 | 96 | 97 | 98 | 99 | 100 |

Fig. 10.2. Use of a peep-hole card on the hundreds chart

Typically during game play most students never actually subtract but find a difference by counting up or adding on from the smaller number to the larger. Because the rules are flexible, any student can feel free to use cubes or the hundreds chart (though not the calculator or paper and pencil) to help find a difference. Most students soon recognize that doing a calculation mentally is faster, even though the element of chance in the card draw keeps the more capable teams from always winning.

While observing the students today, Mrs. Jasper listened as Janie and Cassie interacted during one round of the game. Janie had drawn 68 from the deck of two-digit number cards. According to the game's rules, their task was to work out mentally how many more than 50 the number is.

Janie reached for the hundreds chart and placed the peep-hole card over 50. Then she counted on: "50, 60—and 8 more to 68. So it's 18 more than 50, Cassie."

"I'd do it a different way, but I think you're right. Let me check."

Cassie put the peep hole over 50, moved down two rows to 70, and said, "That's 20 more." Then she counted back two: "69, 68—so that's 2 less than 20. I get 18, too. It checks." As Mrs. Jasper moved to another group, she saw Janie add 18 to the 70 of her calculator display and announce her new score of 88. □

**Note how Mrs. Jasper designed the game to accommodate various learning needs. More capable children could carry out the calculations mentally and use the hundreds chart merely to check. Janie, who needed more time adding and subtracting with the peep-hole card on the hundreds chart because of her short-term memory difficulty, could do so comfortably.**

**Game Rules for "100 Wins It!"**

Materials needed for each team of two players: Calculator, hundreds chart and peep-hole card, ten-trains and loose cubes, deck of two-digit number cards

**Playing the Game:**

*GOAL:* Work collaboratively with a partner to decide how much more or less than **50** a number is. A team wins when its calculator score reaches 100.

1. Each team enters 70 in its calculator as a starting score.
2. Mix the deck of two-digit number cards and put them face down in a pile.
3. At the teacher's signal, each team—
   - draws a card and sets it aside (face up);
   - works out **how many more or less than 50** the number is. (Use available materials to help or check as needed.)
4. If more than 50, add that many more to your calculator score. If less than 50, subtract that many from your calculator score.

Fig. 10.3

## Extending the Special Education Thrust

"100 Wins It!" highlights how Mrs. Jasper encouraged Janie's use of the peep-hole card on the hundreds chart to help her learn. The teacher might interact in other ways with children having different learning needs.

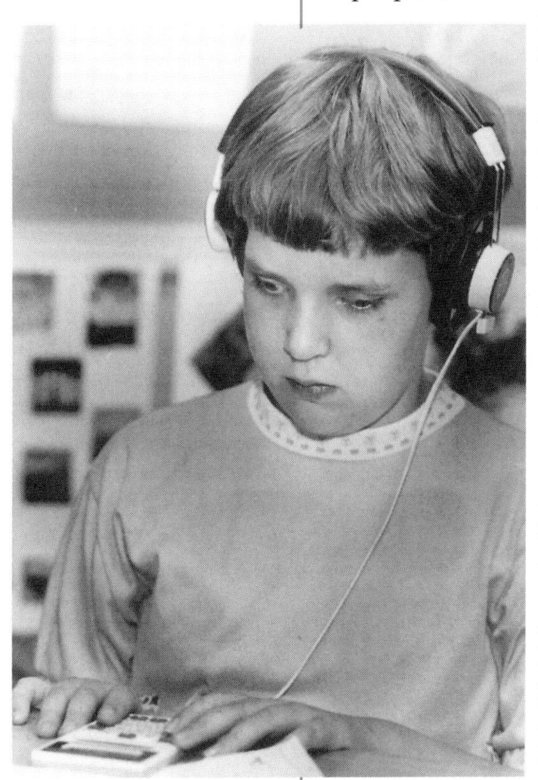

If a student with a visual deficit were in the class, Mrs. Jasper might have supplied a larger-sized number card deck, a personalized enlarged hundreds chart, and a calculator with enlarged keys for the "100 Wins It!" game. Depending on the deficit, she might use bold, black writing on both the number cards and the hundreds chart to give maximum visual contrast. Or she might use color to visually enhance the materials.

Tracing glue over written numerals is helpful for some students because it results in raised digits, which accommodates multisensory input. If the visual deficit is severe, a calculator with a voice synthesizer may be necessary. Collaboration with special educators in the school system is a crucial starting point for understanding and better accommodating the special needs of children with visual deficits and others with special learning needs.

This collaboration may prove invaluable should Mrs. Jasper have students with attention or behavior disorders in her class. She may learn that these students benefit from a program where their teacher's expectations are consistent and any unacceptable behaviors they may exhibit result in logical consequences.

In the context of the "100 Wins It!" lesson, Mrs. Jasper might anticipate that a student with an attention or behavior disorder may have difficulty following the rules and procedures of the game and remaining organized. To minimize disruptions, it is crucial that such a child clearly understand the game's rules. To this end Mrs. Jasper might involve the child in playing both "teacher" and "the class" to model the procedures and expected behaviors of the game.

It may be necessary also to stress a child's own self-management. This can be accomplished by supplying a chart labeled "Am I following game rules?" which requires the child to evaluate personal behavior by entering a check in the "Yes" or "No" column at a regular interval. If they are available, the teacher might provide a printing calculator, so that conflicts can be resolved by checking the printout.

Students with behavior disorders have particular difficulty with peer relationships and often become involved in conflicts with others over even minute details. The teacher can accommodate these students by pairing them with good "reverse" role models—children with rather cheerful, low-keyed dispositions who nonetheless are independent thinkers.

It may be helpful to seat a highly distractible student and the partner in a quiet area of the classroom, with the distractible student facing a blank wall rather than the activity of the classroom. A panel board or carrel may further minimize distractions and help a student with a very high level of distractibility pay attention during the game.

## Extending the Mathematics

Mrs. Jasper's lesson highlighted the value of mental addition and subtraction in the setting of "100 Wins It!" This game challenged students to work collaboratively with a partner to calculate the difference between a two-digit number and the target number 50. The game invites extensions in a variety of ways both at and beyond the grade 2 level, including those outlined below. In each instance, the calculator score determining a "win" should be modified to match the new constraints of game play.

### *Variations of "100 Wins It!"*

***Variation 1:*** Select a different target number, especially a number that is not a multiple of 10 (e.g., 62, 175, 1226), and a game deck with larger numbers.

> *Sample play:* Target number is 175. Start at 100; win at 400.
> Terry draws 215, mentally calculates the difference as 40, and adds this to his calculator score.

***Variation 2:*** Use a target number of 5 and a game deck that contains only two-digit decimal numbers from 0.0 to 9.9.

> *Sample play:* Target number is 5. Start at 7; win at 10.
> Adrienne draws 3.9, mentally calculates the difference as 1.1, and subtracts this from her calculator score.

***Variation 3:*** Use an Explorer calculator, a target number of 1, and a game deck that contains fractions between 0 and 2.

> *Sample play 1:* Target number is 1. Start at 2; win at 3.
> Borris draws 1/8, mentally calculates the difference as 7/8, and subtracts this from his calculator score.
>
> *Sample play 2:* Target number is 1. Start at 2; win at 3.
> Martina draws 1 8/9, mentally calculates the difference as 8/9, and adds this to her calculator score.

***Variation 4:*** Select a target number and provide a game deck that contains both positive and negative integers.

> *Sample play 1:* Target number is 0. Start at 100; win at 200.
> Zina draws –28, mentally calculates the difference as 28, and subtracts this from her calculator score.
>
> *Sample play 2:* Target number is 50. Start at 150; win at 250.
> Stefan draws –32, mentally calculates the difference as 82, and subtracts this from his calculator score.
>
> *Note:* In this last example, some children might think of their body as 0, with negative integers on one arm and the positive integers on the other. Starting at –32, one mentally "travels" to 0 and then to 50, verifying the difference of 82. Alternatively, one could think of starting at 50, moving to 0, and then on to –32, being equivalent to a move of 82.

### About How Many Halves?

Over lunch Mrs. Kraft overheard Joe telling some friends about his weekend trip to his grandfather's. Joe was mildly afflicted with cerebral palsy, and Mrs. Kraft was pleased to note how his social skills had grown. She stayed on to listen.

Joe told the group that for the return trip they needed more gas. Because they were a little short of money, his dad had wanted to know about how much gas they needed to drive the 105 miles back home.

Joe told how he always kept the mileage and gas checklist for his father and remembered that the car averaged 29.5 miles per gallon on the highway. "Dad and I were happy to make an estimate, but my grandfather insisted that we should use the exact figures. We all came up with about the same answer in the end." The bell rang and cut off further conversation, but Mrs. Kraft had an idea for a math activity that afternoon.

**The teacher sees an opportunity to engage Joe in a significant way in a mathematics activity.**

When the class returned to the room, they noticed what Mrs. Kraft had written on the board (fig. 10.4). Mrs. Kraft invited Joe to describe his dilemma. When he had finished, Mrs. Kraft turned to the class and said, "How do you think Joe and his dad estimated the number of gallons needed? Talk it over with your partner." Several students then shared their thinking.

**Mrs. Kraft capitalizes on Joe's verbal skills by having him relate his story, thereby setting the focus for the lesson.**

Fig. 10.4. Joe's dilemma

Mercedes said, "We noticed that 29.5 was almost 30, and 105 was close enough to 90. 90 ÷ 30 is 3, so it would be more than 3 gallons—really closer to 3 and a half."

Cam's approach was a little different: "Angus and I decided it would be better to have too much gas, so we rounded 105 up to 120 and divided by 30 to get 4 gallons. This is more than they would need."

In response to Mrs. Kraft's invitation, Joe explained how he estimated. "I did it like Mercedes, and Dad said my estimate was good. Grandfather divided by paper and pencil and grumbled because his answer wasn't much different but took so much longer to get."

**Although Joe had some physical disabilities, he was extremely capable intellectually and was very articulate.**

Mrs. Kraft decided that it was an opportune time to use the problem to work on understanding division involving decimals, a topic she always found difficult to approach in a relevant way. Her goal was to get students to illustrate the problem situation verbally, with a model, and with a number sentence (see fig. 10.5).

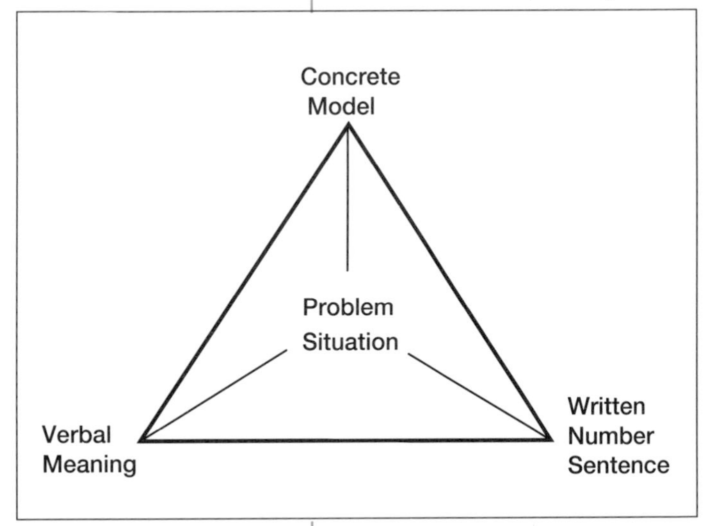


Fig. 10.5. Problem solving: The "hub" of the experience

She first asked for a number sentence that matched Joe's problem and readily got 105 ÷ 29.5 = _____. After being satisfied that the students understood that division was the appropriate operation, she noted that they easily related the number sentence to the estimates that had been discussed earlier. Most estimates were between 3 and 4, or "roughly 3.5."

Mrs. Kraft followed through by challenging her sixth-grade students to check the estimate in different ways. Some multiplied to check and saw that 3.5 was a good estimate; some used base-ten blocks (as in fig. 10.6) and got a similar result; one solved the number sentence on a calculator and got 3.559. Indeed, all the results were very close.

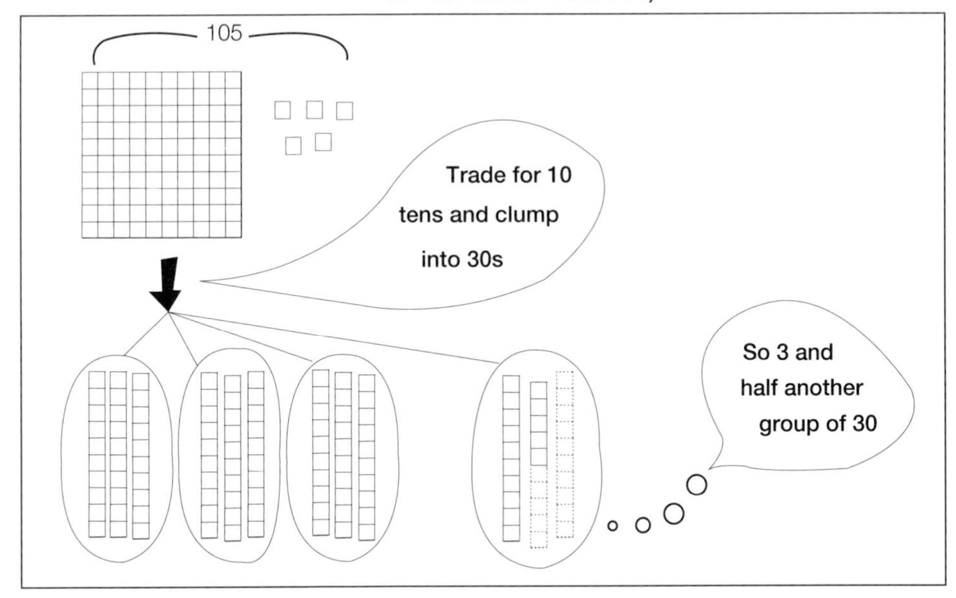


Fig. 10.6

To extend the discussion, Mrs. Kraft shared information from a newspaper clipping that talked about the last cross-country run. One official was stationed at every half mile along the eight-mile course, including the finish line, to record the runners' times. Mrs. Kraft asked the students how many officials were needed. Carl said, "I don't know what the answer is, but we could draw the course and put the officials at every half mile."

Two students drew the eight-mile course on the board and marked it off into half-mile segments as Carl had suggested. When Mrs. Kraft asked for the number sentence that matched the diagram, Carl suggested $8 \div 0.5 =$ ____. Mrs. Kraft inquired, "What does this number sentence mean?"

Jim volunteered: "That means divide 8 in half—that's about 4."

Joe objected. "It can't be 4, because $8 \div 2$ is 4."

Mrs. Kraft said, "Eric, why don't you work with Joe to draw a picture for $8 \div 2$ on the board."

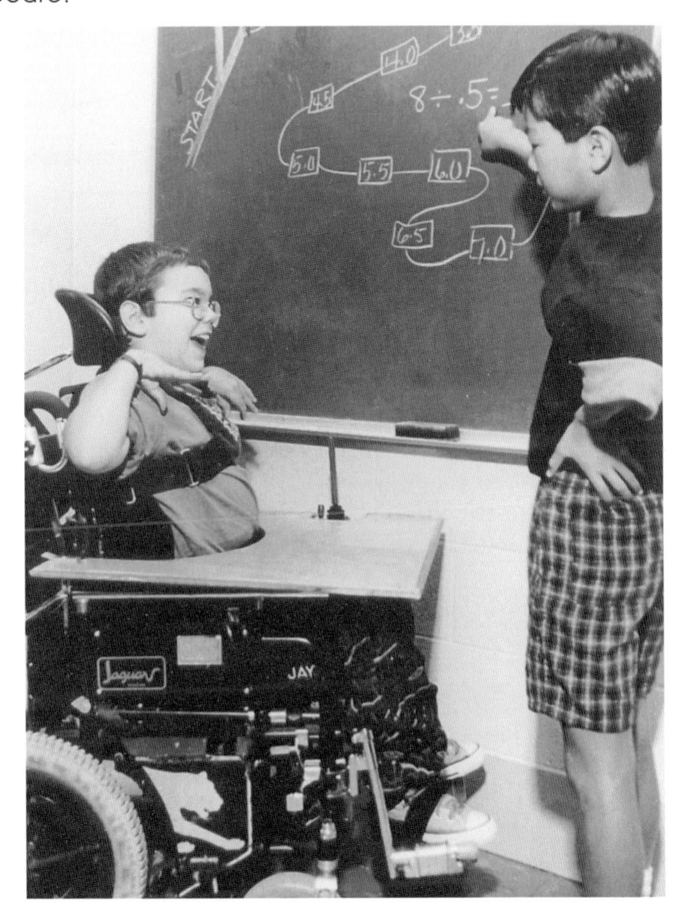


**Eric was Joe's "buddy for the week," so it was natural for Mrs. Kraft to have the two of them cooperate on the task, since Joe's writing skills were limited. At his seat, Joe typically used a nonskid, rubberized mat for steadying paper and books and used a light wrist weight to make writing easier.**

After they drew the picture for $8 \div 2 = 4$, Eric commented, "Our drawing shows there are four 2s in 8."

Joe noted that their drawing was just like Carl's, except there were 2 miles between officials, not just 0.5 of a mile.

"Look," said Carl. "There are more 0.5s in 8. There're 16." Everyone agreed. Then, at Mrs. Kraft's suggestion, the children copied the drawings into their journals, wrote the number sentences for each, and added their own descriptions of each number sentence.

As Mrs. Kraft moved among the students, she could be heard to ask, "If this is true, what does $8 \div 0.25 =$ ____ mean?" □

## Extending the Special Education Thrust

Students with abstract reasoning difficulties as well as those with long- and short-term memory deficits will benefit from the different ways Mrs. Kraft encourages her students to find and check their problem solutions. In particular, the emphasis on verifying or illustrating what is said with base-ten blocks or chalkboard illustrations will help these students establish and retain important conceptual understandings.

To further nurture their learning potential, some students with learning disabilities may need to copy on a large pad what is being illustrated by others. Using colored chalk at the board and providing easy access to colored pencils to highlight critical aspects of an illustration or list is typically helpful. Color coding also benefits students with visual perception strengths, including those with auditory deficiencies who frequently rely on the visual modality to learn.

In the "About How Many Halves?" lesson, many students with learning disabilities would require additional concrete experiences. For example, apples might be sliced in half to demonstrate that $1 \div 0.5 = 2$, $2 \div 0.5 = 4$, and so on. As in the lesson, it is important that these students see or write the number sentence each time and verbalize or "write the equation in words" as well to ensure understanding. An interesting activity could be based on illustrating written problems or equations with pictures or objects as well as with words.

Mrs. Kraft uses the students' questions as a basis for her lesson. Concern about the amount of gasoline required for a trip of 105 miles represents a problem that is relevant to the students' life experience. Students with conceptual and behavioral difficulties will more readily learn and stay attentive when a math lesson relates directly to their own lives.

A further use of questions will benefit students who need conceptual clarification. For example, the teacher might have asked, "If the car traveled about thirty miles on a gallon of gas, about how many gallons would it need to travel sixty miles?" The use of such specific, clear, and repeated questioning will benefit students with a range of special learning needs.

Because this lesson largely involves students communicating with one another, students with auditory-acuity and auditory-processing deficits will require specific accommodations. If such children are in the class, the teacher would wisely collaborate with special educators to plan appropriate intervention techniques. Technological aids, such as amplifiers and (for certain topics) computer-assisted instructional software, are available to assist with these problems. State special education offices can offer suggestions.

The teacher may ask the students themselves about the ways they have been helped by other teachers. They will be able to offer their views on the aspects of the lesson they found most difficult as well as describe specific procedures that were beneficial.

For example, some students with auditory deficits may benefit from a seating location that screens out extraneous noise and allows them to focus their attention on what is relevant. Others may benefit from seeing important information written on the chalkboard. The fact that Mrs. Kraft asked students to *write* both the number sentence and its meaning in words under each chalkboard drawing would also typically help students with auditory impairments.

## Extending the Mathematics

In this lesson Mrs. Kraft capitalized on an interesting situation generated by one of the children in order to give meaning to division involving decimals. Because decimal division rarely occurs, the situation with the gas mileage stimulated the use of division and also promoted the value of estimating.

Variations of this theme can be created using real-world contexts and estimation to develop computation sense, especially where the meanings are difficult for children to understand. The following examples based on percent and the multiplication of fractions further illustrate the strategy exemplified by Mrs. Kraft.

### *Using contextual examples and estimation to develop computation sense*

***Example 1:*** At Toby's Pizza, the cost of a pizza is usually $12.50. They always give Scouts a 5% discount. Luigi's Pizza regularly costs $12.00, but there is no special discount for Scouts. Which is the best deal?

> *Jeremy thinks:* "10% of $12.50 is $1.25, so 5% is a little more than 60¢. That makes Toby's pizza under $12, so it would be the best deal."

***Example 2:*** To try to get Pennsylvania Avenue from a player who is short of money during a game of Monopoly, one player decides to offer 60% of the price. If Pennsylvania Avenue costs $320, what was the player willing to pay?

> *Mariel, the player in trouble, thinks:* "Let's see ... 60% is close to 50% That's half of $320, which is $160. 10% more is another $32. That's not even $200 in all. I think I should hold out for $250."

***Example 3:*** For the school play, each of the ten students in the chorus needs 3 1/2 yards of red cloth for a costume. Juanita was in charge of costumes. The best price she could find for the cloth was $2.90 a yard. If Juanita's mom volunteers to make the costumes, how much should each person pay for the costume material so her mom won't have any additional costs?

> *Shingo says:* "3 yards is about $9. Half a yard is about $1.50 more. That's $10.50."
>
> *Gabriella says:* "I worked it out more exactly. 3 yards is about $9, actually it's 30 cents less. So that's $8.70. And another half yard is exactly $1.45. So the total really is $10.15."
>
> The girls decide that $10.25 will be a fair price because it gives Juanita's mom an extra dime from each student for thread and snaps.

## What Makes Sense?

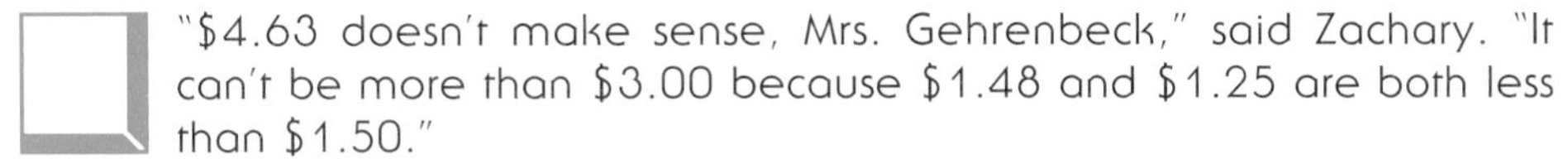

"$4.63 doesn't make sense, Mrs. Gehrenbeck," said Zachary. "It can't be more than $3.00 because $1.48 and $1.25 are both less than $1.50."

Terry interrupted. "That's right! But Miriam and I worked it out a different way. There are 2 one dollars and not enough cents to make a third dollar. So it's less than $3.00 and certainly not $4.63."

**Terry has tunnel vision and relies heavily on oral communication. For him, *to verbalize is to internalize*—a hidden reason for Mrs. Gehrenbeck to promote discussions like this.**

Mrs. Gehrenbeck beamed at the spontaneity of their responses and the clarity with which Terry presented the approach he determined with his partner. She had worked hard to design problems for which she provided answers and challenged her students to mentally check whether her answers were reasonable.

Besides focusing on applications of addition and subtraction in realistic settings, the activity established the importance of estimation and mental math and set an environment that encouraged flexible thinking and a variety of strategies.

It was early in the school year, and so far she had not engaged her fifth graders in any paper-and-pencil computation. From previous experience Mrs. Gehrenbeck had learned that children, including those with special learning needs, would be more likely to test the reasonableness of their own answers in paper-and-pencil computations if they first developed a repertoire of skills and strategies for estimation and mental computation.

Having just finished the lesson warm-up based on checking whether the total costs of several fast-food items were reasonable, Mrs. Gehrenbeck organized the students into groups of three for the major activity of the day.

"Next Sunday is the Clinton Art Festival. The school has received some flyers advertising the festival, and I've listed some of the articles that will be sold and their prices in the chart on the board." (See fig. 10.7.) At the same time, Mrs. Gehrenbeck handed one of the flyers to Terry so that he could read it comfortably. "Terry, I've marked all the items that are in the board chart."

**Terry has no peripheral vision and limited frontal vision. He reads as if through a pinhole, short words or even letters at a time, and needs to bring material close to his eyes. The teacher gives him the flyer she had marked during lesson planning (which really took no extra time to prepare).**

Fig. 10.7

Turning to the class, she continued: "Today your task is to use data from the chart to write four questions based on buying one or more items. Each question should invite a different solution method. I'll write them on the board so you can remember them." She wrote the following:

- One question for which an estimate of an answer would do
- One for which you would find an exact answer mentally
- One you would choose to answer using a calculator
- One you would choose to answer using paper and pencil

**Buddy systems work well for students with special learning needs. To preclude the need for extensive copying by peers, Mrs. Gehrenbeck could use an overhead projector and simply give Terry a Xerox copy of prepared transparencies.**

Miriam, Terry's buddy for the week, copied the list for Terry.

"It's all right to work together to create the four questions and answer them, but each of you should write up your version of the questions and their solutions. Also make clear *why* you chose to solve the problem as you did. Indicate by number which is your special role:

#1 will be the recorder for the group today.

#2 should check that the solution method is appropriate and that the answer obtained is reasonable.

#3 will present the group's questions and solutions to the class."

**Note how unobtrusively Mrs. Gehrenbeck moves to Terry's group to make a role assignment in line with his special learning needs.**

Knowing Terry's strength and his need to verbalize in order to learn, Mrs. Gehrenbeck moved to Terry's group and suggested that he assume role #3 for the group.

When all the groups were ready, they presented their questions and solutions. Mrs. Gehrenbeck made certain that both the reasons for choosing a calculation method were appropriate and that the answers obtained were reasonable. She was so pleased with the kinds of problems that the students created that she made them part of a class display (see fig. 10.8). ❑

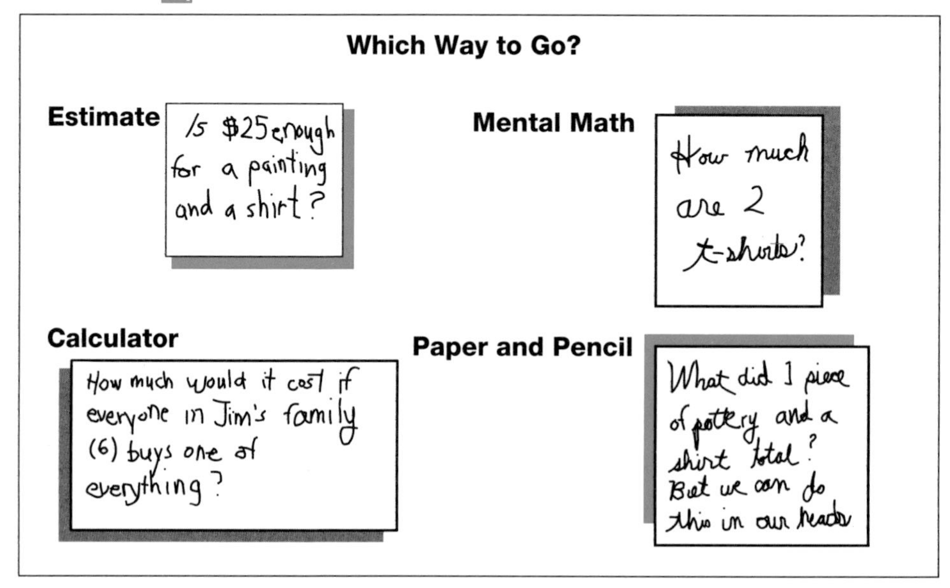


Fig. 10.8

## Extending the Special Education Thrust

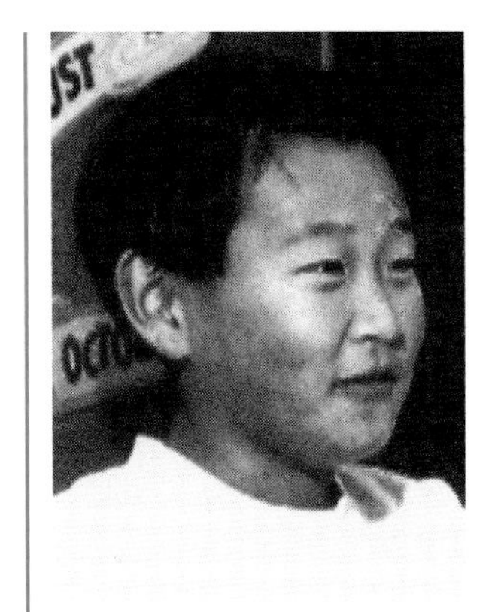

When children are paired for partner or small-group work, as in this lesson, students with learning deficits often learn and gain peer respect in ways that otherwise might not have been possible. They may not initially know or be able to contribute to a discussion, but often they can learn enough in the exchange to present an idea during subsequent class discussions.

Grouping students does require some thought. We may invite frustration for all concerned when very weak and very capable students are teamed. But both types of students can be grouped with "average" students in the class. Occasionally more talented students might be grouped together to create or carry out an extension of a problem; two nonverbal students might be paired to force open communication.

Because this lesson demands that students interact with peers in a positive manner, students with behavior disorders may have particular difficulty. It may be necessary to assign these students to a partner rather than have them work with a larger group of students. The more important factor is that teacher and student agree on a clearly specified brief set of ground rules.

The teacher might design these rules on an individual basis and involve the student in monitoring his or her own on-task behavior (e.g., with a system of checks). At the end of the class period the teacher might determine the extent of compliance and give the student a previously agreed-on reward. A compliment or "high five" exchange might do, or a token toward free-choice time or other items might be preferred. In general, students with behavior disorders have particular needs for structure, for regular schedules, for clearly established limits, and for a clear definition of rewards and consequences.

Because this lesson largely involves students communicating with one another to solve problems, those with communication disorders may also have particular difficulties. Like Terry, these students will benefit from being assigned special roles—ones they *can* assume. For example, they may have the skills to be the recorder or the checker, but they may not have the skills to be the presenter.

A teacher might ask the students to advise them on the role they would be most comfortable assuming or on any special materials they might find useful in carrying out a role. It may be that a child with a language or communication disorder would be willing to present the group report if the proceedings could be written in advance on a transparency and just shown to the class with little or no further comment. Alternative ways of making presentations, like these, should be encouraged.

When group reports are given, teachers might occasionally initiate interaction with questions like, "What did your group decide?" "Did any group do it a different way?" "What did the group think about that?" The student reporting therefore *shares* the responsibility with group members for incorrect thinking (and thus "saves face") as well as the credit for correct approaches.

The second question above deserves further comment. Used frequently, this type of question sets a positive tone both for student work and follow-up discussion. When students know that a teacher expects and encourages

diverse solutions, talented and gifted students are encouraged to stretch to their limits; all students appreciate the risk-free environment and know that "it's OK if I do it differently."

## Extending the Mathematics

In situations requiring computation, children should be encouraged to choose from a variety of solution strategies. Mrs. Gehrenbeck encouraged her students to estimate, to mentally calculate, to use a calculator, and when appropriate, to use paper-and-pencil procedures. A powerful way of getting students to reflect on "which strategy to use when" was modeled in her lesson by having the children construct problems to match different approaches.

For a lesson like Mrs. Gehrenbeck's, it is important to provide follow-up activities that reinforce what has been developed. Several ideas for seat work which illustrate or extend the thrust of different solution approaches are given below.

***Extension 1:*** For a problem set on a given page, have students do only those problems with answers less than $X$ (a target number you select).

> *Mrs. Gehrenbeck says:* "You don't have to do all the problems in this set. Just work those with answers less than 20. It's all right to do more, but I'll check only those with answers less than 20."
>
> *Note:* For talented students (who do not need as many practice problems) and for children with learning, physical, or sensory deficits that indicate that a smaller practice set is necessary, Mrs. Gehrenbeck can consult the answer key and assign different target numbers.

***Extension 2:*** For a problem set on a given page, have students do only those problems that cannot be readily computed mentally.

> *Mrs. Gehrenbeck says:* "We need to agree on which problems in this set can be calculated mentally. Work with your partner to decide." After a few minutes, children report during a whole-class share session. Mrs. Gehrenbeck establishes on the board a list of problems that can be done mentally. She then calls on students to explain different strategies for mentally computing solutions to some of these problems.
>
> *Note:* Although taking the time for this teacher-pupil interaction decreases the amount of time for written computation, it pays big dividends in terms of nurturing overall "computation sense." Sessions like this further reinforce mental approaches to mathematical computation.

***Extension 3:*** For a problem set, have students create a problem whose solution lies between the answers to the first and the last problem in each row.

> *Mrs. Gehrenbeck says:* "For your assignment, turn to Problem Set 28. You don't have to write answers to any of these problems. For each row, just write and solve a problem of your own whose answer lies between the answers to the first and the last problem in that row."

***Variations for example 3***

- Create a story problem whose answer lies between the greatest and the least solution in the row.
- Create a story problem whose answer is greater (less) than the greatest (least) answer in the row.
- Create a story problem whose answer is greater (less) than half the greatest (least) answer in the row.

## CREATING WINDOWS OF OPPORTUNITY

Many able people have been disadvantaged in mathematics because of school programs that were dominated by written computation rather than problem solving. Such programs have created an enduring mind-set that calculations can be executed only with paper and pencil in hand. As a result, a high proportion of people lack the confidence and skills to use mental approaches, estimation, and even calculators for computation. In reality these people are "handicapped" because they lack operation and computation sense.

This chapter has focused on the need to make connections between real-world situations, the meaning of the operations, and the use of a variety of computation strategies. It highlights the need to support the development of mental calculation skills through the use of appropriate visual and tactile experiences. Opportunities like these free students from the constraints imposed by a program dominated by written computation and empower them with a pervasive *computational sense* that will enable them to deal confidently and competently with the quantitative data of their daily experience.

## REFERENCES


Bley, Nancy S., and Carol A. Thornton. *Teaching Mathematics to the Learning Disabled.* 3rd ed. Austin, Tex.: Pro-Ed Publishers, 1994.

Burns, Marilyn. *Math and Literature (K–3).* Sausalito, Calif.: Math Solutions Publications, 1992.

Carpenter, Thomas P., Deborah A. Carey, and Vicky L. Kouba. "A Problem-Solving Approach to the Operations." In *Mathematics for the Young Child*, edited by Joseph N. Payne, pp. 111–31. Reston, Va.: National Council of Teachers of Mathematics, 1990.

Cobb, Paul, and Graceann Merkel. "Thinking Strategies: Teaching Arithmetic through Problem Solving." In *New Directions for Elementary School Mathematics*, 1989 Yearbook of the National Council of Teachers of Mathematics, edited by Paul R. Trafton, pp. 70–81. Reston, Va.: The Council, 1989.

Coburn, Terrence G. *How to Teach Mathematics Using a Calculator.* Reston, Va.: National Council of Teachers of Mathematics, 1989a.

———. "The Role of Computation in the Changing Mathematics Curriculum." In *New Directions for Elementary School Mathematics*, 1989 Yearbook of the National Council of Teachers of Mathematics, edited by Paul R. Trafton, pp. 43–56. Reston, Va.: The Council, 1989b.

Hope, Jack A., Larry Leutzinger, Barbara J. Reys, and Robert E. Reys. *Mental Math in the Primary Grades.* Palo Alto, Calif.: Dale Seymour Publications, 1988.

Hope, Jack A., Barbara J. Reys, and Robert E. Reys. *Mental Math in the Middle Grades.* Palo Alto, Calif.: Dale Seymour Publications, 1987.

———. *Mental Math in the Junior High.* Palo Alto, Calif.: Dale Seymour Publications, 1988.

Irons, Rosemary Reuille, and Calvin J. Irons. "Language Experiences: A Base for Problem Solving." In *New Directions for Elementary School Mathematics*, 1989 Yearbook of the National Council of Teachers of Mathematics, edited by Paul R. Trafton, pp. 85–98. Reston, Va.: The Council, 1989.

National Council of Teachers of Mathematics. *Curriculum and Evaluation Standards for School Mathematics.* Reston, Va.: The Council, 1989.

———. *Estimation and Mental Computation.* 1986 Yearbook, edited by Harold L. Schoen. Reston, Va.: The Council, 1986.

Payne, Joseph N., Ann E. Towsley, and DeAnn M. Huinker. "Fractions and Decimals." In *Mathematics for the Young Child*, edited by Joseph N. Payne, pp. 175–200. Reston, Va.: National Council of Teachers of Mathematics, 1990.

Rathmell, Edward C., and DeAnn M. Huinker. "Using 'Part-Whole' Language to Help Children Represent and Solve Word Problems." In *New Directions for Elementary School Mathematics*, 1989 Yearbook of the National Council of Teachers of Mathematics, edited by Paul R. Trafton, pp. 99–110. Reston, Va.: The Council, 1989.

Rathmell, Edward C., and Paul R. Trafton. "Whole Number Computation." In *Mathematics for the Young Child*, edited by Joseph N. Payne, pp. 153–72. Reston, Va.: National Council of Teachers of Mathematics, 1990.

Stumpf, Jean S., and Carol A. Thornton. *Math and Popular Literature Resource Card Set: Addition and Subtraction, Grades K–2.* Lincolnshire, Ill.: Learning Resources, 1994a.

———. *Math and Popular Literature Resource Card Set: Multiplication and Division, Grades 2–4.* Lincolnshire, Ill.: Learning Resources, 1994b.

Thornton, Carol A. *Basic Number Facts: Addition and Subtraction.* (A teacher resource with card sets.) Normal, Ill.: Illinois State University Press, 1994a.

———. *Basic Number Facts: Multiplication and Division.* Normal, Ill.: Illinois State University, 1994b.

———. "Strategies for the Basic Facts." In *Mathematics for the Young Child,* edited by Joseph N. Payne, pp. 133–51. Reston, Va.: National Council of Teachers of Mathematics, 1990.

Thornton, Carol A., and Diana Behnke. *Mental Warm-Ups: Today and Every Day—Grade 1.* Normal, Ill.: Illinois State University Press, 1990.

Thornton, Carol A., and Anita Bohn. *I Can Number the Ways.* Lincolnshire, Ill.: Learning Resources, 1993.

Thornton, Carol A., and Jan Brown. *Mental Warm-Ups: Today and Every Day—Grade 3.* Normal, Ill.: Illinois State University Press, 1990.

Thornton, Carol A., Wendy Burns, Janet Barnard, Myrna Peterson, and Rebecca Lockett. *Mental Warm-Ups: Today and Every Day—Grade 4.* Normal, Ill.: Illinois State University Press, 1990.

Thornton, Carol A., Graham A. Jones, and Kevin M. Hill. *Ways to Number: Ones, Tens, and Hundreds.* Lincolnshire, Ill.: Learning Resources, 1994.

Thornton, Carol A., and Leigh Warner. *Mental Warm-Ups: Today and Every Day—Grade 2.* Normal, Ill.: Illinois State University Press, 1990.

Tucker, Benny. "Seeing Addition: A Diagnosis-Remediation Case Study." *Arithmetic Teacher* 36 (January 1989): 10–11.

**Carol Thornton** is a Distinguished University Professor in the Mathematics Department at Illinois State University. She works with special education and regular class early childhood, elementary, and middle school in-service and preservice teachers, has written widely in the fields of mathematics and special education, and has a special interest in staff development for nurturing children's mathematical thinking.

**Graham Jones** is a visiting professor of mathematics education in the Mathematics Department at Illinois State University. Formerly he was pro-vice-chancellor and professor at Griffith University, Gold Coast, Australia. He is interested in children's thinking in number, probability, and data analysis.

**Paula Crowley** is an assistant professor of special education in the Specialized Educational Development Department at Illinois State University. She teaches courses on methods of teaching students with learning and behavioral disorders. She taught such children and adolescents in the public schools and in a residential treatment center. She is interested in the accomodations general education teachers can make for students with disabilities in general education classrooms.

**Paula Smith** is a professor of special education in the Specialized Educational Development Department at Illinois State University. She teaches courses related to learning disabilities and is interested in models of service delivery to students with learning and behavior problems, instructional adaptations for such children, and restructuring schools.

11

# Measurement for Young Children

*Helene J. Silverman*
*Phyllis Paris*

THIS chapter demonstrates how young children can be engaged in activities that help them understand what it means to measure. It emphasizes the involvement of children in the processes of measuring, including—

- selecting attributes of measure;
- making comparisons based on a selected attribute;
- making and using estimates of measure;
- selecting and using nonstandard units of measure;
- solving problems based on everyday measurement situations;
- (eventually) making and using measuring instruments calibrated in standard units.

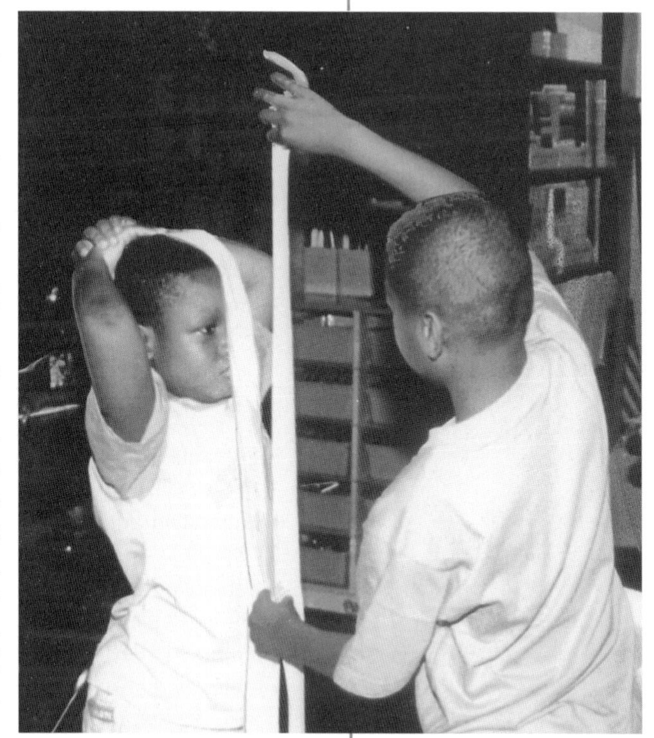

According to the research and recommendations for teaching in a whole language environment (e.g., Routman 1991), young children should be encouraged to share their investigations and discoveries through listening, speaking, reading, and writing as they express their ideas about measurement through meaningful language developed in a rich learning environment. There should be no rush to using standard measuring instruments. For each attribute explored, young children need a variety of informal experiences, such as those described in figure 11.1.

### Measurement of Capacity, Weight, and Time

- Estimating and then pouring water from one container to another to determine whether the water in one container can fill another
- Estimating and then determining how many small cups of sand can be filled from a pail of sand
- Calibrating a large cup to make a graduated container for measuring

..................

- Estimating and then counting to discover how many buttons will balance a toy dinosaur
- Making a balance scale and dough-ball weights

..................

- Ordering the sequence of events for a day
- Estimating then counting how many beads can be strung on a shoelace in a minute

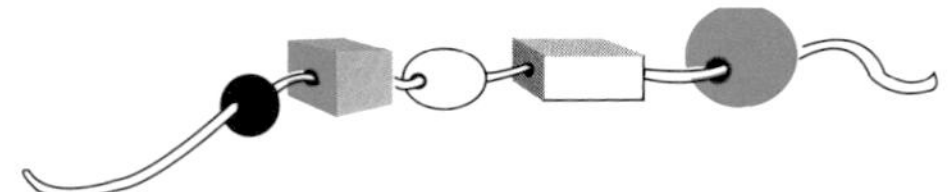

- Making a birthday book
- Keeping track of the passing of days on a calendar

### Measurement of Length

- Aligning two clay "worms" to see which is longer
- Finding and then sorting objects to determine which are shorter than, about as long as, or longer than a straw

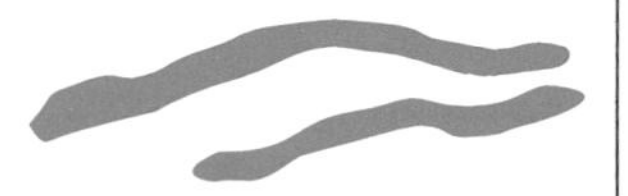

- Cutting a piece of yarn equal to the length of a straw
- Estimating and then finding how many double unit blocks it takes to go around another child
- Estimating and then counting how many children it takes to go around a designated space
- Selecting the adding-machine tape that represents personal height
- Making a body-measurement book

### Informal Measurement of Area

- Estimating and then pasting squares on a strip of cardboard to determine how many will fit
- Estimating and then counting how many rectangles are needed to make a place mat
- Estimating the number needed and then covering a box with stickers and counting to check
- Estimating and then determining if there is enough paper to cover a box
- Estimating and counting how many cubes fit in a box

Fig. 11.1. Measurement experiences for young children

## ILLUSTRATIVE CLASSROOM EPISODES

A pervading theme of the *Curriculum and Evaluation Standards for School Mathematics* (National Council of Teachers of Mathematics 1989) is that mathematical learning should emerge from problem situations and the context of children's "real world" experiences. It is in this spirit that the following three classroom episodes, focusing on aspects of measurement for young children, have evolved.

The activities and discussions that are recorded in the following episodes occurred in three early childhood classrooms of five- to eight-year-olds, of whom large numbers had physical disabilities. Professionals, including teaching aides, a nurse, a speech teacher, a physical therapist, and an occupational therapist, worked together with the children and the special education teacher at different times during the day, so their presence is a natural part of the interaction that is captured in the episodes. Much of this interaction was recorded as the result of the children's activity in measurement learning centers.

These centers and the classrooms themselves had been carefully arranged to accommodate the wheelchairs and walkers used by many of the children. The teachers, with the help of the occupational therapist, had designed special trays for several of the children to use with wheelchairs or alongside floor area tasks. The episodes illustrate how naturally other children worked with these children using the trays.

Following each episode, ideas for extending both the special education intervention and the mathematics are presented. Throughout, the emphasis is on assessing and building on the children's informal knowledge and experience to reach and stretch their mathematical thinking through problem solving and language.

### Footprints

*Children explore length measurement using a personal footprint as a nonstandard unit of measure and basis for direct comparison.*

Mrs. Johnson and her aide had just helped each of the six children at the Measurement Center trace a footprint onto newsprint. Even before all had cut out their tracing, she noted that the children were spontaneously comparing the different sizes, commenting that some were "bigger" or "fatter." She saw the aide record Inez's statement on adding-machine tape: "My foot is longer than your foot."

**Direct work on vocabulary and sentence building is necessary for many of the children in the group.**

The children made a game of finding someone with a shorter footprint and saying, "My foot is longer than your foot." To comfort Nila, the child with the shortest foot, and to highlight different measurement vocabulary, Mrs. Johnson suggested that the game rule be changed to "My foot is *shorter* than your foot."

Again, one child was left out. The children reasoned that because he had the "giantest" foot of all, Peter wouldn't be able to find anyone to whom he could say, "My foot is shorter than your foot." Luis saw a way to include Peter in the game. He removed the teacher's sneaker from her

The children in these classes learned that the physically handicapped children had to position themselves first. Each physically handicapped child was assigned a special position to use at each center, and some of these children had learned to move from wheelchairs to walkers with the help of an aide and other children.

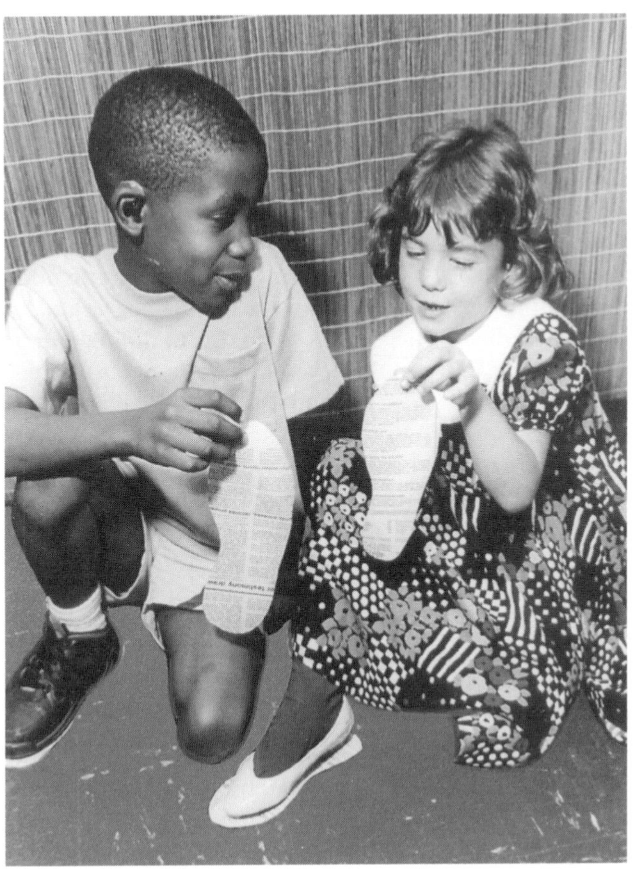

foot and compared it to Peter's footprint, telling his teacher, "You have the giantest foot now."

Then the children got a doll and some stuffed animals and insisted on tracing their feet to find the "babiest" foot of all. Following the lead the children had initiated, Mrs. Johnson used this opportunity to introduce the vocabulary *largest* and *smallest*.

The teacher reflected briefly on how motivated the children were and, after bringing closure to their foot-tracing and comparison activity, gathered them around her to listen to *How Big Is a Foot?* (Myller 1991). The story was a natural lead-in for assigning students to either the Writing Center or the Manipulatives Center. The children knew that eventually they would have a turn at both centers. Mrs. Johnson and other professionals in the room were active in their supportive roles as the children turned to their tasks in the centers.

After some time the aide working with Patricia at the Writing Center signaled to Mrs. Johnson. The child beamed as she told the teacher, "I've written a book about my footprint." Patricia then explained that they had made extra cutouts of her footprint, which she had then used to measure other parts of her body—her arm, leg, hand, and thumb. Her findings were in her book. Each page was illustrated with a sketch of the body part and a footprint (fig. 11.2).

The teacher has planned several special activities for Patricia, who seems to easily learn the new measurement vocabulary for length and can easily read back stories that she creates using invented writing.

The full text of her book is as follows:

> This is Patricia's foot.
> This is Patricia's arm.
> Patricia's foot is shorter than Patricia's arm.
> This is Patricia's foot.
> This is Patricia's leg.
> Patricia's foot is shorter than Patricia's leg.

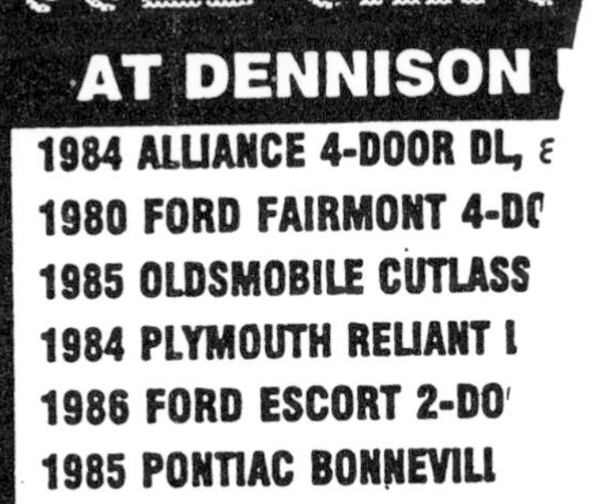


Fig. 11.2. Page from Patricia's book

**To avoid difficulty in audience situations such as a turn in the Author's Chair, the speech therapist worked individually with Jane prior to the oral reading of her book, capitalizing on her receptive language skills (i.e., listening and reading) to help her overcome difficulties with the formation of specific sounds.**

This is Patricia's foot.

This is Patricia's hand.

Patricia's foot is longer than Patricia's hand.

This is Patricia's foot.

This is Patricia's thumb.

Patricia's foot is longer than Patricia's thumb.

This is Patricia's foot.

During her turn at the Author's Chair, Patricia read her book to other children. Looking at Mrs. Johnson, she also made a point of telling how she had estimated first, then measured.

Smiling, Mrs. Johnson said, "Nice job!" and volunteered to laminate the cover and bind her book so it could be added to the class library.

Jane, who suffered an accident in early childhood that affected her hearing, was the next child ready to take a turn in the Author's Chair. The speech teacher assisted her so that she could read her book to the other children.

Mrs. Johnson listened to Jane's story, then moved to the group working at the Manipulatives Center. Here the children were selecting objects from a junk box and comparing them to their footprint. To help them organize their discoveries, Mrs. Johnson had provided sorting mats on which children could place objects to show the results of their measurements (fig. 11.3).

| longer than my foot | about the same as my foot | shorter than my foot |
|---|---|---|
| | | |

Fig. 11.3

Joanna, who is paralyzed from the waist down, used her walker to gain access to the clay at the Manipulatives Center. She and Timara proceeded to roll long and short "worms" out of clay. Then Timara attached all the worms to make a "monster worm, the longest on earth—and for sure longer than my foot!" □

## Extending the Special Education Thrust

The sorting activity in the preceding episode is an interesting, motivating one for children. Yet, special considerations may need to be employed to meet special needs. If a child's disability includes serious lack of motor control and the sorting mat is placed on a tray, for example, weights may be used—on the tray or on the child's arms—to steady the tray or the child's movement during sorting. Higher desks may be needed by children who find it easier to sort while standing.

Students who are easily distracted by sound or motion may have to complete their sorting in a "quiet area" or alone so that they can remain focused. Others with figure-background problems may require dark tactile paper or felt to define their work areas during the sorting task.

At the Measurement Center, hearing impaired children would need to face the teacher to better observe the teacher's lip movements and facial and hand gestures. When books are read or when they participate in the telling and retelling of stories, these children may require a copy of the book or time with a speech or resource teacher to work on the language patterning and story comprehension.

Shy students and children for whom these experiences in the center are new can be paired with other children who can serve as guides. Additionally,

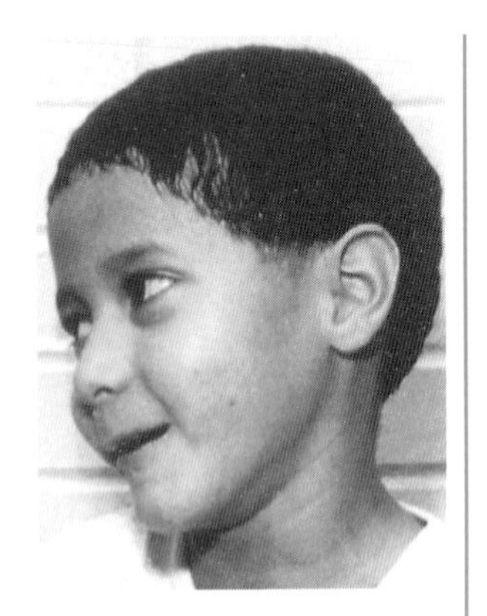

the teacher or aide can physically position these children so that they are easily included. Most children in the group will require additional, similar experiences to assist with the generalization of the concepts and the application of the new vocabulary.

### Extending the Mathematics

The book *Measure Me* (Ziefert 1991) can be used to extend the concept of iterating a unit. For example, children might be asked to predict what would happen if different children used copies of their footprints to measure Patricia's arm. "Will all results be the same?" As children explain their predictions and then measure to check, they may gain greater insights about the fact that different-sized units (footprints) yield different results in a measurement.

Children also could be invited to make chains of their foot cutout to determine the lengths of designated distances marked off in the classroom or other nearby locations. In relation to dinosaur and animal units, the students might represent the parts of animals with string or blocks to show relative length.

Some children may be ready to model a scale replica of the classroom using standard or nonstandard units to represent the dimensions. They can then cut figures from construction paper, newspapers, or magazines to fit in their model.

Other children may collect data from measuring the distance for how far a paper cup can be tossed or the distance between two paper clips dropped from waist high. They might classify leaves and twigs by their measurements or compare the heights of children versus parents on a graph.

Older students from other classrooms might be invited to analyze the data collected in these measurement experiences. They might, for example, determine the range of the results or compare the mode (most frequently recorded result) with the median (middle score when results are ordered) or with the arithmetic average. Such a project invites students to organize and display information using different types of graphs, to describe verbally or write reports of their analyses, and to state predictions, implications, or conclusions that naturally flow from the data.

The philosophy of *experience-based learning* modeled in this unit can also be extended to work with older students who might, for example, lay out a garden for planting. Garden plots as small as $4' \times 4'$ might be investigated. Students could research the growing space, time, planting, and watering requirements. They might also consider the appropriate or aesthetic placement for certain varieties of plants. As an end product students might use graph paper to furnish a sketch of their garden as part of a written report. It may be possible to actually plant the garden on the school grounds.

## Is Biggest Always Heaviest?

*Children explore the relative weight (mass) of objects, sometimes in comparison to size.*

After being introduced to *The Enormous Watermelon* (Parkes and

Smith 1986), a group of children in Ms. Sakalian's early childhood class tied strings to a balloon and pulled it. They had not gone far when they decided that the balloon was not *heavy* enough, and told the teacher that they wanted to make an enormous ball.

The teacher supplied the materials and showed them how to make a papier-mâché ball. While they worked, the teacher suggested that the children brainstorm on things that are heavy. Lynette watched, listened, and made sketches of some of the things. "I want to make a book of heavy things on my next turn at the Writing Center," she explained.

Ms. Sakalian prodded the group to reflect about the weight of some of the things they had suggested. "You named lots of heavy things. Are any heavier than the 'enormous ball' you are making?" she asked.

"I'm heavier," laughed Lakesha—and all agreed.

"And you're bigger than the ball, too, I notice," smiled the teacher.

"Are any of the heavy things you named *smaller* than your ball?"

Several items fit the bill: the paperweight on Ms. Sakalian's desk, the iron Lynette's sister uses, the concrete brick Steven's family use to hold down the lid on their garbage can.

"Hmm," the teacher mused. "Are the bigger things like your ball always *heavier* than smaller things?"

**The teacher uses the children's discoveries to build vocabulary and sentence structure, since many of the children have vocabulary and language structure below the expected level for their age group.**

The children exchanged knowing glances and went about their work.

When the large papier-mâché ball was finished, Lakesha and James cut out paper figures representing the characters in *The Enormous Watermelon* and tied them to the ball. At the Clay Center, Jeanette and Irvin made a large watermelon out of clay and several "baby" watermelons from leftover clay.

The teacher placed the large watermelon on Irvin's outstretched left hand and a "baby" watermelon in his right hand. Irvin said that the large

**A risk-free environment has been created in which conversation about activities is encouraged. The teacher and the other adults interact with the children, encouraging them to talk about what they do, stimulating reflection and further investigation.**

**When Arthur joined the group at the balance center, the children adapted their work so that his tray could become the work area. Because of teacher-modeling, the children naturally moved to include Arthur in the activity.**

**The occupational therapist obtained a chair that was small and close to the ground so that Peter could participate in the activities without taxing his spine.**

**Nora, who has several health problems, is very familiar with the nurse's office. She connected the experience with the balance scale to being weighed by the nurse.**

watermelon was "pushing" him down. The teacher introduced the relationship, "The enormous watermelon is heavier than the baby watermelon," and wrote the sentence on a strip of adding-machine tape.

Then she handed Jeanette a bucket for each outstretched hand and asked the children to think about what would happen if the enormous watermelon was put into one bucket and the baby watermelon put into the other. The children tested the clay watermelons in the buckets several times.

The teacher then brought out a balance scale and placed the watermelons in a pan, one on either side of the center fulcrum. After discussing the relative weights of the melons, the children tested several dolls and stuffed animals to see which were heavier. Jeanette said that the balance worked like a seesaw and they could put two "eensy" monkeys on one side to play with the large bear on the other.

Arthur joined the group. Because he was confined to a wheelchair, the children positioned the balance scale so that it would rest on his tray. The teacher introduced the relationship, "The little monkeys are lighter than the bear." Jeanette added, "But two monkeys can make the bear balance."

In the Manipulatives Center, several children sorted juice cans filled with objects of different weights to determine which cans were heavier. They used a sorting mat like the one in figure 11.4. Then Steven asked to try the cans on the balance scale. The aide asked the children to show with their arms what they thought the balance scale would do for each pair of cans. Peter had to bring his special chair over to the group because he felt tired.

| Lighter | Heavier |
|---|---|
| | |

Fig. 11.4

At the Clay Center, Manette and Albert rolled dough balls of approximately equal size. The teacher gave them a balance scale and some toys. She showed them how to tell whether there were enough clay balls in the pan to balance the toys. They argued over whether they needed four or five dough balls to balance the large Lego block. Manette insisted that she could use only four. To record her decision, she drew the four complete balls—and some small pieces.

Nora and George joined the group. Ms. Sakalian gave them another scale. Nora said that she wanted the nurse to weigh her dough balls. The aide led them to the nurse's office where they examined the scale and watched the aide move the marker along the bar as they were weighed. George was convinced that there were clay balls inside the scale that made the numbers move! ❑

## Extending the Special Education Thrust

Collaborative work such as that done in activity centers often must deal with special needs. If a group making the papier-mâché ball is composed of children with poor social skills or those with behavior problems, for example, these students may have to take specific turns applying their strips of newspaper to avoid conflict. When they work at the sorting mat classifying the cans, these students may need to be limited to a single pair of cans at a time.

To insure accountability for working, students should be encouraged to keep a journal by drawing or writing about what is done. This approach is particularly important for those who have difficulty staying "on task." If some children cannot personally carry out this task because of paralysis or other disability, the services of a teacher's aide or a rotating buddy system in which a partner draws what these students suggest can be used.

To deal effectively with the new vocabulary, some students with language or memory difficulties may need to label the weights and parts of the scale. The visual stimulation of the words on the sorting mat and the strategy of writing verbal observations in complete sentences on adding-machine tape (as in the classroom episode) will be particularly valuable for these children.

Students with low vision may need assistance reading the numbers on the nurse's scale and on other scales in the classroom. They may need to feel the movement of the balance to determine which side has the heavier object.

### Extending the Mathematics

Children's literature books such as *Heavy Is a Hippopotamus* (Schlein 1954) can be read to extend the measurement objectives of this episode. A discussion of the relative weights of the different animals and objects highlighted in the story is a natural follow-up. In dramatic play, the children can mimic the walks of heavy animals like the elephant and lighter animals such as the rabbit. They also might draw pictures of the animals or objects and order them by weight.

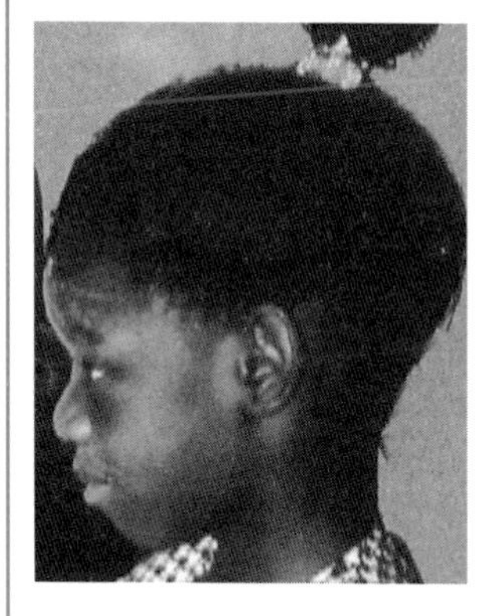

Older students need experiences with a variety of scales, such as the spring balance and the force-arm balance, as well as one-pan and two-pan balances. They can use commercial weights to approximate their hand-made clay "weights," then use the latter to find the weight of various objects.

Some older students may wish to explore the effect of mass on buoyancy in the air and water. Others may explore using balance scales to estimate the weight of large quantities of small objects by sampling. These students should be invited to write about both their estimate of the total weight and their strategy of weighing samples to determine that estimate.

## Empty and Full

*Children re-create scenes from a literature selection as a medium for exploring vocabulary and concepts of capacity. Other explorations stem from this one.*

After Ms. Rodriquez read *Mrs. Wishy Washy* (Cowley 1980), a group of children role-played the story at the Water Play Center. They named a series of rubber puppets after the characters, which they put into the water until the tub was "full," and repeated some of the text. Then they emptied the tub and dried the puppets with paper towels. The teacher's aide

**The use of adding-machine tape helps the children to see the connection between sentences just used and the written word. The tape can also be easily manipulated as a "label" for the product from a task, as a model to copy, and as a stimulus for use in role-play activities.**

**Corey has a limited space in which he can work because of limited mobility. The teacher helped him transfer the boxes and cubes to a work tray and helped him to position his walker and chair so that he could work at a table.**

**When Jane took her turn serving the juice, the teacher's aide helped her to handle the cups. Both Victor and Jane worked together, drawing the circles as a record of their work.**

conversed with them to be sure that they used the words *full* and *empty* within the context of their play. Then Teesa filled cups with water and served "snacks" to each of the puppets. John put the cups in a row and placed a puppet in one-to-one correspondence with each cup.

Haydee filled a kidney basin with clay, repeating, "Oh, beautiful mud." She then drew characters from the book, cut them out, and stood them up in the "mud." The aide repeated the sentence, "The basin is full of mud," and wrote it on adding-machine tape.

At the Manipulatives Center, Kenny, Jason, and Corey filled small boxes with Unifix cubes. After filling each box, Corey emptied out the cubes and linked them together. He showed Ms. Rodriquez that the jewelry box held more cubes than the paper-clips box. He said the "rope" he made was longer for the jewelry box. The teacher showed him how to paste squares to represent the number of cubes to keep a record of his experiment. He asked to trace the bottom of the box so that he knew which was which and asked the teacher to write *more* and *less* on the appropriate parts of the record that he had made.

Angel and James each took fists full of Unifix cubes. They linked the cubes together, aligned the linked cubes, and decided that James's fist held more cubes because his cube train "stuck out more." The teacher repeated the sentence, "James's fist held more cubes than Angel's fist." Then she wrote the relationship on adding-machine tape. The children pointed to the words and read in chorus, "James's fist held more cubes than Angel's fist."

During snack, Jane and Victor served the juice. They counted how many small cups the teacher filled from the large bottle. They guessed that the second bottle would fill more cups than the first bottle did; they then drew circles to keep a record of the number of cups as they were filled. They were surprised to see that the same number of cups were filled from the second bottle! ☐

### Extending the Special Education Thrust

A baby's bathtub was adapted to become a Water Play Center for those children who had to work directly on the floor. If this idea is replicated and used with children having poor coordination, it will be useful to cover the floor with newspapers and have the children wear plastic aprons.

Water play is very difficult for children who need to use walkers or stands to position themselves. To avoid frustration with water play, these children need a carefully chosen buddy and often the direct assistance of an aide.

Students with low vision benefit from the opportunity to *feel* the depth of the empty containers, the individual objects, and the wetness from any overflow. They also need to feel the linked cubes to experience the differences in the volumes of the boxes.

Children who lack the motor coordination for distributing a juice snack can still participate by keeping records as the juice is distributed by others. They can also role-play the experience when the cups are empty. Additionally, activities like the careful filling of small boxes with cubes and recording on paper are helpful for developing fine-motor skills.

### Extending the Mathematics

The water-play activity can easily be extended to experiments with buoyancy, surface tension, and displacement (e.g., adaptions from Wiebe [1990] or McKibban et al. [1987]). Children can be encouraged to observe, classify, and discuss results and then make a graph for easy comparison.

A large container can be calibrated to make a tool for measuring liquid volume. The graduated scales on the measuring cups can be determined by actually filling the instruments with previously measured liquid.

Some children may be able to estimate the number of cubes that fill a box and relate the number in one layer to the actual measure. Others might use cubic inch and cubic centimeter blocks to determine if the boxes were made to customary or metric dimensions and then classify them. Some children may also be able to fold and cut their own boxes to fit specified dimensions.

## CREATING WINDOWS OF OPPORTUNITY

This chapter examined the teaching of measurement in early childhood classrooms having large numbers of physically handicapped children. The children worked in mathematics activity centers, each reflecting developmentally appropriate measurement activities in a whole-language environment. A risk-free, student-centered environment characterized the episodes.

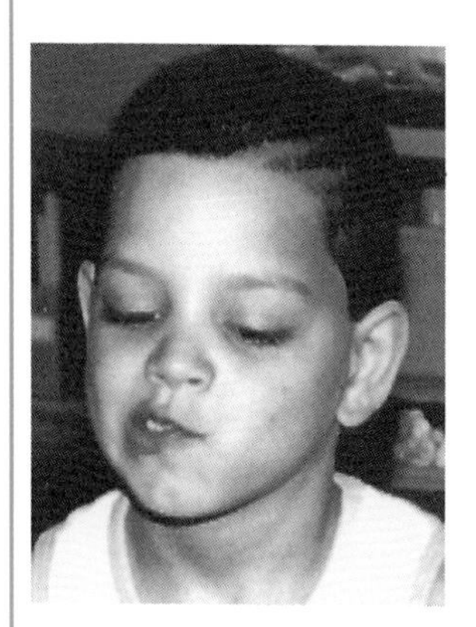

Because of the age range of the children involved—five to eight years—heavy emphasis was placed on selecting the measurable attribute, observing and classifying objects according to the measurements, estimating and then directly comparing measurements, and counting nonstandard units to carry out different measurement tasks. Children were encouraged to record their work in a variety of personally meaningful ways. They used both commercial and homemade measuring instruments as they solved everyday problems and made connections to other contexts.

Experiences were planned so that the children learned to apply new vocabulary to a variety of contexts. For example, relational language was emphasized as the students made comparisons based on the selected attribute. They were then often given opportunities to apply or practice newly learned vocabulary through open discussion and in the books they created. These measurement experiences indeed opened windows of opportunity to mathematical thinking and decision making.

## REFERENCES


Cowley, Joy. *Mrs. Wishy Washy.* 1980. Bothell, Wash.: The Wright Group, 1990.

McKibban, Mike, Kathleen Landon, Walt Laidlaw, and David Lile. *Floaters and Sinkers.* Fresno, Calif.: AIMS Education Foundation, 1987.

Myller, Rolf. *How Big Is a Foot?* New York: Dell Publishers, 1991.

National Council of Teachers of Mathematics. *Curriculum and Evaluation Standards for School Mathematics.* Reston, Va.: The Council, 1989.

Parkes, Brenda, and Judith Smith, retellers. *The Enormous Watermelon.* Australia: Nelson, 1986. Distributed by Rigby, Crystal Lake, Ill.

Routman, Regie. *Invitations: Changing as Teachers and Learners K–12.* Portsmouth, N.H.: Heinemann Educational Books, 1991.

Schlein, Miriam. *Heavy Is a Hippopotamus.* New York: W. R. Scott, 1954.

Wiebe, Ann. *Soap Films and Bubbles.* Fresno, Calif.: AIMS Education Foundation, 1990.

Ziefert, Harriet. *Measure Me.* New York: Harper Child Books, 1991.


## FOR ADDITIONAL READING


Bruni, James V., and Helene Silverman. "Let's Do It: An Introduction to Weight Measurement." *Arithmetic Teacher* 23 (January 1976): 4–10.

———. "Let's Do It: Developing the Concept of Linear Measurement." *Arithmetic Teacher* 21 (November 1974): 570–77.

Hirsh, Elizabeth S. *The Block Book.* Washington, D.C.: National Association for the Education of Young Children, 1982.

Inskeep, James E., Jr. "Teaching Measurement to Elementary School Children." In *Measurement in School Mathematics*, 1976 Yearbook of the National Council of Teachers of Mathematics, edited by Doyal Nelson, pp. 60–86. Reston, Va.: The Council, 1976.

Jensen, Rosalie, and David R. O'Neil. "Let's Do It: Meaningful Linear Measurement." *Arithmetic Teacher* 29 (September 1981): 6–12.

Liedke, Werner. "Measurement." In *Mathematics for the Young Child*, edited by Joseph P. Payne, pp. 229–49. Reston, Va.: National Council of Teachers of Mathematics, 1990.

Neufeld, K. Allen. "Body Measurement." *Arithmetic Teacher* 36 (May 1989): 12–15.

**Helene Silverman** is a professor in the graduate and undergraduate programs in early childhood and elementary education at Lehman College, City University of New York. She teaches mathematics methods and advanced curriculum courses where she emphasizes approaches to working in urban classrooms. A mentor for master's thesis projects, she guides students as they create change in their classrooms.

**Phyllis Paris** is an assistant principal at Public School 85, School District 10, in the Bronx, where her major responsibility is the supervision of early childhood classes. Her major areas of interest include the development of literacy, the use of learning centers, mainstreaming students with multiple handicaps, and parental involvement in schools.

12

# Geometry and Spatial Sense

*James V. Bruni*
*Susan Rovet Polirstok*

GEOMETRY for students with special learning needs? That hardly seems like a high priority. After all, in a regular elementary school classroom, geometry is usually assigned low priority or treated as an enrichment topic. With the extra attention that students with special needs require to learn number concepts and skills, geometry may seem like a luxury. But is it? This chapter presents three examples of geometry experiences that make the case that geometry is clearly not a "frill" but an essential and invaluable component of mathematics programs for *all* children—with extraordinary dividends for the special-needs learner!

The National Council of Teachers of Mathematics (NCTM) *Curriculum and Evaluation Standards for School Mathematics* (1989, pp. 48–50, 112–15) emphasizes the importance of geometry in elementary school mathematics. The *Curriculum and Evaluation Standards* presents a comprehensive vision of geometry—one that goes far beyond identifying geometric shapes. It recommends a wide range of first-hand experiences that will involve children in making, examining, and transforming two- and three-dimensional shapes and exploring the space around them, as they achieve "insights and intuitions" about geometric shapes and develop "spatial sense."

The activities described in this chapter are examples of the kinds of geometry investigations proposed by the *Curriculum and Evaluation Standards* (NCTM 1989). In implementing these activities, a teacher has the opportunity to observe students demonstrate their ability to do the following:

- Describe, compare, and contrast properties of two- and three-dimensional shapes

The authors wish to extend special thanks to the faculty of the Birchwood School in West Nyack, New York (Regina Baxter, Terence Burke, Nita Fierro, Fran Friedman, and Marcia Frey), and their principal (Art Jakubowitz) for their generous cooperation and sincere interest, which made the development of this manuscript possible.

- Make and describe figures created by combining shapes
- Recognize two- and three-dimensional shapes from different perspectives (e.g., how two-dimensional shapes look in different positions on a plane; how three-dimensional shapes look from above, below, and the sides and how they are formed from flat patterns)
- Describe movement from one place to another in space
- Give instructions (using directional terms) for moving from one place to another on a map
- Relate geometric shapes to shapes in their environment and the relationship of form to function

Many special-needs learners are often categorized as poor mathematics students on the basis of arithmetic scores from standardized assessment measures. However, these scores are usually based primarily on arithmetic ability and are not reflective of their spatial abilities, which in many instances are quite remarkable. Geometry investigations offer these students an opportunity to experience an entirely different form of mathematics and the success that has eluded them. Also, as indicated by Morrow (1991), "Because geometry provides opportunities for intuitive, holistic thinking, it serves as an approach to other math topics for children who are less likely to quickly grasp ideas presented sequentially, symbolically, and through logic" (p. 20).

Other special-needs learners may manifest an array of perceptual deficits in areas such as the following:

- Spatial perception—the ability to discriminate right from left, top from bottom, front from back, and so on
- Visual discrimination—the ability to discriminate similarities and differences when comparing objects, letters, or numbers
- Visual memory—the ability to remember visually one object from an array or a series of objects in a specific order
- Visual object constancy—the ability to identify an object regardless of its position in space
- Visual figure-ground—the ability to identify an object from a background of other objects

Deficits in these skills make it difficult for students to perform well on standard school tasks involving reading and writing alphabet letters or numerals, using space on a page appropriately for computation, using rulers and other measuring devices, understanding diagrams, and following oral or written instructions that involve directional terms. For these students, geometry and spatial-sense experiences can provide important opportunities to improve performance in these problematic areas.

## ILLUSTRATIVE CLASSROOM EPISODES

Three geometry teaching episodes follow that demonstrate the reactions and interaction of children in rich, problematic experiences. These lessons actually took place in a special education day school committed to a small-group, pull-out mode of instruction for students with severe learning and behavior disorders.

## Thinking Geometrically Using Tangrams

Ms. Montez attended a math workshop and really enjoyed creating different shapes with the tangram pieces. She even played with the tangram pieces at home and was thrilled to be able to arrange all seven tangram pieces into a square (fig. 12.1). She thought this might be wonderful material to use with Jose, Mike, and Jessy as a way of giving them some valuable geometry experiences.

**Ms. Montez learns the potential of the tangram pieces by first using them herself.**

As she used the tangram pieces herself, she realized that tangram investigations presented excellent informal opportunities to help her students visualize shapes in different orientations, combine shapes to make other shapes, examine special features of shapes, compare and contrast properties of different shapes, and so on. The more she reflected, the more opportunities she saw for helping her students think geometrically using the tangram pieces.

Fig. 12.1

She thought about how she might structure a series of activities to meet the special needs of these six- and seven-year-old children who were functioning at a readiness level. They were easily frustrated, they were impulsive, they had difficulty following directions, and their language development was delayed (although their mathematics performance was better than their language performance on standardized tests).

Ms. Montez decided to start with a story and then allow them each to use a set of tangrams to create shapes and describe what they made. She came to class with sets of plastic tangram pieces and paper versions of them. She also located a square plastic shape that was the same size as the tangram square. The next morning she asked the three children to sit with her at a table and told them a story.

*Ms. Montez:* This is a story about a little Chinese boy named Tan who loved to collect things. One day he found this very special tile. (*She holds up the square shape.*) He was so excited! He thought the tile was so special. (*She pauses.*) What's so special about this tile?

*Jessy:* It's blue!

*Jose:* It's smooth!

*Ms. Montez:* Yes. It *is* a lovely blue. And it feels smooth. (*She gives each of them a chance to feel it. Then she runs her finger along the edges of the tile.*) What kind of a shape does it have?

**The teacher makes learning experiences multisensory to compensate for the learning channel weaknesses and perceptual difficulties of the students in her group.**

| | |
|---|---|
| *Mike:* | It's square. |
| *Ms. Montez:* | I'd like each of you to stay in your seats and look around the room to see if you can see something that has a square shape like this tile. (*She gives them a chance to look around.*) |
| *Mike:* | The door window! |
| *Jessy:* | The table! |
| *Mike:* | The table's not square. It's too long? It's a *rectangle*! |
| *Ms. Montez:* | Why does Mike call the shape of the table top a rectangle? |
| *Jose:* | It has long sides and short sides. A square has all short sides. |
| *Ms. Montez:* | Well, let me tell you what Tan did with this special tile! He put the tile into a bag and then into a large envelope. (*She puts the tile into a plastic sandwich bag and then into a brown envelope that already contains another bag with a set of tangram pieces.*) Then he ran home to show it to his mother. But, as he ran home, the envelope fell. (*She drops the envelope.*) Can you guess what happened? |
| *All:* | It broke! |
| *Ms. Montez:* | Yes. Tan looked into the envelope and saw that the tile had broken into pieces. He ran home crying and showed the pieces to his mother. (*She takes the pieces out of the envelope and places them on a sheet of paper in front of the children on the table.*) Tan's mother looked at the pieces and said, "Why that's amazing! Those pieces are very special!" (*She pauses to have the children look at the pieces.*) What's so special about those pieces? |
| *Mike:* | How did you do that? Where's the square tile? |
| *Jessy:* | Those pieces were in the envelope! |
| *Ms. Montez:* | You're right, Jessy. You discovered my secret! Now, let's look at these pieces. What's special about them? |
| *Jose:* | They have triangles. |
| *Mike:* | There's a little square! |

**This is an illustration of how a child with a learning disability (or any child) can do "surface learning." By probing, one finds that the necessary underlying concept may not have been internalized.**

Ms. Montez led a brief discussion about the shapes, asking them what was special about a triangle. She then had the students compare it to a square and locate all the pieces that were shaped like a triangle. The children realized that a triangle has three sides and a square has four sides, but when Ms. Montez asked them to count the sides of the square Jose simply ran his finger along one side as he counted to four, and Jessy counted the corners, or vertices, of the shape.

She asked them each to take their square piece and pretend that it was the floor of the room. She asked them to show with their fingers where they would go if they went to a *corner* of the room and then where they would go if asked to go to a *side* of the room. From their actions, it appeared this helped clarify the vocabulary. She realized from this discussion that Jessy did not see the difference between the square and the parallelogram pieces. But she knew she had to keep the discussion brief and made a mental note to revisit these ideas. She gathered the students around her.

*Ms. Montez:* Getting back to our story... Do you know what Tan did? When he realized how special those pieces were, he began making all sorts of shapes by putting the pieces together, like this. (*She models making a house with the pieces.*) What do you think he made with these pieces?

**Successful activities with special-needs learners often require modeling by the teacher.**

*All:* A house!

*Ms. Montez:* I'm going to give you a chance to make things out of these pieces just like Tan did. These pieces have a special name. They're called *tangram pieces*. I'm going to give you each a set of tangram pieces and a cardboard tray to put the pieces on. I want you to keep the pieces in the tray while you work. Do you agree to keep all the pieces on the tray and not make them fall onto the floor? Will you do that?

**Reinforcement of the rules for handling manipulatives is important.**

Everyone agreed, and Ms. Montez distributed the sets of tangram pieces on trays. She had the children open their bags, take out the pieces from the bags, and check to see if they had the same pieces and the same number of pieces as Tan's (those pieces were still on a sheet of paper in the middle of the table). This turned out to be a good way to see how well the children could compare shapes and recognize congruent shapes. The children began making all sorts of shapes with the tangram pieces.

**Whenever a new material is introduced, children need a reasonable amount of play time to familiarize themselves with the material. Shortening this time can produce frustration and anger.**

*Mike:* I made a lamp!

*Jessy:* Look at my one-legged lady!

Ms. Montez encouraged them to describe their creations. When she tried to have them make other designs with the tangram pieces, they were reluctant to destroy what they had made. So she told them she had a way they could save their designs and share them with others. She now showed them the paper tangram pieces and how she made a design with these pieces and saved it by pasting them on colored paper. Ms. Montez gave each child a sheet of colored paper, a glue stick, and a set of paper tangram pieces and challenged them to copy their designs onto the colored paper. She asked them questions about how they knew where to put the different pieces.

**Not only did this activity provide excellent hand-eye training, but being able to preserve their designs gave the students a true sense of "ownership" and helped sustain their attention.**

With the children so motivated to make designs with tangram pieces, she took advantage of this opportunity for language development by writing the children's descriptions alongside each of their creations (fig. 12.2). When she suggested that they might like to make a book of these creations with their descriptions, the children greeted the suggestion enthusiastically. Ms. Montez knew that geometric activities using tangrams would be an ongoing project in her classroom for weeks to come! □

**The mathematics activity provided a natural opportunity for meaningful language development.**

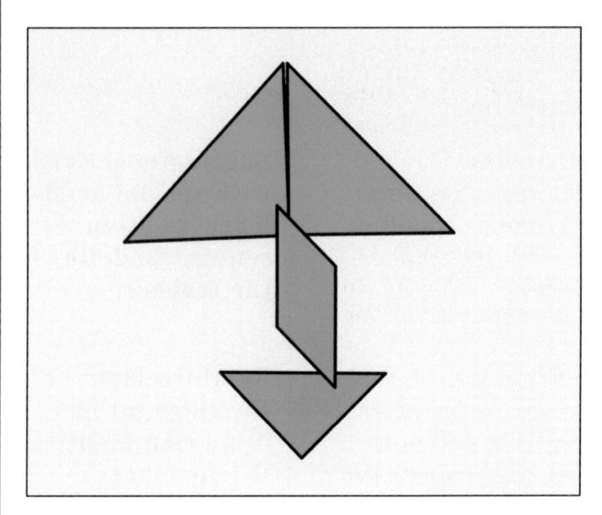

The Diamond Lamp

This lamp is very valuable and costs a lot of money. This lamp can break easily. This lamp belonged to an old man.

Mike

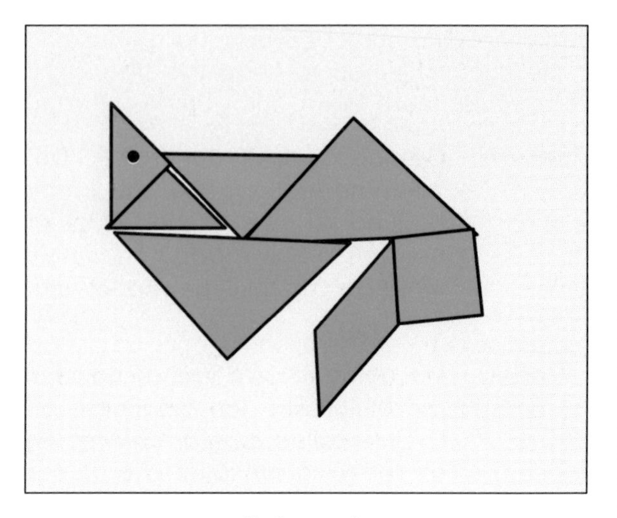

A Spaceship

The piece with a dot is a wing with a weapon. You throw it like a Ninja turtle.

Jose

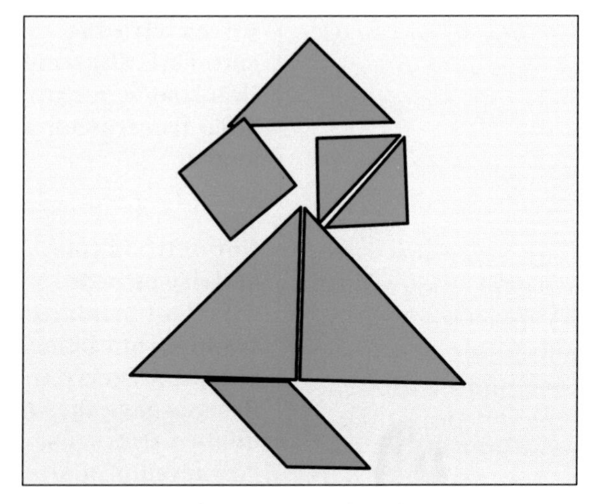

A One-Legged Lady

The lady has one leg because there wasn't another piece for a leg.

Jessy

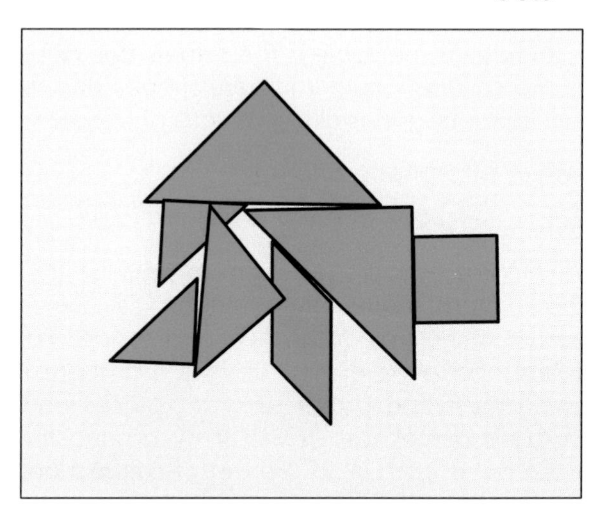

The Trapper Spaceship

Aliens travel in this spaceship and trap people. Aliens are not nice.

Mike

Fig. 12.2

## Extending the Special Education Thrust

The tangram activity that Ms. Montez used can be further modified for older students who have listening and spatial deficits. A highly motivating extension is to challenge students to make a design with the tangram pieces and then describe that design to other students *so they can attempt to replicate it without seeing it.*

One student can be designated the "design maker," who gives specific directions about how to place each piece in the design. The design maker stands behind the group, to eliminate the right-left opposite problem inherent in direction giving, and can use a magnetic board and magnetic tangram pieces. The other students, using their own tangram pieces on a mat or tray, must listen carefully, interpret the directions, and place the pieces so their designs will match that of the design maker.

Not only does this task emphasize the importance of giving specific directions and listening carefully to directions, but it requires the use of directional words and spatial concepts. To get the activity started, it is useful for the teacher to assist in modeling the activity and then to assist the design maker by asking such questions as, Which piece should we start with? (Possible response: the biggest triangle.) How do we place the longest side of the triangle? (Possible responses to be encouraged: horizontally, vertically, pointing upward toward the right, pointing upward toward the left.) Some students may require the use of directional prompts on their mats or trays—R for right, L for left, V for vertical, H for horizontal, and D for diagonal, along with arrows pointing in the appropriate direction (fig. 12.3).

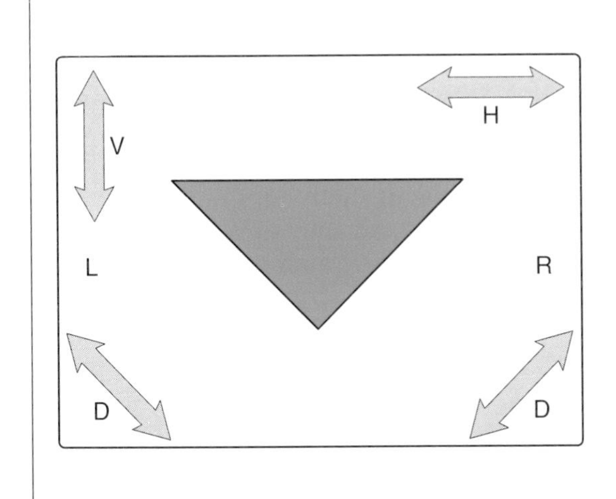


Start with the biggest triangle. Put it on the mat so that its longest side is horizontal and the opposite corner is pointing down.

Fig. 12.3. A tangram piece is placed on a mat having directional prompts.

As a writing activity, students can make a list of the sequence of directions they will use and may use a walkie-talkie from outside the classroom to "broadcast" the directions. Again, the teacher can model the direction writing, using a "cloze" format, with students supplying key words to complete the directions. Over time, the teacher can increase the number of words to be supplied until direction writing becomes an independent activity. Analyzing the children's directions—which ones conveyed the correct information and which ones caused confusion—is not only an important problem-solving activity but also an important communication skill.

### Extending the Mathematics

Numerous tangram puzzles at various difficulty levels are available (see, e.g., Martschinke 1990) for nurturing the development of spatial sense (see the examples in fig. 12.4). As they use the tangram pieces and compare and contrast the various shapes, children develop informal understandings of the basic properties of triangles, squares, and parallelograms.

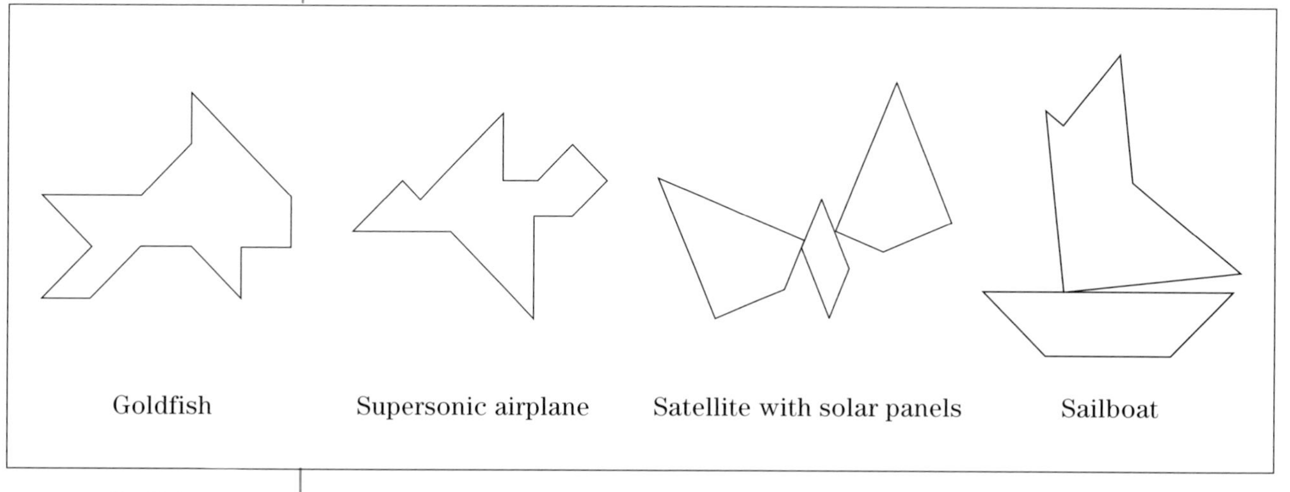


Fig. 12.4

By comparing tangram shapes with the shapes of objects in their environment (e.g., rectangles, circles, ellipses, rhombuses, trapezoids, other quadrilaterals, other kinds of triangles, etc.), they move from identifying and describing the overall shape of objects to looking at specific properties of shapes (like the relative lengths and positions of their sides and their angles or the amount of opening between the sides), and they develop an intuitive understanding of the basic properties of different shapes as an important basis for more formal definitions.

## From Maps to Logo

Mr. Brooks wanted to find ways to meet the special needs of two of the students in his fifth-grade class, Peter and Maria. Both of these students repeatedly demonstrated spatial difficulties, indicated by such behaviors as their inability to describe the location of things or to give directions for how to get from one place to another. They had short attention spans and memory problems, and they did not follow directions well. Their records also consistently indicated poor social skills and low self-esteem.

He wanted to organize some activities that could help them develop their spatial sense and, at the same time, be motivating and help them experience success. Mr. Brooks had participated in some Logo workshops during the summer and felt that "turtle geometry" would be a great way to help them develop spatial skills. But he needed to find a way to introduce it so that Maria and Peter could be successful.

One day, a wonderful opportunity arose. As part of a science lesson, his class was talking about how robots are used in so many different ways in industry. Peter seemed interested in the topic. He told Mr. Brooks how he heard on television that they used a robot to remove an explosive from a building. Mr. Brooks seized this opportunity! He decided to plan some activities that would challenge them to teach a robot how to move around the school and maybe even lead to their being able to program movement using Logo on the computer.

**This mapping activity used an area with which the students were familiar to reduce the level of abstraction of the task.**

The next day he asked Peter and Maria to work on a special project with him. He showed them a model of a robot he had made with a pencil and pipe cleaner and a large sheet of grid paper on which he had drawn the outline of the corridor of the main floor of the school building with key locations indicated (fig. 12.5). He explained that the pencil was supposed to be a robot and asked what they thought the paper represented.

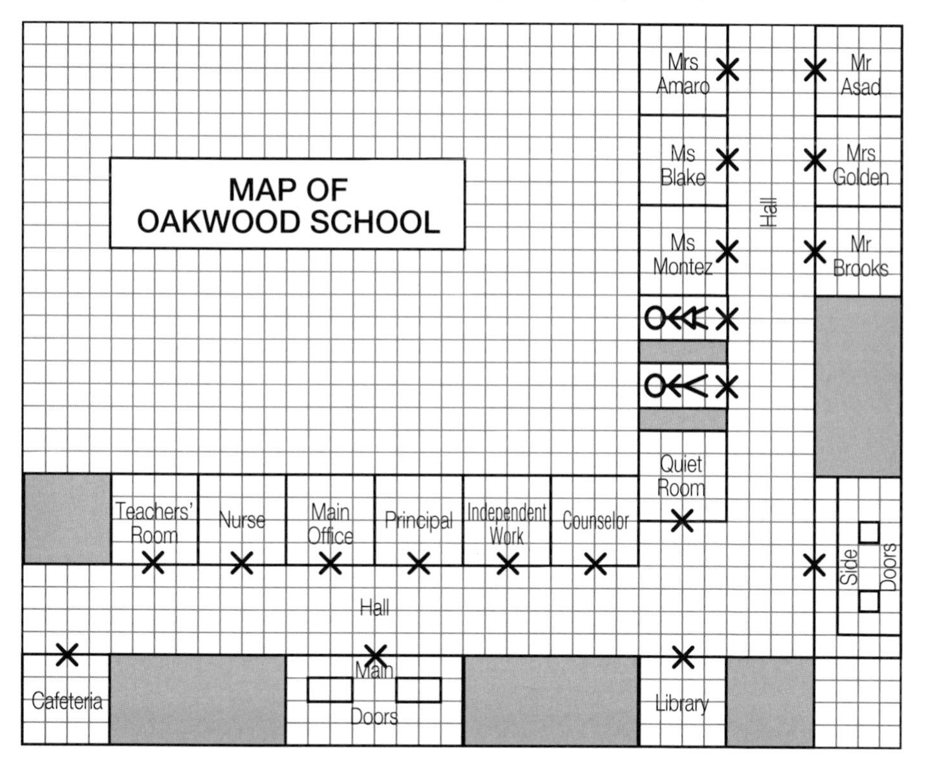


Fig. 12.5

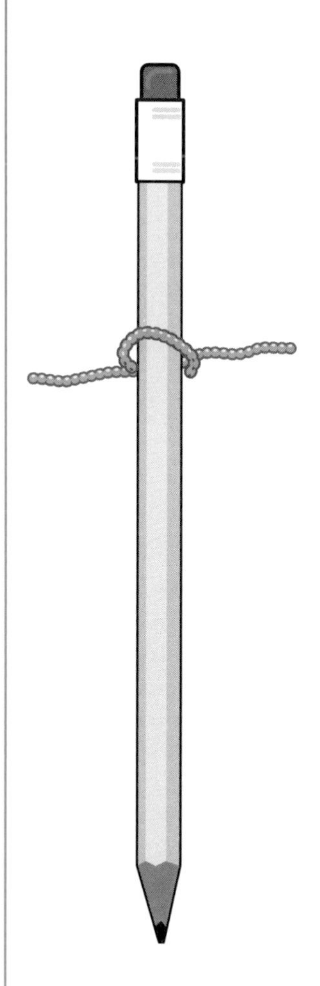

The students recognized that this was supposed to be a map of the rooms in the school and pointed out the main office and the cafeteria. Mr. Brooks asked them which other rooms on the map they could name. Peter and Maria identified several—the principal's office, the boys' room,

**The walk was a concrete, visual, sequential memory-training activity and helped students make the connection to the more abstract diagram of the arrangement of the rooms.**

**This is an example of backward chaining. The correct completion of the map each time reinforced visual memory by providing prompts for revisualization.**

**Mr. Brooks actively involved the children in using directional words for a specific purpose and could determine their understanding directly by the correlation between the words they used and their actions as they moved the pencil robot along the map.**

the girls' room, the teachers' room, a few of the classrooms, and the library. They were unsure of some of the rooms. Mr. Brooks decided to take them for a walk, and together they verified the location of each room on the map as they walked along the hall until the entire map was completed and Mr. Brooks was confident that they could relate it to the classrooms in the building.

As a reinforcement activity, Mr. Brooks presented the map with one room unnamed and asked the students to supply the correct room label. He did this several times, leaving out one room, two rooms, four rooms, and so on (more rooms each time), until the map had no labels and the students had to revisualize the entire schema.

With the map completed, he took the pencil robot, placed it at the main entrance door on the map, and introduced a problem.

*Mr. Brooks:* Suppose this robot was at the main entrance of the school. In what direction would it have to move to go to the library?

*Maria:* It would have to go that way (*pointing down the hall*).

*Mr. Brooks:* Yes, but we would have to give it clear directions. It would have to move *forward* into the hall (*he models the action with the pencil robot*) then ... would it have to *turn right* or *turn left* to face the direction of the library?

*Maria:* Right?

*Mr. Brooks:* What do you think, Peter?

*Peter:* (*Hesitatingly*) Right. But how can a robot understand what we're saying?

*Mr. Brooks:* You're right, Peter. That would be a big problem. So let's teach it some very simple words. When we want the robot to go forward, we'll just use the word *forward.* (*He has each of them practice moving the pencil robot along the corridor of the school map when he gives the command "forward."*)

Mr. Brooks then challenged them to teach the robot how to move from one place to another on the map by posing problems like, "Suppose the robot is at the door of the library—what commands should we give it to have it get to Mr. Asad's room?" Together they discovered that the most useful commands were "right" for "turn right," "left" for "turn left," and "forward" for "go straight ahead."

He had them verbalize the commands and wrote them for the students as they acted them out. Maria and Peter realized that they also needed to tell the robot *how far* to go forward. Mr. Brooks asked them how they could do that. Peter suggested that the boxes on the grid paper could be used to tell the robot how far to go forward. Using Peter's idea, they practiced using commands like "forward 20" to move the robot. But sometimes they overestimated how many boxes forward were needed. So they taught the robot how to go backward and invented the "backward" command.

Mr. Brooks wrote the four commands on cards—*forward* ____, *backward*____, *right, left*—and gave each of them a chance to move the pencil robot along the corridor on the map using the commands to get from one room to another. The only commands that posed a problem were "right" and "left," because the students were sometimes confused by them. Mr. Brooks suggested that they draw a face on the eraser part of the robot (to personalize it) and then that they put a red pipe cleaner bracelet on the right arm of the pencil robot to help them remember which direction was "right." Peter preferred to think of the pipe cleaner as a flashlight. Mr. Brooks thought that was a fine idea.

**Prompts and cues are necessary for success with many students (especially those with directionality or reversal difficulties).**

As a follow-up to this activity, Mr. Brooks prepared copies of the large map on grid paper so that each of them could have their own maps. He gave Peter and Maria a chance to make their own pencil robot models, and then he challenged them to figure out where the robot would end up after following a sequence of directions (fig. 12.6). Peter and Maria each modeled the movements using their pencil robots along their maps and were delighted to find out that they ended up at the same place. As they followed the directions to move the robot forward a large number of boxes, Mr. Brooks encouraged them not to just count by ones but to count by twos, fives, or any number they chose.

**The students used multiples of a number and the multiplication tables for a real purpose, to move more quickly along the map.**

START AT MAIN DOOR
GO FORWARD 2 STEPS
TURN RIGHT
GO FORWARD 10 STEPS
TURN LEFT
GO FORWARD 4 STEPS

WHERE IS THE ROBOT?

START AT DOOR OF LIBRARY
FORWARD 2
LEFT
FORWARD 16
RIGHT
FORWARD 4

WHERE IS THE ROBOT?

Fig. 12.6. Sample sequence of directions for pencil robot

Mr. Brooks then challenged Maria and Peter to make up a sequence of commands to move the robot from one place to another along the corridor of the school by suggesting they choose a starting point and figure out the commands needed to get to the place they wanted the robot to end up. He worked through a set of commands with them (fig. 12.7). The students then made up sets of commands for each other and tried to figure out where the robot was going in each case.

Maria and Peter swapped sets of commands and enjoyed trying to figure out where the commands were leading the robot. Mr. Brooks thought he could take this activity a step further and have the children use these commands with Logo on the computer. He asked them whether they had ever used Logo. Peter hadn't, but Maria remembered she had used it last year and commented, "It's *very* hard." Mr. Brooks assured Maria and Peter that because they could

START AT DOOR OF ______________
FORWARD ______ BACKWARD ______
LEFT RIGHT
FORWARD ______ BACKWARD ______
LEFT RIGHT
FORWARD ______ BACKWARD ______

WHERE IS THE ROBOT?

Fig. 12.7. Sample open-ended sequence of directions

teach the robot how to move around, they would have a good time with the Logo "turtle."

The next day Mr. Brooks brought in an overhead transparency he had made from the map of the school. He placed the transparency on the computer monitor (where it attached easily because of the static electricity from the screen) and moved the turtle to the location on the screen that coincided with the front entrance of the school, using the coordinates of that location (fig. 12.8). He then invited Peter and Maria to try to see if they could figure out how to move the turtle from the main entrance to different parts of the map.

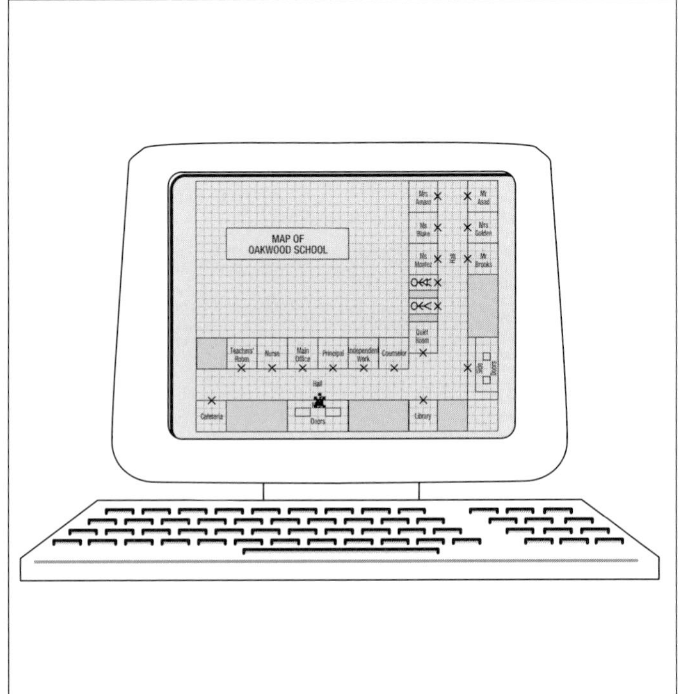


Fig. 12.8

Mr. Brooks taught the students how to make the turtle go forward by typing "F" followed by a space and the number of "turtle steps" forward they want it to go. He also taught them the "R 90" command for turning right, the "L 90" command for turning left, and the "B" command for moving backward. To make it less confusing, he started with the turtle not drawing a path. With lots of trial and error and Mr. Brooks's supervision, Maria and Peter took turns attempting to move the turtle from one place to another on the map of the computer screen. They realized that the turtle steps were not the same size as the side of a box on the map they had made, and through trial and error they discovered that it took about ten turtle steps to move one box length. Both Maria and Peter were successful in directing the turtle to their desired destinations on the map. Maria was overjoyed—she could do it! And last year it was so difficult! Mr. Brooks felt the pencil robot activities really set the stage for this success.

## Extending the Special Education Thrust

This mapping activity has direct applications for special education, because reading a map and locating one's position on a grid are skills necessary for daily living. Mr. Brooks's use of the floor plan of the school as the training map increased the possibility of success because it was a familiar environment.

A key to success with this activity was being able to discriminate right from left. Students with learning disabilities often experience spatial and temporal confusion, and teaching students to compensate by using prompts available in the environment is one way to deal with this confusion. For example, a student may be encouraged to wear an article of jewelry on one hand to differentiate it from the other. Key landmarks in a neighborhood might be associated and serve as prompts for right and left (e.g., at the McDonald's, turn right).

Mr. Brooks chose to have the students put the red pipe cleaner on their robots as a prompt to help discriminate right from left. He used a backward chaining technique to make sure that students could successfully revisualize the locations of all the sites on the map. Using a grid, he began to teach students to use simple commands to reach coordinate points. This activity not only provided spatial training but reinforced direction giving (both oral and written) as well as direction receiving (auditory sequential memory training).

The use of the map to develop the language of Logo commands was an excellent pre-Logo preparatory experience. With preparatory experiences involving movement on a map, students can use Logo more successfully. They can focus their attention on the challenges of being able to estimate distances traveled by the turtle on the screen and using right and left commands from different spatial orientations to successfully develop programs for moving the turtle from one place to another on the map.

This mapping activity and its Logo extension can be carried out with students who have varied types and degrees of disability. Even students with physical disabilities who do not have the use of their arms can use a head wand to give the necessary Logo commands, once the pretraining activities have been experienced. Blind students can participate in Mr. Brooks's school map training as long as the grid is a raised one and has the appropriate Braille labels. Similarly, if the acetate screen has a raised grid and appropriate Braille labels and the keyboard has Braille letters, using Logo can be accessible to these students.

### Extending the Mathematics

The map of the school can easily lead to considering other maps, including maps of the community and road maps of the area. Students can discover the value of coordinate grids for locating specific streets or other places and create their own maps.

Through extended work with Logo, they can explore what happens when they use the R or L command with numbers other than 90 and thus develop an understanding of "amounts of turn" and "angles." Students can also learn how to place the turtle anywhere they want on the map by using coordinates. Their work with Logo leads naturally to skill in sequencing instructions and creating programs for teaching the turtle to make a specific design or create a desired path. The world of Logo is open to them, offering a whole new kind of mathematical thinking.

## Making a City

Martin and Brian are very active nine-year-olds in Mrs. Amaro's classroom and pose a special challenge for her. Their records describe each of them as hyperactive, aggressive, easily frustrated, and having low self-esteem. With their test scores below grade level and with the speech problems they exhibit, they have been classified learning disabled.

Mrs. Amaro has noticed that one of the activities they like most is to build with the blocks she has in an activity center. Most of the blocks are either cubes, rectangular solids, or cylinders. She has decided to introduce some other shapes and to see if she could use a building activity they enjoy to introduce them to ideas about two- and three-dimensional geometric shapes and the associated language.

Today Mrs. Amaro brought in a few special sets of small wooden geometric solids (fig. 12.9). She had placed each set in a bag and invited Martin and Brian to join her because she had "a special set of blocks for them to use." To begin the activity, she took out a cone from the bag.

*Mrs. Amaro:* What's special about this block?

*Brian:* It's pointy.

*Martin:* It looks like an ice-cream cone!

*Mrs. Amaro:* (*Taking out a cylinder*) How about this one?

*Martin:* It rolls. Like a can.

*Mrs. Amaro:* How are these two blocks the same? (*Mrs. Amaro let the students take turns holding the solids and examining them.*)

*Brian:* Round bottoms.

*Mrs. Amaro:* What does Brian mean, Martin?

*Martin:* They both can roll.

*Mrs. Amaro:* And how are these two shapes different?

*Brian:* That one's not pointed. See (*he picks up the cylinder-shaped block*), round bottom, round top.

**To understand a concept, one must learn what it *is* as well as what it *is not* (attributes versus nonattributes).**

Mrs. Amaro took out two more blocks and attempted to follow the same type of questioning procedure to help them focus on the characteristics of these geometric solids. But both Brian and Martin began building with the blocks already taken out and wanted to see the rest of them. She decided to take one complete set out, have the children build with them, and talk about their special properties as they were building.

**Acknowledging the students' need to explore the use of the blocks on their own, Mrs. Amaro incorporated block building into the lesson framework and reduced the likelihood of confrontation.**

She had prepared a photocopy of the blocks and had them match each block to the photocopy as she removed it from the bag (to see if they could recognize the two-dimen-

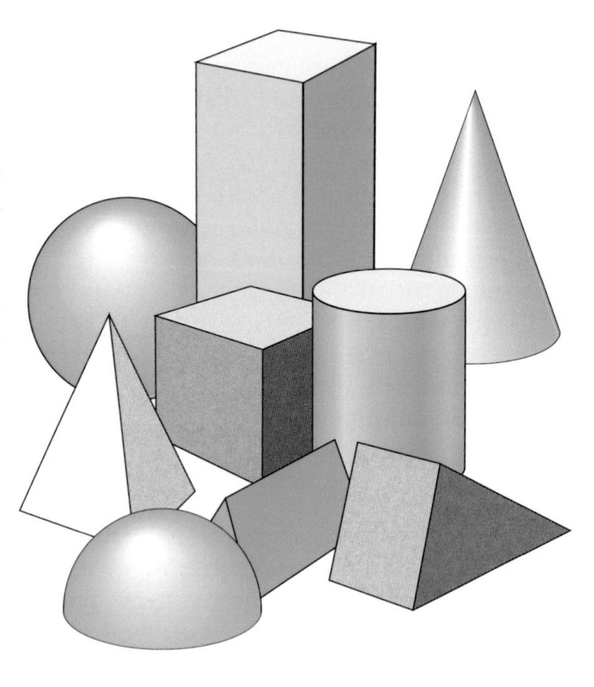

Fig. 12.9

sional representations of these three-dimensional shapes). When there was a difference of opinion over which picture represented which shape, she had them tell what clues they saw that made them think the picture represented a certain shape. In that way, the children came to agreement on the correct matching.

After Mrs. Amaro took one complete set out of the bag, she gave each student a large tray. She explained that when she gave them their blocks, they were to build on that tray and be careful not to let the blocks fall to the floor. Of course, each boy wanted to use all the blocks from the bag they had been examining. So Mrs. Amaro got a second bag of blocks and decided to use this as an opportunity to have the boys verbalize the characteristics of the shapes.

She asked the boys to take turns describing a block from the set of blocks they had been using. At each description, the other boy helped Mrs. Amaro find that shape in the second bag. In the end the boys agreed that this was a good way to be sure that it was fair—that each had the same set of blocks.

Satisfied that both bags were identical, each boy began building a variety of structures. When the structures got too high, Mrs. Amaro warned them that they might fall off the table. Watching them create these structures gave her the idea that the boys might enjoy using their structures to make a model of a city.

So Mrs. Amaro got a large sheet of white paper and asked the students if they would like to use their blocks to make a city. The boys were very enthusiastic about that prospect. She put down the paper and asked them where the main road should be on the city and then used their response to draw it on the sheet. She had them name the road and then had them take turns putting in other roads and naming them. Then she encouraged them to make buildings for the city. As they made different buildings, she talked with them about their buildings.

*Mrs. Amaro:* What happens in this building?

*Brian:* It's a bowling alley. See the bowling ball on top? (*He points to the sphere-shaped block and smiles.*)

*Mrs. Amaro:* This building looks like a Toys Я Us store!

*Martin:* Yeh. It has the same shape. It has those pointed things in front that look like ice-cream cones.

*Mrs. Amaro:* Yes. These pointed shapes you put in are called cones.

Mrs. Amaro let them continue creating their city and was amazed with the complexity of the structures they created. She used every opportunity to have them describe their structures and the features of the blocks they used. She was delighted that they were on task for such a long period of time (which was uncharacteristic of them) and awarded each the maximum number of points that could be earned in their behavior modification system for a given period. The boys were extremely happy and felt successful. In fact, they were so involved in this activity that they did not

**Mrs. Amaro realized how important it is to clarify rules for using the materials and the value of providing individual block sets and work spaces to minimize potential conflict between students.**

**The teacher discovered a purpose for describing the geometric characteristics of the blocks that has real meaning for her students while responding sensitively to their need for fairness.**

**Mrs. Amaro recognized that city planning and model building provide a meaningful context for language development, as well as for making connections across curriculum areas and for relating geometry to their real world.**

**Allowing students to make decisions regarding the choice, composition, and naming of structures and streets confirmed their ownership of the project and consequently enhanced their self-esteem.**

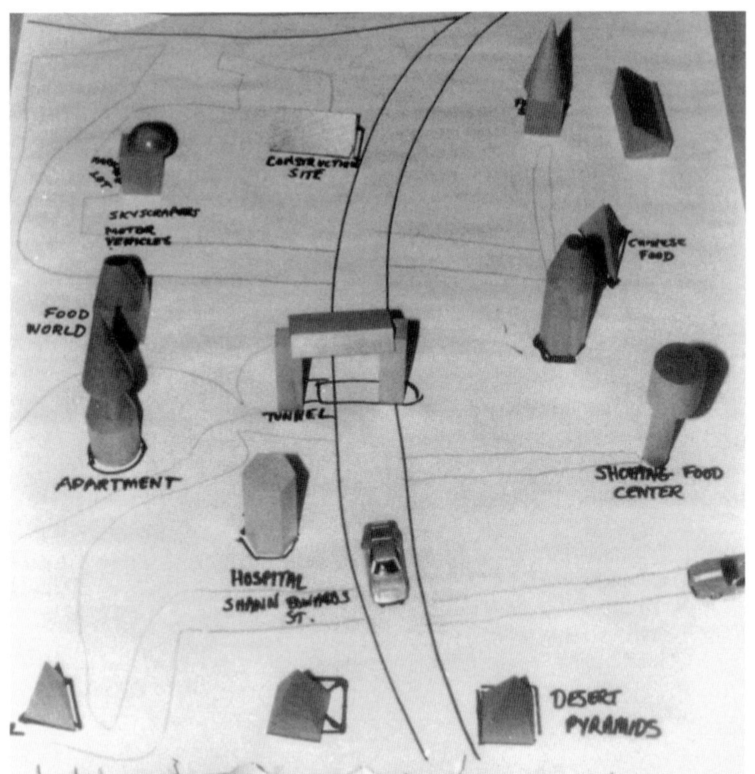


want to stop. Mrs. Amaro was delighted with their involvement and decided this would be an ongoing activity in her classroom. ❑

## Extending the Special Education Thrust

The use of geometric solids to build a city in a cooperative group provides excellent opportunities for language development and interfaces with science and social studies curricula and fosters development of appropriate social skills. The process of building a city requires that students work together, respect each other's ideas, and ultimately make decisions about what needs to be included in the city. Through this process, a teacher can make many clinical observations and interventions to improve a student's interactive skills and enhance the efficiency of the group as a working unit.

For students with learning and behavior disorders, small-group interaction time can be very difficult because of impulsive behavior, hyperactivity, and marked social skill deficits, including aggressive behavior. Using a behavior modification point system and a structured agenda that pinpoints objectives to be accomplished for each session, students with learning and behavior problems can be successful in the group interactive process. The group talk can also provide teachers with an opportunity to address the students' need for self-esteem and to actively listen as they verbalize what is important to them.

From a language-development point of view, the oral discussion of what elements and structures should be included in the city can be valuable in terms of sentence development and the acquisition of new vocabulary. Social skills, such as negotiation, can be enhanced when students cooperatively have to decide what should be included in the city. Once some of the key decisions about the city are made by the group, a myriad of writing projects can evolve.

Depending on the cognitive level of the students, the scope of the task can be modified. Instead of building a city, students might be asked to build a street they know, a school they would like to go to, a special classroom they would like, or some other structure. If the level of aggression is such that students cannot work successfully in a group, then they might be asked to work with a partner that the teacher has determined to be compatible. If cooperative activities are not possible, each student could be assigned a particular section of the city to develop independently. The project's culminating activity would then be to unite all the sections into one completed city. Thus, each student can feel ownership for the project, even though he or she worked on only one specific part.

One of the clear advantages of the making-a-city activity is that students found it to be fun. When students with learning and behavior problems perceive an activity to be enjoyable and nonthreatening, the possibilities for teaching and learning are greatly enhanced.

### Extending the Mathematics

The students' desires to save the structures that they build provides an exceptionally fine opportunity to help them explore ways to represent the structures they have created. Taking photographs of these structures from different spatial orientations (from above, from the front, from the back, from the sides) can help them visualize what they made from different perspectives and can lead to valuable discussion as they describe what the photographs represent.

Similarly, students can attempt to sketch the basic blocks that are used for each structure on separate sheets and label them so that the structure can be recreated (they can write a kind of "recipe" for making their structure). These activities can lead into the use of geoblock activities for developing an understanding of how three-dimensional structures look from different orientations and also the relationship of three-dimensional blocks to their two-dimensional faces (fig. 12.10). Using isometric drawing and understanding actual blueprints of structures is another potential extension, where basic ideas about ratio, proportion, and scale can be an excellent connection between mathematical concepts and their real world.

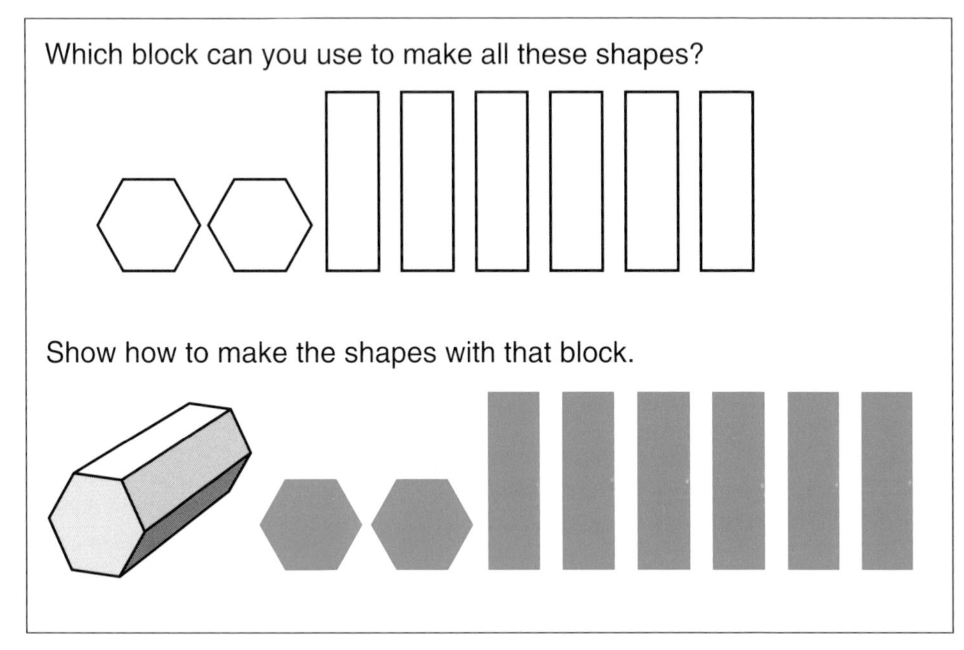


Fig. 12.10

## CREATING WINDOWS OF OPPORTUNITY

The teachers described in the three classroom episodes of this chapter discovered that the geometry activities they developed provided a unique way to help their special-needs learners. Far from being a "frill," geometry provided these students with extraordinary opportunities to experience success with mathematics. These hands-on, multisensory investigations proved very valuable not only for developing many perceptual abilities but also for enhancing oral and written language skills as well as interpersonal and social development.

Geometry for students with special needs? Absolutely!

## REFERENCES


Battista, Michael T., and Douglas H. Clements. "A Case for a Logo-Based Elementary School Geometry Curriculum." *Arithmetic Teacher* 36 (March 1988): 11–17.

Bruni, James V., and Roslynn Seidenstein. "Geometry Concepts and Spatial Sense." In *Mathematics for the Young Child*, edited by Joseph N. Payne, pp. 203–27. Reston, Va.: National Council of Teachers of Mathematics, 1990.

Burger, William F. "Geometry." *Arithmetic Teacher* 32 (June 1985): 52–55.

"Developing Mathematical Processes." *Topic 8: Three Dimensional Shape.* Nashua, N.H.: Delta Education, 1974.

Jensen, Rosalie, and Deborah C. Spector. "Geometry Links the Two Spheres." *Arithmetic Teacher* 33 (August 1986): 16.

Lerner, Janet W. *Learning Disabilities*, 5th ed. Boston: Houghton Mifflin, 1989.

Mann, Philip H., Patricia A. Suiter, and Rose Marie McClung. *Handbook in Diagnostic-Prescriptive Teaching.* 3rd ed. Boston: Allyn & Bacon, 1987.

Martschinke, Judi. *Tangramables—a Tangram Activity Book.* Deerfield, Ill.: Learning Resources, 1990.

Mercer, Cecil D. *Students with Learning Disabilities.* 4th ed. New York: Macmillan, 1991.

Morrow, Lorna. "Implementing the Standards: Geometry through the Standards." *Arithmetic Teacher* 38 (August 1991): 21–25.

National Council of Teachers of Mathematics. *Curriculum and Evaluation Standards for School Mathematics.* Reston, Va.: The Council, 1989.

Rowan, Thomas E. "Implementing the Standards: The Geometry Standards in K–8 Mathematics." *Arithmetic Teacher* 37 (June 1990): 24–28.

Sgroi, Richard J. "Communicating about Spatial Relationships." *Arithmetic Teacher* 37 (June 1990): 21–23.

Van de Walle, John A. *Elementary School Mathematics—Teaching Developmentally*, pp. 266–304. White Plains, N.Y.: Longman, 1990.

Watt, Daniel. *Learning with Logo.* New York: McGraw-Hill, 1983.

**James Bruni** is a professor and coordinator for the graduate program in elementary education at Lehman College, City University of New York. While teaching mathematics methods courses, he works in urban classrooms to nurture a leadership group of teachers attempting to implement the NCTM Standards. He was chair of the Editorial Panel of the *Arithmetic Teacher* and a member of the Editorial Panel of the newly established *Mathematics Teaching in the Middle School.*

**Susan Polirstok** is an associate professor and coordinator of the graduate programs in special education at Lehman College, City University of New York. She teaches special education courses involving the theories that underlie students' learning and behavioral problems and the strategies, both cognitive and behavioral, that teachers can use to maximize performance. Her interests include peer tutoring, self-evaluation strategies, whole-language applications for special education, and parent training.

13

# Discovering High School Geometry

*Melfried Olson*
*Subhash Jani*
*Judith Olson*

*in collaboration with*
*Jan Klippert*

GEOMETRY, an avenue to spatial experiences and logical reasoning, is essential for providing a rich high school mathematics curriculum for *all* students, including those with special learning needs. Even more than arithmetic or algebra, geometry can build on environmental situations to help students formulate convincing arguments and engage in real-world problem solving. The National Council of Teachers of Mathematics (NCTM) in its *Curriculum and Evaluation Standards for School Mathematics* (1989) describes the desired components of the secondary school geometry curriculum. Suggestions (p. 157) include the study of geometry in two and three dimensions so that all students can—

- interpret and draw three-dimensional objects;
- represent problem situations with geometric models and apply properties of figures;
- classify figures in terms of congruence and similarity and apply these relationships;
- deduce properties of, and relationships between, figures from given assumptions.

Elaborating how these ideas should be explored, NCTM (1989, p. 157) offers the following guidelines: "The 9–12 geometry strand should provide experiences that deepen students' understanding of shapes and their properties with an emphasis on their wide applicability in human activity.... Students should have opportunities to visualize ... in order to develop spatial skills fundamental to everyday life and to many careers. Physical models and other real-world objects should be used to provide a strong base for the development of students' geometric intuition so that they can draw on these experiences in their work with abstract ideas."

In relation to the teaching and learning of geometry, the van Hiele model (see, e.g., Burger and Culpepper [1993]; Woodward and Hamel [1990]) is often referenced. This model of geometric understanding characterizes major thinking processes and suggests that the acquisition of knowledge of geometry proceeds through a series of sequential "levels" of geometric thought, namely, visualization, analysis, informal deduction, formal deduction, and rigor. The first four levels are briefly described below:

- *Visualization.* At this stage students see only entities around them. Objects are viewed as whole objects rather than as having component parts.
- *Analysis.* At this stage students begin to focus on the properties of a figure. For example, a quadrilateral is considered a rectangle because it has right angles in addition to looking like a rectangle. However, students cannot understand the relationship of properties.
- *Informal Deduction.* At this stage students understand class inclusion, knowing, for example, that a square is also a rectangle. Students can establish relationships among the properties of figures, such as that having two equal angles in a triangle that implies two sides are also equal.
- *Formal Deduction.* The role and function of deductive thinking is understood. Relationships among such advanced concepts as theorems, undefined terms, and postulates are comprehended.

Although students progress developmentally through these levels in a sequential order, advancement in geometric understanding is more a function of the content and experiences provided during instruction than of the student's age. Each level has its own vocabulary and its own system of relations. If a student is functioning at one level and instruction is at a higher level, learning will be difficult at best and may not occur at all.

Unfortunately, much instruction in high school geometry is conducted at the formal-deduction level, whereas many students, principally because of inadequate attention to geometry in the elementary and middle grades, are functioning at lower levels. They lack the prerequisite knowledge and experiences to understand and successfully use the relationships required for proving theorems. This situation causes a major mismatch between the learner and the content and helps to explain why many students resort to memorization in order to successfully negotiate their way through high school geometry.

To address the methods and organization of geometry instruction for all students, but particularly for those with special learning needs, teachers need to be sensitive to where students are on the continuum toward formal deduction. The following points are useful to keep in mind:

- Begin new topics with informal geometric explorations by developing activities that move students through the levels of geometric understanding—with emphasis on visualization and analysis before informal and, finally, formal deduction are required.

- Establish a risk-free environment that emphasizes concepts and active learning processes (conjecturing, explaining, justifying, etc.) over content coverage alone. Although many formula relationships such as surface area and volume or the Pythagorean theorem are useful, it is also important that students comprehend at some intuitive level the relationship a formula describes.

- Move from concrete to abstract ideas, making connections between models and symbols. Starting with concrete activities alone will not suffice. The connection between the concrete and the abstract must be explicit and purposeful.

- Furnish examples of the uses of geometry in daily life to keep learners motivated and aware of the relevance of geometric concepts.

- Use problem solving as a means as well as a goal of instruction. Students should learn geometry through problem solving as well as apply it to solve problems.

Geometry should be a hands-on, minds-on experience that leads all students to make generalizations and informal or formal deductions from explorations they have performed. Nurturing a positive, accepting attitude toward active-learning approaches, basic to this experience, is a necessary condition for success (Smith 1991; Schloss, Smith, and Schloss 1990).

As was discussed at the beginning of chapter 12, some special-needs students have perceptual or spatial deficits—visual discrimination, visual memory, visual-object constancy, or visual figure-ground problems—that affect their academic progress. Rich geometric experiences are possible for these students, particularly when abstract notions are introduced tactilely and concretely and when the students are given the opportunity to interact with their peers in refining justifications into logical and convincing arguments.

Geometric activity structured in this way is consistent with recommendations in the special education literature for a large percentage of special-needs students because it promotes active engagement at a student's level of challenge (Baroody 1991; Hallahan and Kauffman 1991; Deshler, Alley, Warner, and Schumaker 1981). This approach helps students compensate for their learning deficits, which frequently stem from poor visual perceptual and information-processing abilities, a lack of experience with geometric ideas and the resulting deficiencies in prerequisite geometric understandings, or both.

## ILLUSTRATIVE CLASSROOM EPISODES

The following three episodes describe how different teachers have created geometry experiences that begin with lower levels of geometric understanding and gradually move students to make generalizations and construct logical, convincing arguments that support their thinking. The first episode demonstrates how the study of similarity can begin with the use of pattern blocks and build to more abstract concepts. The second has students draw line segments, find midpoints, and measure segments to determine the relationship between a segment joining the midpoints of two sides of a triangle and the length of the third side.

The last episode illustrates how students were challenged to move from thinking of a rectangle in standard form to locating all rectangles that might be drawn with a given segment as a diagonal. Each episode is followed by suggestions for adapting the lesson for students of differing special needs as well as by ideas for extending the mathematics in natural or exciting ways.

## Similarity

As a warm-up to the day's lesson on similarity, Ms. Sage's geometry students were busily engaged with a partner in examining the pattern blocks, comparing and contrasting them on the basis of their properties. It was the first time the group had used the blocks this year, so Ms. Sage felt that a brief period for exploration might be useful.

**These concrete experiences get Mike motivated and actively participating. At the same time, the structure of the written task is useful in keeping him on task.**

As she moved among her students, Ms. Sage noted that her requirement to list "four or five observations about the blocks" gave a structure to the exploratory time that seemed to be working well for Mike, who had an attention deficit disorder (ADD). Mike typically had difficulty following directions and remaining on task, and he tended to rush through assignments. Ms. Sage waited a few minutes longer, then closed the warm-up by having one student from each team report on *one* likeness or difference they noted.

**Having Mike record for the group further ensures his full, on-task attention. It also enhances the potential of his remembering key points in the discussion.**

Mike had been appointed to synthesize comments under the "How Alike?" or "How Different?" columns of a chalkboard chart. An impressive list was soon formed! Under differences, for example, student teams noted that some were regular polygons (square, equilateral triangle), whereas the others were not; that all but the trapezoid had sides *one* unit long and the trapezoid's longest side was *two* units in length; that all but the hexagon and equilateral triangle were quadrilaterals; that most had

60-degree or 120-degree angles, or both, but the square (all 90-degree angles) and the tan rhombus (two 15-degree angles) were exceptions. Some teams investigated symmetries and reported the differences they found.

Likenesses, too, were reported. Mike had recorded that all were polygons, all had straight sides, and all blocks the same color were congruent.

When Myra reported that "one way the blue rhombus and the red trapezoid were alike was their angle measure," Mike and several others seemed surprised.

"How did you and Mario determine that the angles were alike, Myra?" asked Ms. Sage.

Demonstrating, Mario replied, "We just put the blocks on top of each other and matched up the corners."

"Did anyone else reach the same conclusion using a different approach?" she pressed.

Indeed, several students had. Gustavo and Naoko, for example, used the fact that each angle of an equilateral triangle measures 60 degrees to determine that the rhombus (consisting of two equilateral triangles separated by the shorter diagonal) and the trapezoid (which could be partitioned into three equilateral triangles) both had two 60-degree and two 120-degree angles.

Satisfied that the students had noted many important relationships among sides and angles, Ms. Sage was ready to begin the major phase of the day's lesson. "Today we will use the pattern blocks to learn about similarity; we will also investigate ratio and proportion. These are important concepts used by engineers, botanists, and scientists. We will use the blocks to make similar shapes, then we'll compare the different measurements of the shapes we have made."

Then Ms. Sage displayed a transparency that outlined the first investigation: "Work with your partner. Using the green triangle as the unit of area and its side as the unit of length, can you make with the pattern blocks a larger equilateral triangle with sides twice as long as the green block? Can you make a triangle with sides three times as long? Four times as long? Can you predict how many triangles will be needed each time? Complete a chart like the one on the chalkboard [see fig. 13.1] to record the findings of your investigation."

| Side Length | Number of Triangles Used | | Perimeter of Design |
|---|---|---|---|
| | Prediction | Actual | |
| 1 | | 1 | 3 |
| 2 | | 4 | |
| 3 | | | |
| 4 | | | |

Fig. 13.1. Expanding triangles

Ms. Sage knew that some teams would finish before others. As they did, she suggested that they create a similar chart for squares—each time

**Furnishing a written as well as an oral outline of the task is more likely to get Mike to attend to following directions. The transparency serves as a visual reminder throughout the activity, increasing the potential of his on-task participation.**

**Ms. Sage's instructions are brief as she involves student teams in collaborative problem solving. The average-ability, low-keyed but conscientious individual she selected as Mike's partner serves as a positive role model for not rushing through the tasks.**

forming a larger square by using multiple copies of the orange square—for sides of lengths 1, 2, 3, and 4.

When all the groups had finished the chart for triangles, Ms. Sage suggested that they examine it with their partner and make note of any patterns they noticed.

During the wrap-up on this activity, Joy reported for her team, "We noticed that the larger triangles had the same shape as the smaller triangle."

"Can you explain what you mean by 'same shape'?" Ms. Sage asked.

LaTray, Joy's partner, responded, "Because the angles are the same."

"Will equal angles in triangles always assure that the triangles have the same shape?" asked Ms. Sage.

"Yes, but they may not be the same size," LaTray commented.

When this discussion drew to a close, Ms. Sage asked the students who had worked on the squares-investigation chart to share their results and compare them with those presented for triangles. These students drew attention to the lengths of the sides and the perimeter measures.

Chang said, "When each side length is doubled, so is the perimeter. When each side length is tripled, so is the perimeter."

"What do you think the perimeter would be if each side length were one hundred times the original length?" queried Ms. Sage. "Tell your partner what you think."

The students' interaction on the question she had posed needed no follow-through. Ms. Sage could *hear* the message "one hundred times bigger" clearly enough!

Rather than push for a generalization just yet, Ms. Sage posed a similar investigation using the blue rhombus as the starting block. "Using only blue rhombuses, work with your partner to see if you can build a larger, similar rhombus with the same shape as the blue block and with each side two units long. Complete a chart like that on the board [see fig. 13.2] and build larger rhombuses as well—with sides three units long and then four units long. Compare what you find to your work with triangles."

| Side Length | Number of Rhombuses Used | Perimeter of Design |
|---|---|---|
| 1 | | |
| 2 | | |
| 3 | | |
| 4 | | |

Fig. 13.2. Expanding rhombuses

When the chart was completed, Ms. Sage asked for observations.

Trissa said, "I noticed that the angles of each rhombus we made were the same as in the original rhombus."

"Will equal angles in figures always ensure that they have the same shape?" asked Ms. Sage.

Since no real consensus emerged from the group, Ms. Sage suggested that they explore further by making a variety of block shapes with equal angles. Mike thought they could form a rhombus shape with six small rhombuses and tried to make one for his partner, Malcolm (see fig. 13.3). He soon realized that although the larger, new figure had angle measures equal to those of the original blue rhombus, it did not have the same overall "shape" as that rhombus. In fact, since all the sides were not of the same length, it was not a rhombus at all!

"That just means it takes more than equal angles to form larger shapes the same size as the start figure," offered Simone. "Pedro and I think you have to account for the lengths of the sides, too."

"We saw that when the sides were twice as long, the perimeter was also twice as long and when the sides were three times as long, the perimeter was also three times as long," said Khedra.

"We saw that, too," said Kim, "but when the side was twice as long, we used more than twice as many pieces. Why?"

"The number of pieces used is really the square of the side length," noted Zak. The group readily agreed with Zak and recognized that the "number of pieces used" was really the area, with the blue rhombus as the unit of area.

Ms. Sage used this opportunity to explore the relationship of the areas of similar shapes. "Decide with your partner: Does that same relationship between the square of the side length and area—the number of pieces needed for the shape—always hold?"

Student pairs studied the charts they had made and decided that Zak's observation always held. Next Ms. Sage challenged them to formalize their observations by using proportions. Mike wrote the group's conclusions on the chalkboard: that *similar* shapes have equal angles and correspond-

**Learning is richer in settings like this when examples meeting some but not all of the conditions of similarity are explored. Without the intervention, Mike's misconception that surfaces in this task may have gone undetected.**

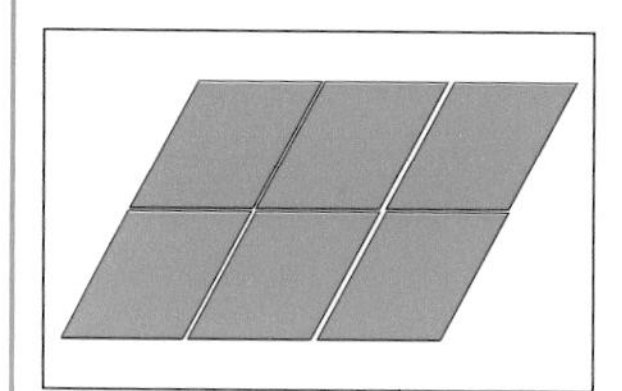

Fig. 13.3

**Mike's role as group recorder helps meet his personal needs to synthesize the main points of the session.**

ing sides are in proportion. Further, the areas of two similar shapes are proportional to the squares of their linear measures.

In the time that remained, Ms. Sage asked the student pairs to try to make and compare both the linear and area measures of larger, similar trapezoids formed from the red blocks and to write up their observations. ☐

## Extending the Special Education Thrust

The previous episode on similarity illustrated how Ms. Sage involved Mike, a student with Attention Deficit Disorder. Different adaptations would be necessary for visually impaired students like Rhodes, who is in another of Ms. Sage's geometry classes. Although legally blind, Rhodes has some residual vision that allows him to discern light, form, and color. Ms. Sage tries to meet Rhodes's special needs in at least three different ways: through her classroom organization, through tactile-kinesthetic means, and by increasing the size or magnification of the visual materials she supplies to Rhodes. Some of the adaptations Ms. Sage has made are described below. Similar ideas might be useful for working with other students having visual deficits.

In Rhodes's case, planning was critical. After Ms. Sage organized her classroom at the beginning of the year, she met with Rhodes before school started to help him map pathways through the room. She seated Rhodes centrally, away from windows that might cause glare or shadows. When the school year began, she established a rotating buddy system to assist Rhodes with transit needs and to facilitate note sharing for assignments and the important points of lessons.

For cooperative group work, used frequently during the class, Ms. Sage involves Rhodes in roles he can handle successfully, like doing calculator work (with headphones and a voice synthesizer) or orally summarizing discussions at different points throughout the work period, so these summaries can be tape recorded for his future study.

For Rhodes and many students with visual deficits, pattern block activities like those of the lesson above provide important kinesthetic-tactile experiences useful for learning. The blocks can be finger-traced to trigger recognition, and the color coding further enhances discrimination. A student who wishes to examine two particular pattern block figures more carefully might be given access to a larger demonstration set (which can be made in house by volunteers in shop classes) or be invited to refer to the images of pattern blocks as they are projected onto a wall or chalkboard. It may be useful for some students to use broad-tipped markers to trace the projected block shapes onto large newsprint sheets as they compare and contrast the properties.

If the Logo extension of this lesson suggested below is used, students with visual deficits might be given a larger-sized screen that accommodates larger font sizes and the magnification of graphics. Larger keyboards, also available, can be useful for some visually impaired students who have minimal keyboarding skills and need to look at the keys. If necessary, braille symbols can be taped over the keys. Enlarged keyboards also are valuable for students with physical disabilities for whom manual dexterity is a problem.

As a self-organizer and learning aid, students themselves might be involved in creating or adding to a summary notebook or audiotape of important geometry ideas. This can be a separate notebook or tape or a specially designated section of the regular mathematics notebook. Such notebooks and tapes have proved useful not only to visually impaired students like Rhodes but also to students with various learning disabilities and behavior or attention deficits. These notebooks and tapes might include, in the student's own words, a description or visual illustration of such key terms as *similar shapes* or important conclusions or theorems that are central to a lesson. Students who otherwise might miss important information through inattentiveness, those who experience memory deficits, or those needing additional time to process verbal information all find that such a personal resource is particularly helpful.

## Extending the Mathematics

### Counterpoint

To reinforce the ideas of similarity, students might consider rectangles $JKLM$ and $NOPQ$ (see fig. 13.4). All the angles are equal to 90 degrees, but what is true about the ratios of the sides? The ratios of sides $JK$ and $NO$ and sides $JM$ and $NQ$, 10/6 and 3/2, are not equal. Thus, the rectangles are not similar even though the corresponding angles are equal.

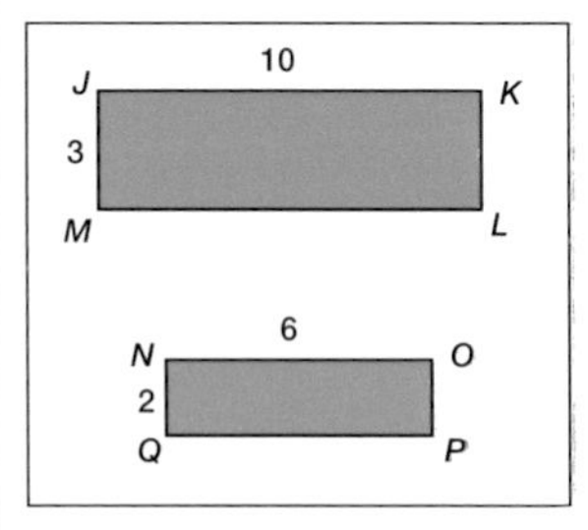


Fig. 13.4

Next have the students examine polygons $RSTU$ and $VWXY$ (see fig. 13.5), in which the ratios of sides $RU$ and $VY$ and sides $RS$ and $VW$ are proportional,

$$\frac{2}{4} = \frac{3}{6},$$

but the polygons are not similar. The foregoing elaboration revisits an issue raised during the lesson and further demonstrates to students that two polygons are similar if and only if the corresponding angles are congruent and the corresponding sides are proportional.

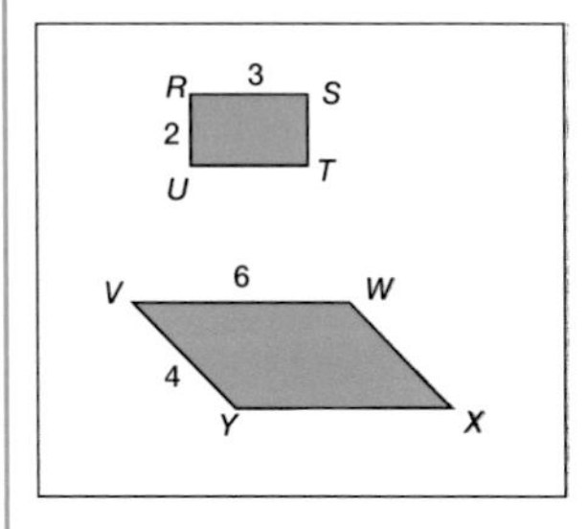


Fig. 13.5

### A Special Case: Triangle Similarities

Students can also be invited to study similarity applications for triangles more closely. They might, for example, construct a triangle $ABC$ and a larger triangle $DEF$ such that angle $D$ equals angle $A$ and angle $E$ equals angle $B$. What conclusion can be drawn about the relationship of angles $F$ and $C$ and about the triangles themselves?

Students should readily state that angle $F$ equals angle $C$, since the measures of the three angles of a triangle must add to 180 degrees. By measuring the lengths of the sides of both triangles, they will find that

$$\frac{AB}{DE} = \frac{AC}{DF} = \frac{BC}{EF}.$$

Repeating this procedure with other pairs of triangles leads to the relationship that if two angles of one triangle are congruent to two angles of another triangle, then the triangles are similar. Drawing from their experiences in the lesson above, students should expect to find that the ratio of the perimeters of the two triangles is the same as the ratio of two corresponding sides.

Because there is a relationship between the perimeters of similar triangles and those of other pairs of similar shapes, students can explore to determine the relationships among all linear measures, such as between corresponding altitudes in similar triangles, corresponding medians, or corresponding angle bisectors in similar triangles. Work with similarity and right triangles can lead to the derivation and use of trigonometric functions. Interesting applications of similarity like this, including relations to body size and scaling, can be found in *Similarity* (Apostol 1990).

### *Using Similarity to Measure Inaccessible Heights*

Similarity has many uses. For example, similar triangles can be used to calculate the height of objects that cannot be measured directly, such as the height of a tall tree. If a shadow is cast by the tree and the length of the shadow can be measured, similar triangles can be used to find the height of the tree.

On a sunny day student teams might go outside and gather the measurements needed to calculate the height of a tree. A person's height and shadow will form a right triangle that is similar to the right triangle formed by a vertical tree and its shadow. This proportion can be used:

$$\frac{\text{person's height}}{\text{shadow}} = \frac{\text{tree's height}}{\text{shadow}}$$

Students should notice that the tree's height is the only value in the proportion to be calculated; the other measurements are known. This indirect method can be used to determine the heights of other tall objects as well.

Logo investigations further enable students to explore similarity. Data disks containing pattern block procedures like that in figure 13.6 can be prepared in advance for students to use. The procedures were written with variables so that shapes of various sizes could be made by students.

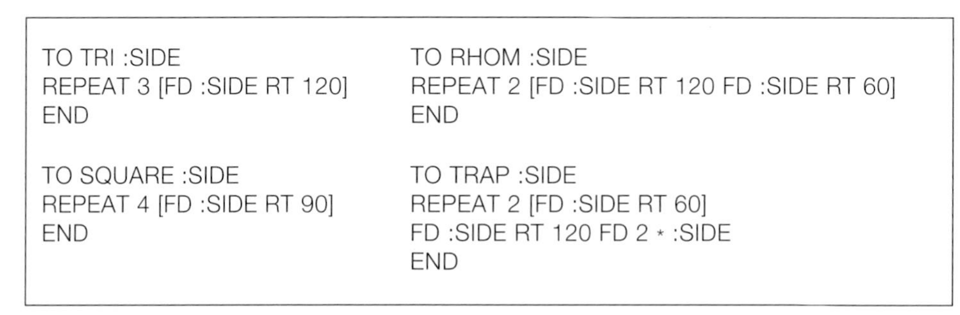

```
TO TRI :SIDE
REPEAT 3 [FD :SIDE RT 120]
END

TO SQUARE :SIDE
REPEAT 4 [FD :SIDE RT 90]
END

TO RHOM :SIDE
REPEAT 2 [FD :SIDE RT 120 FD :SIDE RT 60]
END

TO TRAP :SIDE
REPEAT 2 [FD :SIDE RT 60]
FD :SIDE RT 120 FD 2 * :SIDE
END
```

Fig. 13.6

Students might be asked to use the procedures to draw SQUARE 20 and SQUARE 40 or draw TRIANGLE 20 and TRIANGLE 40 and to compare the measures of the sides, perimeter, and area of the two pairs of shapes. Students might be asked also to record the ratios of the corresponding measures and to write a paragraph summarizing their conclusions.

When the REPEAT command is understood, the students can write their own procedures to explore similarities of other figures, such as pentagons, hexagons, and 13-gons. Using variables will allow many similar figures to be drawn on the screen. Logo gives students a tool for making conjectures and conducting explorations in a nonthreatening environment.

## Midpoints and More

The students filed past Mrs. Adams as they came into the classroom, taking the sheet she had offered. Jose looked at the paper (see fig. 13.7) and grinned at his teacher. "Not much to go on, Mrs. Adams," he said.

Mrs. Adams smiled back and said, "But it is a start, Jose—all you'll need." She elaborated only to herself, "By the end of the period that piece of paper should have you concluding that a segment joining the midpoints of two sides of a triangle is parallel to, and one-half the length of, the third side." Out of the corner of her eye she saw Jose settle in beside his partner, Tino. Jose was getting out the materials Mrs. Adams had listed on the chalkboard and Tino was examining the handout more closely.

As soon as the bell rang, Mrs. Adams greeted the group and said, "Today we will develop a simple midpoint theorem that architects and engineers use in building designs and construction. I'd like you to begin by using your ruler to draw and label triangle *ABC*, using the line segment *AB* on your paper as one side. Try to draw a different-shaped triangle from those of others around you."

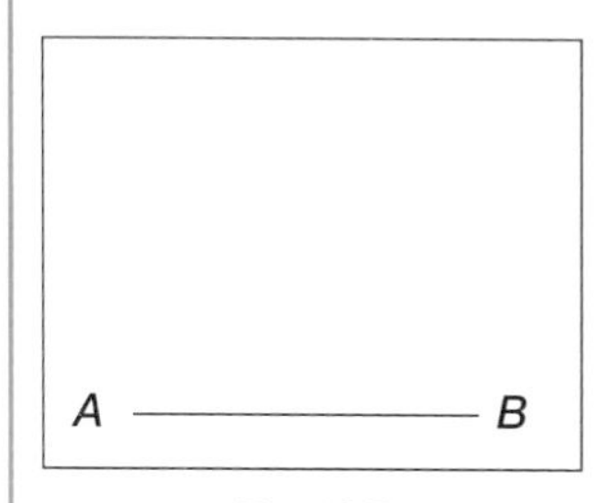


Fig. 13.7

**Tino is a student with visual-perception and memory deficits. It helps that he can familiarize himself with the visual handout prior to using it in class.**

**Like many other students, Tino will benefit from the verbal explanation of the activity, with its emphasis on the real-world relevance of the day's lesson. During the lesson, group work is the norm. Jose will help Tino check his measurements but without measuring for him.**

**Mrs. Adams is making an attempt to develop the geometry in intuitive ways. This approach will help Tino understand and remember the term *midpoint*.**

**Mrs. Adams is asking a question that, in addition to making a point of seeing something "in the mind's eye," challenges Tino's visualization skills.**

As individual students finished, Mrs. Adams suggested they write the ways in which their triangle differed from others around them. When all had finished, Mrs. Adams spoke to the group. "Side *AB* of your triangle [see fig. 13.8] was drawn for you. Next I'd like you to find the midpoint of each of the other two sides. Where will these midpoints be on the sides?" Shauna responded, "In the middle."

"Yes," said Mrs. Adams. "A midpoint is in the middle. It is really halfway. Find the halfway point, or midpoint, of segment *CA* and label it *P*. Then find the midpoint of segment *CB*, and label it *Q*."

After the students located points *P* and *Q*, Mrs. Adams asked, "Can you 'see' segment *PQ* even though it has not been drawn?" She then directed the students to measure segment *PQ* in millimeters and compare their measure with those of others near them. She prepared a table on a separate transparency (see table 13.1) to record the measures obtained by the students.

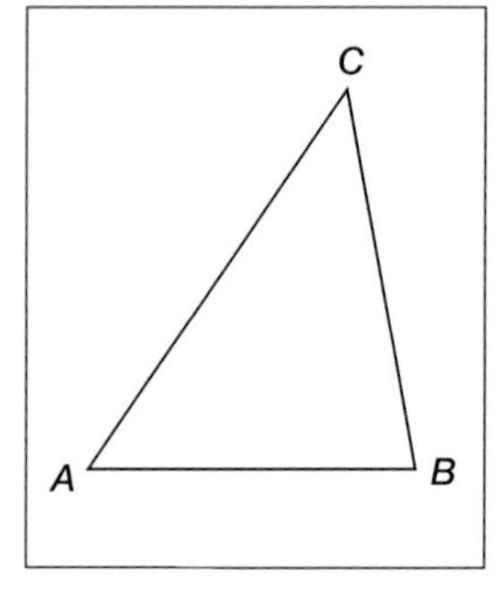


Fig. 13.8

Table 13.1

| Name | *PQ* |
|---|---|
| Jim | 50 |
| Dedra | 48 |
| Jose | 51 |
| Tino | 52 |

As Mrs. Adams finished writing the fourth set of measurements, Olga raised her hand and commented, "All the measurements are close to the same number. They all seem to be about fifty millimeters."

The class discussed this coincidence. The students were surprised that no matter how far point $C$ was from segment $AB$, the points $P$ and $Q$ stayed the same distance apart. Following this discussion, Mrs. Adams directed the students to measure segment $AB$ in millimeters. On finding the measure of $AB$ to be 100 mm, they conjectured that their measurements for $PQ$ should be half of 100.

Byrony asked, "Does this mean that if we had started with segment $AB$ of length two hundred millimeters, then the distance between $P$ and $Q$ would be one hundred millimeters?" Mrs. Adams had the class investigate Byrony's question before continuing. As the students worked, Mrs. Adams asked three students to trace their triangle on a blank transparency she provided.

"You noticed that the line drawn through the midpoints of two sides is half as long as the third side." She displayed the transparency the three students had prepared and asked, "Do these triangles suggest any other patterns?" Even before all three students had shown their triangles, Carlos noted, "All segments $PQ$ seem parallel to segment $AB$."

Following this observation, Mrs. Adams asked each student to write what was learned from this activity. As the students worked, Mrs. Adams stopped by Tino's desk. She asked him to finger-trace the two segments as he described the relationships between them, then she colored over the segments with an orange marker.

Before collecting the papers, she called on several students to share what they had written. Mami noted, "The measure of segment $PQ$ was the same for all of us no matter where we placed point $C$. The measure of $PQ$ was one-half the measure of $AB$. Also, segment $PQ$ is parallel to $AB$."

**Tino's retention is strengthened through completing the table and listening to the discussion of others. In-depth investigations with discussions like this one cater to Tino's auditory strengths and are a rich source of learning for him.**

**Paraphrasing either orally or in writing is important for memory. The finger tracing and color coding on the handout and the follow-up summary statements by various students will further enhance Tino's retention.**

Later Mrs. Adams directed the students to the textbook where the theorem—*A segment connecting the midpoints of two sides of a triangle is parallel to, and one-half the length of, the third side*—was located. She invited the students to discuss the similarities between this theorem and the comments they had made about the activity just completed. □

## Extending the Special Education Thrust

The lesson above highlighted the accommodations the teacher made for Tino, a student with perceptual and visual-memory deficits. Different accommodations might have been made for students with cerebral palsy like Ken, who has a condition known as diplegia. Ken is confined to a wheelchair because of the weakness in his leg muscles. He has minimal stiffness in his arm muscles but has difficulty moving his hands accurately. Several adaptations to meet Ken's special needs include the following:

- Providing a special wheelchair desk
- Pairing him with partners who have the patience to allow more time to complete a task
- Supplying a clipboard to secure papers and handouts
- Assigning a buddy to help locate needed materials
- Adapting the task when possible (e.g., allowing Ken to measure in larger units; using centimeters instead of millimeters works well for this lesson)
- Furnishing a larger-sized pencil
- Allowing more time for verbal responses and for substituting verbal activities for some written work

A computer with an oversized keyboard and microcomputer programs like the Geometric Supposer (1985) or Geometer's Sketchpad (1991) would be ideal for this lesson. Such programs would allow Ken to construct the figures without having to draw them manually. It may be useful to pair Ken with another student if the use of the keyboard poses a problem.

Mrs. Adams's structured directions and pairing of students are particularly beneficial to students with behavior disorders. These students have a tendency to seek attention, show off, be disruptive, and annoy others. In addition to employing behavior-management techniques such as those discussed in preceding chapters, Mrs. Adams could try to move these students away from distractions and pair them with individuals who are persistent and analytic and who might help keep them attentive and challenged during group activities.

## Extending the Mathematics

The manner in which the triangle midpoint problem was presented in the classroom episode above invites many extensions. Several of these extensions are summarized below.

### *Midpoint Lines*

To challenge their ability to generalize, students might be asked also to find the midpoint $M$ of segment $AB$, draw segments $MP$ and $MQ$, and list their observations about their modified drawing (fig. 13.9). Students might, for example, observe that—

- three segments join the midpoints and each segment is one-half of, and parallel to, the third side;
- all the five triangles, $ABC$, $PQC$, $AMP$, $MBQ$, and $QPM$ are similar. In fact, this configuration resembles that formed with pattern block pieces in the first classroom episode, "Similarity." From the work with similarity, students can see that triangle $CPQ$ is one-fourth the area of triangle $ABC$.

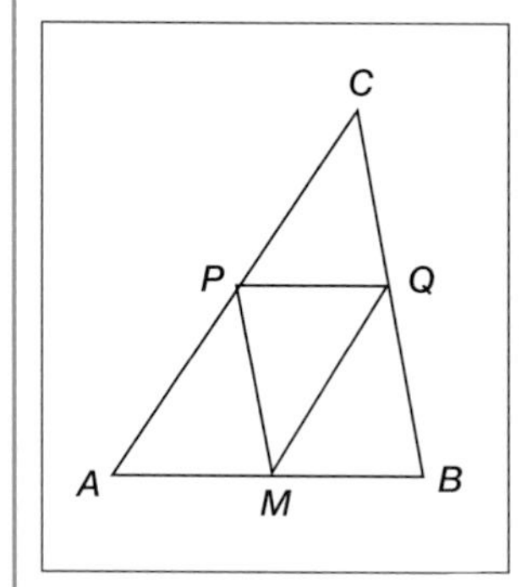


Fig. 13.9

### *"One-Third" Lines*

Instead of finding the midpoints of the sides of a triangle, students could be asked to find "one-third" points $X$, from $C$ to $A$, and $Z$, from $C$ to $B$ and to explore the relationships between segments $XZ$ and $AB$ (fig. 13.10). On concluding that segment $XZ$ is one-third the length of, and parallel to, segment $AB$, the students can compare the area of triangles $CXZ$ and $CAB$. To convince the students that triangle $CXZ$ has one-ninth the area of triangle $CAB$, the teacher might direct them back to the work with similarity and the scaling performed with pattern blocks.

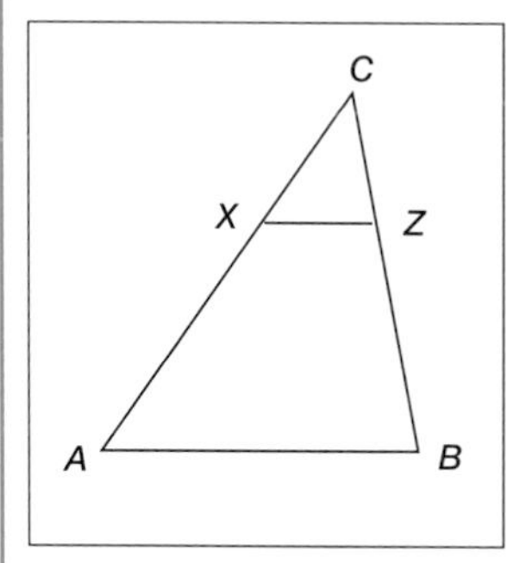


Fig. 13.10

### *Microcomputer Extensions*

Students can be given the opportunity to explore these relationships with the Geometric Supposer (1985) the Geometer's Sketchpad (1991), or the Geometry Inventor (1992). In this instance, similarity could be "verified" by measuring the corresponding angles or by computing the necessary ratios.

### *Three-Dimensional Extensions*

After exploring the concept of similarity, the students could cut out their triangles in figure 13.9, fold them along segments $PQ$, $PM$, and $QM$, and check whether they can fold their triangles into a triangular pyramid (tetrahedron). This activity serves as a nice connection between two- and three-dimensional work in geometry. Students should predict the outcome before doing this activity to enhance their ability to "see in the mind's eye." Some triangles will fold into a tetrahedron, whereas others will not (those students with right or obtuse triangles will not be able to fold their figures into a tetrahedron).

### *Chaos*

For further work related to similarity, the students can play the chaos game (NCTM 1991, p. 22), and observe how it is related to this investigation. Some students may enjoy creating their own variations of this game.

## A Rectangle Becomes a Circle

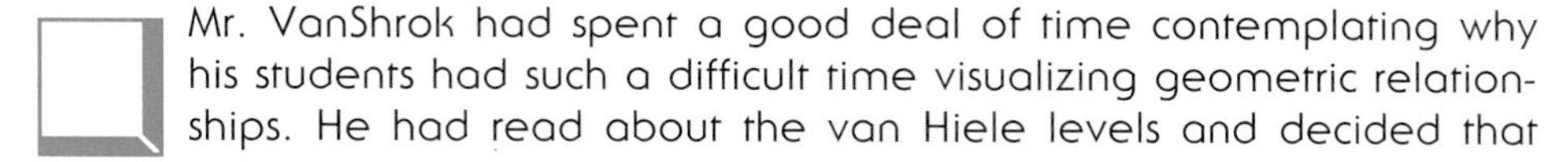

Mr. VanShrok had spent a good deal of time contemplating why his students had such a difficult time visualizing geometric relationships. He had read about the van Hiele levels and decided that perhaps it would help if he could involve the students in more activities that focused on visualization. That was the goal of the locus problem he planned to present today—to challenge the students to apply their knowledge of the properties of a rectangle to visualize and describe the figure formed by all rectangles in a plane having a specific line segment as a diagonal.

Mr. VanShrok began the class by having his aide show the students a three-inch-by-five-inch index card. He asked them to describe what they saw.

Chelese, talented in mathematics, is a self-starter who learns best through assignments, projects, and discussions that nurture her creative and insightful reasoning abilities.

"I see a rectangle," said Minwha, and the others agreed.

Mr. VanShrok asked, "What makes this a rectangle?"

Poanne responded, "Because it has two long sides parallel and two short sides parallel."

Shelton said, "It is a rectangle because the two long sides are equal and the two short sides are equal."

"It is a rectangle because it is a quadrilateral with right angles," said Chelese.

Mr. VanShrok could sense from their reactions that the students favored Chelese's definition, which said it all. "Of course!" he heard Elana say to Harold. "We forgot the *right* angles!"

Mr. VanShrok continued the discussion by asking students to define *diagonal* to their partner and tell the number of diagonals in a rectangle. As they interacted, he saw several students referring to his index card, finger-tracing its diagonals in the air. Mr. VanShrok used this opportunity to discuss what it meant to "see in the mind's eye" without actually sketching with a pencil. After a few minutes he called for wrap-up and obtained a consensus that a diagonal of a polygon joins two "nonadjacent" vertices and that every rectangle has two diagonals.

"Today's task will challenge you to 'see in your mind's eye,'" he said. "Start by using your ruler to draw a three-inch segment near the center of the piece of plain paper that is on your desk." After a few moments he added, "Label the endpoints *A* and *B*."

Mr. VanShrok indicated that the segment was to be considered the diagonal of a rectangle and their task was to draw a rectangle having this segment as a diagonal. He distributed index cards and suggested that they might be used instead of rulers to help get the right angles. Although some students struggled with this task and had to seek help from a part-

ner, most were quickly able to draw one rectangle. Mr. VanShrok asked Felice to explain how she was able to draw hers (fig. 13.11).

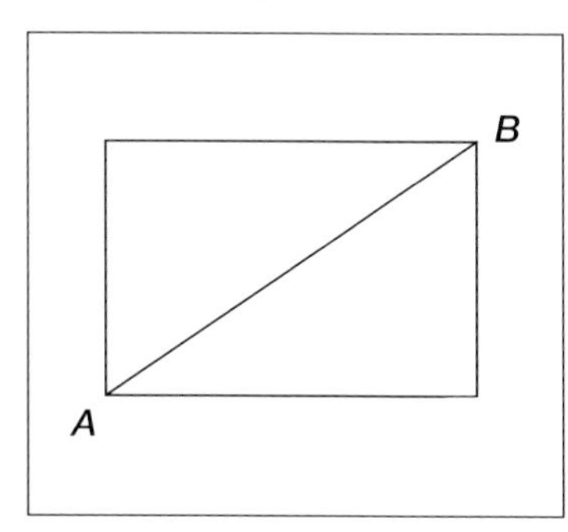


Fig. 13.11

Felice responded by saying, "I rotated my paper so the diagonal line—was diagonal! Then I drew up from point $A$ until it was even with point $B$. Next I connected to point $B$ and did the same thing on the other side."

As several others agreed and showed their papers, Mr. VanShrok became aware of how very limited some students' visual perceptions were. Many could work only with a diagonal that was oblique—making (about) a 45-degree angle with the horizon! They had cleverly rotated their paper so they could draw their rectangle with its longest side "horizontal."

Hoping to broaden their perceptions, Mr. VanShrok asked all the students to show their drawings and suggested that they find and discuss any differences in the rectangles with their partner. He observed that some students noted the different orientations and the different dimensions among the rectangles, but he created quite a stir nonetheless when he directed them to "find *another* rectangle having $AB$ as a diagonal. Place it on the same sheet you are now using. If you quickly find another rectangle, see if you can find several more rectangles."

As he moved about the room, several students had questions: "Do you really mean 'draw another rectangle'?" and "Must it be the exact diagonal,

**Gifted students commit readily to a challenging task. Mr. VanShrok planned a task that would benefit all but also invite Chelese's commitment and challenge her ability to mentally manipulate ideas and images.**

**Mr. VanShrok did not single out Chelese's approach, which did *not* involve rotating the paper. He wanted the students to grapple with the problem themselves. Gifted students who respect the need for others to do so have an understanding that will enhance personal interactions. Students no doubt will look carefully at Chelese's paper when she displays it.**

or can it be part of the diagonal, or can we extend the diagonal?" Mr. VanShrok clarified and encouraged them to "use their mind's eye" and keep trying. After several frustrating minutes, Minwha excitedly claimed, "I found another one!"

"Great!" said Mr. VanShrok. "How did you find it?"

"I used the edges of the index card you gave. I lined them up to touch the ends of the segment and then marked where the corner was. This gave me a right angle. I then made the other side of the diagonal the same," said Minwha.

After Mr. VanShrok had Minwha explain her procedure, Shelton said, "I figured it out another way. I noticed that the length of the short side of the index card was the same length as the diagonal you gave us. I found the midpoint of the segment and marked the midpoint of the edge of the index card. I then laid the index card midpoint on the segment midpoint and found the other corners for my rectangle."

"Excellent!" said Mr. VanShrok. "Does anyone have another way?"

After several more minutes of exploration, the students found other methods that worked, and when they completed the discussion, Mr. VanShrok had several of the students place their rectangles on a single transparency (fig. 13.12). He asked for observations, and Felice said, "It looks like the corners of the rectangles are forming a circle."

**Mr. VanShrok keeps the students involved while promoting divergent thinking.**

"Could we prove that the vertices of all the rectangles that can be drawn with the same diagonal would form a circle?" asked Mr. VanShrok. He continued with another challenge: "If all the rectangles were drawn, would the interior of the circle be covered?"

Mr. VanShrok pursued these ideas until he was able to close by connecting this activity to the relationship that an angle inscribed in a semicircle is a right angle. Several students, including Chelese, had independently concluded this and had moved the discussion in this direction.

**Gifted students often generate creative, innovative solutions. Through his questioning, Mr. VanShrok draws out insightfulness and creativity.**

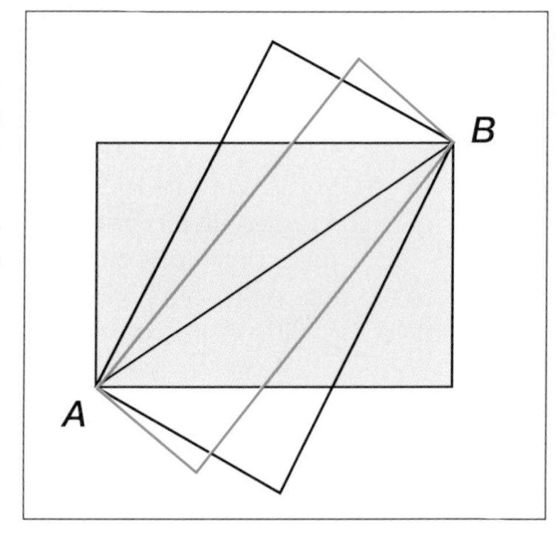


Fig. 13.12

The lesson approached this relationship from an investigative approach and built on the properties of rectangles. The results of the investigation, summarized by the students themselves, led to the statement of the theorem.

Chelese suggested that they think of the rectangle as a baton being twirled in a parade. She answered Mr. VanShrok's question about covering the interior of the circle by noting that as the baton is twirled very fast, what the eye sees is a "circle." "It's like the baton is balanced at the midpoint of the diagonal by the person twirling it."

Shelton recognized that this approach was very similar to his, which involved using the midpoint and the index card to find the other corners of the rectangle.

"Yes," commented Mr. VanShrok. "In essence Shelton was twirling the segment about the midpoint of the diagonal. This would not only form the desired right angles and rectangles, but it would generate a circle while doing so." ❑

**Gifted students often excel at understanding formal relationships. One advantage of a heterogeneous group is that opportunities like this arise, in which Chelese raises the level of thinking by producing a relevant example that generalizes the geometric situation.**

## Extending the Special Education Thrust

Students with abstract-reasoning or receptive-language deficits are frequently mainstreamed. Among other problems, students with these learning and communication disabilities have difficulty with abstract concepts and multiple word meanings. Connecting new information with prior learning is also problematic. One such student in Mr. VanShrok's class is Renee.

To meet the special needs of students like Renee, it is important to involve them in a lesson at times when the concepts and participation in a discussion are within the student's success range—or can be with a little coaching. For example, Mr. VanShrok might rotate the three-inch-by-five-inch index card, and ask Renee specifically, "Is it still a rectangle?"

Similarly, he might follow up on Chelese's analogy to the baton twirler by inviting Renee to think of a friend twirling a baton, balancing it at the center. Much more directly, he can connect this new learning with Shelton's twirling the segment about the midpoint.

In group problem-solving situations, Renee should be placed with a group that has good language role models and where she might initially assume a nonthreatening role. Cooperative learning techniques benefit many types of special-needs students. The group structures that best meet the needs of individual students vary, but generally a teacher looks for compatible personalities and ability levels that are not extreme. Naturally, the teacher must structure and require individual accountability in some way.

To adapt this lesson for students with behavior or attention disorders, Mr. VanShrok might make greater use of an explanation of the activity and structure at the beginning and at major transition points within the lesson. For example, he might have begun the lesson by saying, "We will explore three major ideas related to rectangles today."

To advise the students during the lesson of the teacher's expectations for their performance on the tasks, students might also be told, for example, that at the end of the lesson they will be asked to summarize the three ideas. At relevant times during the lesson, these ideas might be enumerated as part of a chalkboard list that is retained for reference.

### Extending the Mathematics

The exploration in the previous classroom episode allows for many extensions. Several ideas follow.

#### *A Ragged Edge*

Students might be given a paper similar to that in figure 13.13 and again be asked to draw a rectangle. Although the students may still draw a rectangle with "horizontal and vertical sides," this paper invites different orientations.

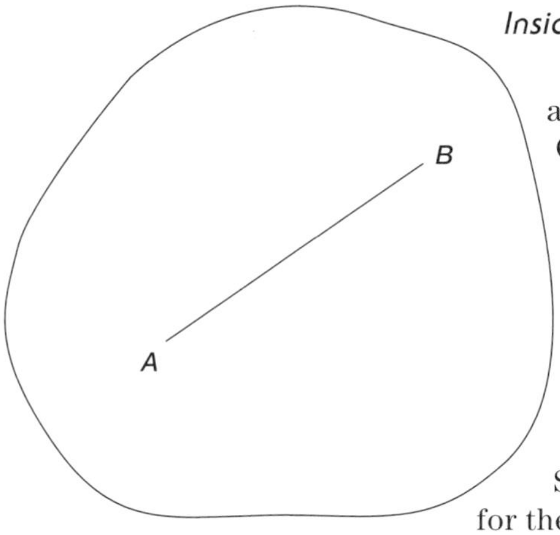

#### *Inside or Outside the Circle*

Given a circle with diameter $AB$, the students then know that for any other point $C$ on the circle, angle $ACB$ is a right angle. Challenge them to predict the size of angle $ACB$ if point $C$ is picked *outside* the circle; if point $C$ is picked *inside* the circle.

By experimenting, students may find that for points *inside* the circle, angle $ACB$ is obtuse. Conversely, angle $ACB$ is acute when point $C$ is chosen *outside* the circle. These conclusions emphasize the notion that a right angle is a boundary between acute and obtuse angles.



Fig. 13.13

#### *Variations on the Original Problem*

Suppose, as in the foregoing episode, we fix segment $AB$ but look for the locus of points $C$ where angle $ACB$ is a specific measure, say 30 degrees. This idea could be explored manually or with the assistance of a computer.

Another interesting variation is to take a three-dimensional look at the problem. Students might, for example, be challenged to draw all rectangles in space that have the same diagonal. In this instance, they should conclude that a sphere emerges. The students might formulate the following theorem: *If AB is a diameter of a sphere and C is any other point on the sphere, then angle ACB is a right angle.* They could also explore the relationships between acute angles and points outside the sphere, obtuse angles and points inside the sphere, and the location of points $C$ in space where the measure of angle $ACB$ is 30 degrees.

## CREATING WINDOWS OF OPPORTUNITY

This chapter has focused on special-needs learners in high school geometry classrooms. These special learners may have greater potential for acquiring higher-level geometry understandings and skills than might be predicted from their test scores. Many of these students have good spatial abilities even though they have low computation or problem-solving scores on standardized tests. Many have demonstrated that they can master the regular content of the class if instruction employs a variety of learning strategies and cooperative learning practices (Schloss, Smith, and Schloss 1990; Kerr, Nelson, and Lambert 1987).

Geometry lends itself to multiple teaching strategies, multiple levels of presentation, and multiple uses of materials. For these reasons, geometry is a natural vehicle for including many levels of learners. It may be that geometry can be taught to an academically and socially diverse group because fewer prerequisites are needed to study geometry taught from an exploratory, problem-solving, constructivist mode. Therefore, geometry may be the most accessible high school course for all students, including a wide range of students with special learning needs.

A broad variety of instructional and organizational practices have been highlighted in this chapter to offer ideas for meeting the diverse special learning needs encountered in a typical geometry classroom. These approaches can enable many students with special learning needs to learn to think mathematically through geometry and to develop important processes, including these:

- Conjecturing and discovering patterns
- Generalizing
- Drawing, making, and using models
- Visualizing and right-brain thinking
- Problem solving, reasoning, and critical thinking
- Convincing, proving, and communicating in cooperative learning situations
- Modeling in measurement, data analysis, and other areas of mathematics

The three teaching episodes and related special education discussions of this chapter demonstrate how active-learning approaches in geometry can benefit students with differing mental and physical abilities. The examples in the episodes focused on similarity, a triangle-midpoint theorem, and locating all rectangles that can be drawn with a given diagonal. The mathematical extensions following the teaching episodes focused on ways in which student thinking about important geometric concepts and ideas might be extended in interesting, fruitful ways.

In the spirit of the van Hieles, the chapter has moved from reinforcing visualization and analysis to an emphasis on making conjectures and justifying conclusions. The teaching episodes and extensions presented examples of problem tasks that can help propel students' thinking and reasoning toward more formal deduction. Helping all students, including those with special learning needs, to "bridge geometric gaps" was the basic instructional goal. It was accomplished by opening windows of opportunity through concrete and intuitive experiences to more logical and abstract levels of geometric thinking.

## REFERENCES

Apostol, Tom M. *Similarity.* Project Mathematics! series. Pasadena, Calif.: California Institute of Technology, 1990.

Baroody, Arthur J. "Teaching Mathematics Developmentally to Children Classified as Learning Disabled." In *A Cognitive Approach to Learning Disabilities*, edited by D. Kim Reid, Wayne P. Hresko, and H. Lee Swanson. Austin, Tex.: Pro-ed, 1991.

Burger, William, and Barbara Culpepper. "Geometry." In *Research Ideas for the Classroom: High School Mathematics*, edited by T. Owens. New York: Macmillan, 1993.

Coxford, Arthur F., Jr. *Geometry from Multiple Perspectives.* Reston, Va.: National Council of Teachers of Mathematics, 1991.

Deshler, Donald D., Gordon R. Alley, Michael M. Warner, and Jean B. Schumaker. "Instructional Practices for Promoting Skill Acquisition and Generalization in Severely Learning Disabled Adolescents." *Learning Disabilities Quarterly* 4 (1981): 415–21.

Geometer's Sketchpad. Berkeley, Calif.: Key Curriculum Press, 1991.

Geometric Supposer. Newton, Mass.: Education Development Center, 1985. Available from Sunburst Communication, Inc., Pleasantville, N.Y.

Geometry Inventor. Scotts Valley, Calif.: Wings for Learning and LOGAL Educational Software, 1992.

Hallahan, Daniel P., and James M. Kauffman. *Exceptional Children: Introduction to Special Education.* 5th ed. Englewood Cliffs, N.J.: Prentice Hall, 1991.

Kerr, Mary Margaret, C. Michael Nelson, and Deborah L. Lambert. *Helping Adolescents with Learning and Behavior Problems.* Westerville, Ohio: Merrill Publishing Co., 1987.

National Council of Teachers of Mathematics. *Curriculum and Evaluation Standards for School Mathematics.* Reston, Va.: The Council, 1989.

Schloss, Patrick J., Maureen A. Smith, and Cynthia N. Schloss. *Instructional Methods for Adolescents with Learning and Behavior Problems.* Needham Heights, Mass.: Allyn & Bacon, 1990.

Shaughnessy, J. Michael, and William F. Burger. "Spadework Prior to Deduction in Geometry." *Mathematics Teacher* 78 (September 1985): 419–28.

Smith, Corrine Roth. *Learning Disabilities: The Interaction of Learner, Task, and Setting.* 2nd ed. Needham Heights, Mass.: Allyn & Bacon, 1991.

Suydam, Marilyn N. "The Shape of Instruction in Geometry: Some Highlights from Research." *Mathematics Teacher* 78 (September 1985): 481–86.

Woodward, Ernest, and Thomas Hamel. *Visualized Geometry: A van Hiele Level Approach.* Portland, Maine: J. Weston Walch, 1990.

**Melfried Olson** is a professor of mathematics education at Western Illinois University, Macomb, Illinois. He teaches undergraduate and graduate mathematics education courses, primarily for elementary and middle school teachers. He works extensively with in-service projects at various levels. With Judith Olson, he codirected a National Science Foundation grant, Implementing the NCTM Standards in Geometry, that trained 300 teachers nationwide.

**Subhash Jani** is a professor of special education at Western Illinois University, Macomb. He teaches graduate and undergraduate courses in evaluation, characteristics and methodologies in special education, and learning disabilities. He pursues his interests in learning disabilities as they affect visual and academic processing through part-time practice.

**Judith Olson** is a professor of mathematics education at Western Illinois University, Macomb. She teaches undergraduate and graduate mathematics education courses, primarily for elementary and middle school teachers. She has interests in educational uses of technology, diagnostic instruction, and the encouragement of women to study mathematics. In addition to the National Science Foundation grant mentioned above, she is also codirector of another grant, Connecting to the Past with the Future: Women in Mathematics and Science.

**Jan Klippert** teaches methods for teaching students with behavioral disorders; in addition, she is the supervisor of field experiences for the Department of Special Education, Western Illinois University, Macomb. She was a special education teacher for eleven years, working with behavior-disordered adolescents in Princeton, Illinois. She spent one year at Macomb High School as a resource teacher for the learning disabled and as the prevocational education coordinator.

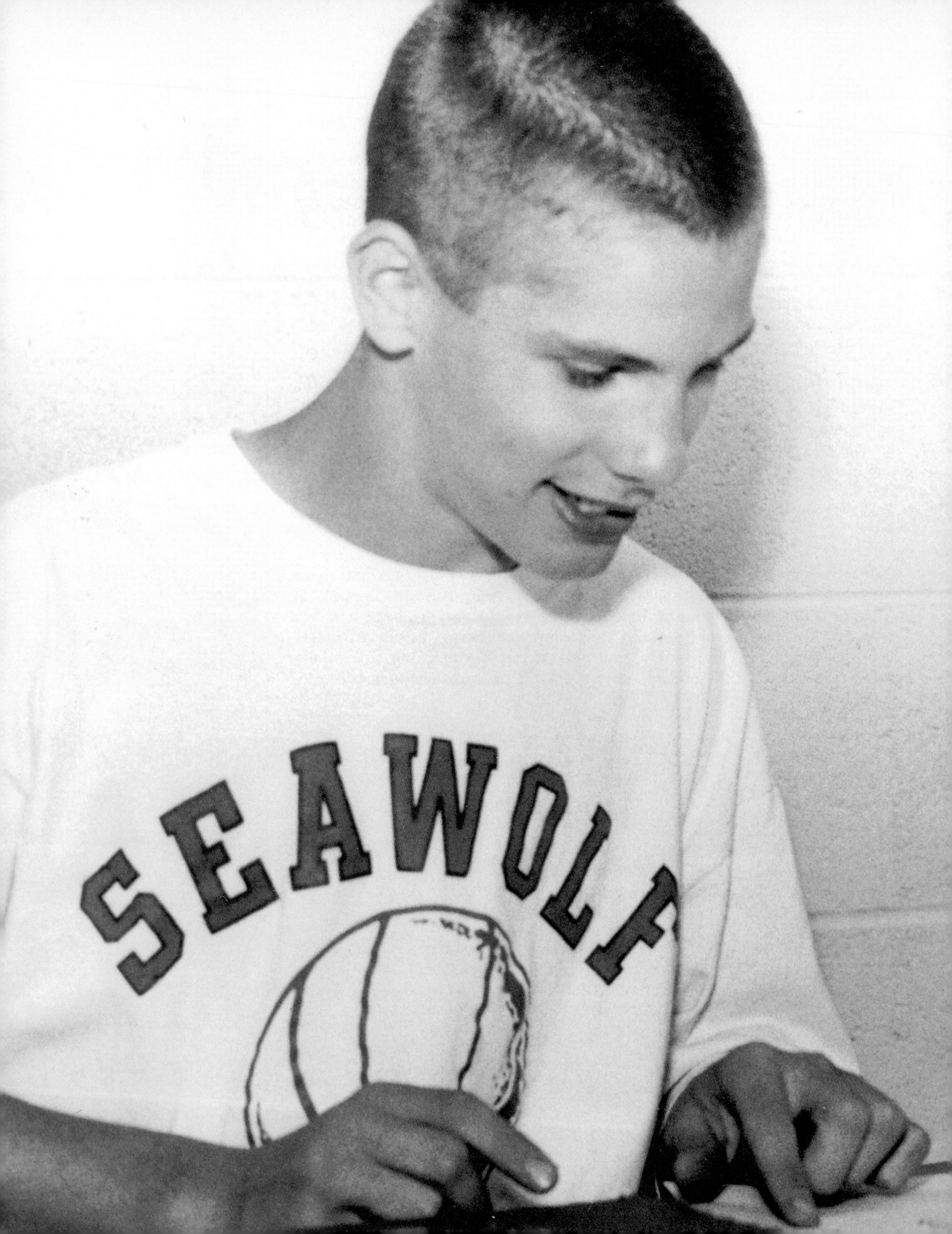
SEAWOLF

14

# Making Sense of Algebra

*Francis J. Gardella*
*David Glatzer*
*Joyce Glatzer*
*Daniel B. Hirschhorn*
*Carolyn Rosenberg*
*Margaret A. Walsh*
*in cooperation with*
*Angela Waters*
*Joan Stanton*

SIMPLY put, making sense of algebra has been a difficult task for the majority of students in the United States. According to 1990 NAEP data (Mullis et al. 1991), 83 percent of 17-year-olds have been exposed to at least one semester of algebra at some level, but only 13 percent of 17-year-olds survive at least two years of algebra to take mathematics courses in third-level algebra, precalculus, or calculus. In the teaching of school algebra, 85 percent of those enrolled in the courses drop out of the mathematics pipeline (70 percent of the total population). Although the high school geometry course may also contribute to this, it is clear that too many students, whether with special needs or not, are bewildered and "turned off" by mathematics during their exposure to school algebra.

There have been two standard explanations for the lack of success of students in school algebra. One is the lack of prior knowledge of algebra the students have coming into an algebra course. As Flanders (1987) noted, "In grades 6–8, the message given by typical mathematics texts is that very little new mathematics beyond arithmetic exists; ... then students encounter an algebra course." Commenting on this, Usiskin (1987) writes, "Students in junior high school are lulled into a pace whereby they expect to learn a new idea only a few times a year and a new skill only once every couple of weeks. Then, in algebra, the new ideas come monthly and the new skills come daily."

The second reason, and the one that is the focus of this chapter, is simply that the content and teaching of school algebra is inappropriate for the majority of students. The combination of a huge catalog of unconnected algorithms being presented to students almost exclusively by lecture has led to the filtering out of too many students in mathematics. Students with special learning needs suffer the most from poor selection of content and poor teaching methods. Yet it is often these students who will have the most need for algebra as they face more complex problems of working and living.

To make sense of algebra, one has to know what algebra is. Algebra, however, is hard to define. A nice working definition is that school algebra is related to variables. But there are many uses of variables. Fortunately, the different uses of variables lead to a varied and rich conception of algebra. The examples in the next paragraphs, drawn from Usiskin (1988), highlight this point.

A first use of variables is to describe patterns. For instance, the sentence "For all $n$, $n \cdot 0 = 0$" describes the pattern that any number multiplied by zero gives you zero. This leads to the conception of algebra as generalized arithmetic. A second use of variables is as values or parameters in functions. For instance, the linear equation $y = 4.9x$ describes the relationship between the quantities $y$ and $x$. When graphed, this relationship can be seen visually as one of constant increase. Variables as function values lead to the conception of algebra as the study of relationships between quantities.

A third use of variables is that of unknowns, such as in the equation $13^2 = 5^2 + x^2$. This leads to the traditional conception of algebra as the study of procedures for solving certain kinds of problems. Finally, a fourth use of variables is simply as symbols to be manipulated, such as in the problem "Prove the identity: $2 \sin^2 x - 1 = \sin^4 x - \cos^4 x$." For this use, algebra is the study of structures, an idea expanded on late in high school and in college.

The *Curriculum and Evaluation Standards for School Mathematics* (NCTM 1989) has strongly supported this more expansive view of school algebra and its exposure in earlier grades. In every single standard for grades 5–8 and grades 9–12 there is patterning, graphing, or the use of variables. Similarly, many of the vignettes in the *Professional Standards for Teaching Mathematics* (NCTM 1991) describe algebraic situations or situations that could easily adapt to algebra. Even in the arithmetic vignettes, the teachers are consistently trying to get the students to describe the patterns they see, to detail the procedures they used, and to picture the relationships between the quantities in the problem being discussed. The teaching methods illustrated in these standards are appropriate and desirable for teaching school algebra, especially to students with special needs.

## ILLUSTRATIVE CLASSROOM EPISODES

The episodes that follow show approaches for school algebra from the first three conceptions of algebra shown above. Yet all are interrelated. The first vignette shows an application of the relationship between quantities using technology. Using graphing calculators allows the students to make sense of the relationship between the variables in both tabular and graphical form. The second vignette shows the study of a procedure using manipulatives. As students explore linear equations with algebra tiles, they learn how to use variables to communicate their ideas to one another. The third vignette shows generalizing a geometric pattern using both manipulatives and technology. As students get deeper into the problem, the knowledge called on spans from arithmetic to geometry to second-year algebra. In all three vignettes students are communicating among themselves, using problem-solving skills, and connecting and extending the algebraic concepts learned.

The three overriding conceptual frameworks of algebra (generalized arithmetic, relationships between quantities, and the study of procedures) open the way for alternative and diverse instructional techniques by which students with special needs can begin and moreover continue to grow in their knowledge of algebra. Given that algebra provides access to many career opportunities to which these students might otherwise be denied, the use of applications, manipulatives, technology, group work, patterning, and problem solving provides the teacher with powerful tools for tailoring instruction and opening windows of opportunity for success in algebra for their students.

## Applying Algebra

Ms. Snyder, a high school teacher, has several classes of at-risk students with a history of low achievement in mathematics. In order to motivate her students and expand their mathematical capabilities, Ms. Snyder uses graphing calculators. Ms. Snyder's room is normally arranged so that the desks are in clusters of four in order to facilitate interaction and dialogue among students.

**The graphing calculator can be used as a tool to investigate relationships and to explore patterns and their algebraic and geometric representations.**

In the lesson that follows, she initiates an investigation of a situation known to all her students—the cost patterns of renting video games. Students were first given the task of investigating the cost of renting video games in local neighborhood stores. Several students reported that Video Palace charges its club members $3.50 to rent a video game. The one-time membership fee is $25.00.

**When students are given a task to do outside of class, they gain "ownership" of the problem, and the discussion goes better. When physically challenged students get data from stores, it should be predetermined if those stores are accessible and helpful to people with special needs.**

Ms. Snyder asked the entire class, "How much does it cost to rent the first game?"

Several replies were heard: $3.50, $28.50, and $25.00.

Ms. Snyder said, "Work this out in your groups."

The discussion continued without another word from the teacher, and soon the entire class agreed that the cost of the first video game was $28.50. The consensus of the groups was that the membership fee had to be included.

One student exclaimed, "Wow, I could have bought that game for that price!"

Ms. Snyder asked, "What is the total cost for renting two games?"

The unanimous answer was $32.

Ms. Snyder then followed up, "This is important. What is the cost at Video Palace of renting *no* games?"

Some students seemed confused, others said it cost nothing, and two others said it cost $25. One of these latter students explained that the $25 was the cost of the membership fee.

Ms. Snyder picked up her calculator (which was attached to an over-

**When using applications, the numbers discussed are in context and thus have a meaning to the students. Often these contexts are fodder for interesting discussion, but here Ms. Snyder quickly returns the class to the task at hand.**

**By using iteration (the process of performing the same operation on the last value received), the calculator may be used to build a table of values. This is analogous to the way a spreadsheet is developed by a computer.**

**By circulating around the room, the instructor is able to monitor students' learning by observing their work. This is a much more powerful technique than calling on a single student for an answer. This strategy provides immediate feedback on the progress of the entire class and allows the instructor to use student thinking or needs to modify the lesson as appropriate.**

head projection system). She said, "Let's enter these data on the home screen of the calculator. Before you rent any games, you have to pay the $25 membership fee." She and each of the students recorded 25 and pressed the ENTER key. "If it costs $25 to rent zero games, what is the cost of one game?"

Pat spoke up, "Just add $3.50."

Ms. Snyder replied, "Let's do just that on the calculator. All we need to do is enter +3.50 and press the ENTER key." The following was displayed on the overhead projection screen:

```
25
                 25
Ans + 3.50
               28.5
```

She continued: "For two games, we just press ENTER again. In your groups, continue to use your calculator to determine the total cost of renting three and four games."

After the groups did this, the display was projected on the classroom screen (fig. 14.1).

"It looks good so far, but how much would it cost to rent the first ten games?" asked Ms. Snyder.

Ms. Snyder then overheard this conversation from one of the groups. "That's easy," said Mary, "It's gotta be $35."

Carlos disagreed, "No way! Remember the $25 to join."

"Oh, yeah," said Mary, "I forgot about that. So $25 plus $35 is $60. Gee, that's a lot of money!"

George said, "I did it another way. All I did was press ENTER six more times."

Ms. Snyder said to the whole class, "There were two ways discussed to get the answer. Let's try it both ways and see what happens." She pressed ENTER six more times. Then she entered 3.50×10+25. The display on the home screen of the calculator is shown in figure 14.2.

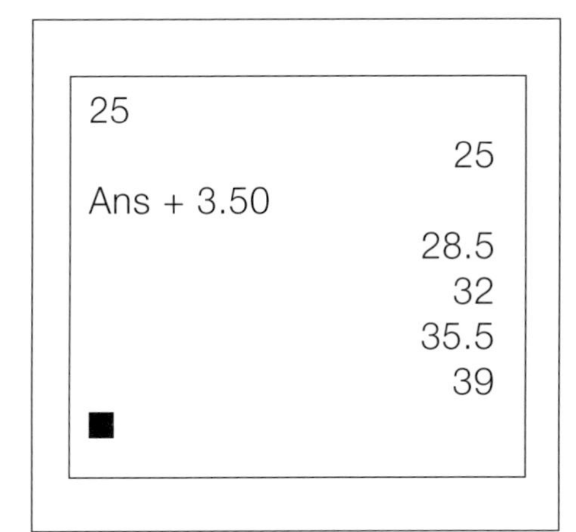

Fig. 14.1

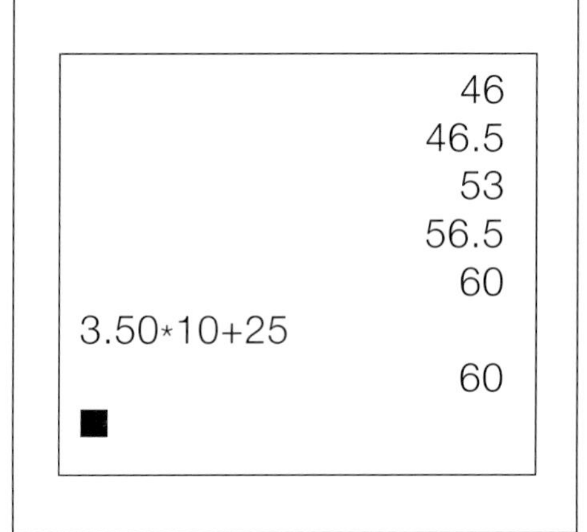

Fig. 14.2

"In your groups, determine the rental cost of twenty, thirty, or fifty video games," she challenged. Ms. Snyder monitored the activities of the work groups by walking around the room and listening to their conversations. She observed that students realized that the second strategy was more practical for large numbers of rentals.

"Jane, tell us what your group did."

"No problem, Ms. Snyder! We took $3.50 times the number of video games rented and then we added $25.00."

With a smile, Ms. Snyder said, "Great! Now, how can we express that thought so that we can find the cost of *any number* of game rentals using the symbols of mathematics?"

Jerry said, "Let's use an $x$. That's what math teachers always do."

The instructor asked, "How can we use an $x$, and what would it mean in this problem?"

Mary responded hesitantly, "Would it be $3.50x + 25$? We've been doing 3.50 times the number of games plus the fee to join."

The class unanimously agreed.

Ms. Snyder said, "Tomorrow we will use the graphing calculators to draw a graph or picture of the cost of video game rentals at Video Palace. Think about how we might do that." □

## Extending the Special Education Thrust

Many students with special learning needs have never mastered computational skills and are traditionally denied the opportunity to learn meaningful mathematics and to develop the problem-solving skills that are encountered in an algebra course. These students can be motivated to mathematical achievement through relevant, real experiences. In addition, the use of a graphing calculator helps these students circumvent computational difficulties, thereby allowing them to focus on, visualize, and conceptualize important mathematical ideas.

There are natural extensions to the mathematics in the episode above. For example, start with the equation $y = 3.50x + 25$. Note that both the slope and $y$-intercept have meaningful interpretations. The $y$-intercept is where you start (the membership fee), and the slope is the cost of each game. In general, for the equation $y = mx + b$, the $y$-intercept $(0, b)$ might be the start, the first point plotted when graphing by hand, and the slope $m$ is the rate of change as you move one unit to the right.

Next, the students could graph the algebraic representation of the cost of renting video games at Video Palace, $y = 3.50x + 25$, in the standard $10 \times 10$ viewing window. Ask if this graph is an appropriate "picture" of the cost of renting games at Video Palace (fig. 14.3). It is important for them to realize that only the first-quadrant values apply to this real-life situation, because the number of games rented and the corresponding cost will always be positive.

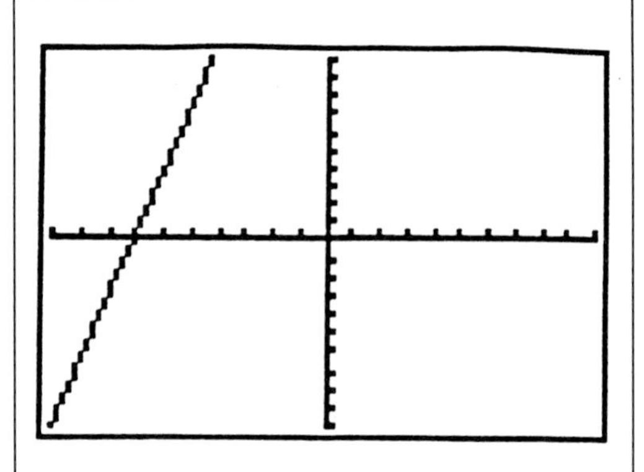

Fig. 14.3

Encourage students to experiment with determining a "friendly" viewing window that is appropriate for this problem. To use the calculator as a "counter" (in order to reflect on the discrete nature of this problem), an appropriate viewing screen must be used. Depending on the graphing utility, a possible viewing window might be as follows:

$$0 < x < 95$$
$$0 < y < 126$$
$$\text{scl} = 10$$

The TRACE feature of the calculator could be used to confirm the cost of renting 3, 4, 5, 10, 20, and 50 video games. Students should note that they are obtaining exactly the same solutions from their graph that they found by building a table of values in the introductory lesson (fig. 14.4). Explicit questioning may be helpful to focus their thinking and help them make this connection.

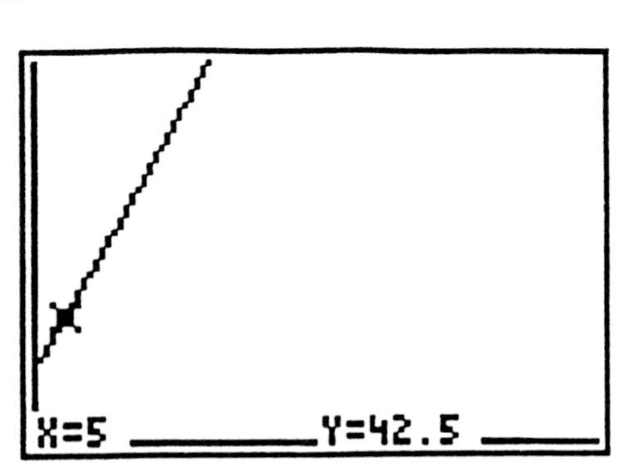


Fig. 14.4

Students should note that they are able to continue to obtain information about the graph of $y = 3.50x + 25$, even though both the graph and the cursor are no longer visible in the viewing window. The TRACE feature of the graphing calculator continues to follow the path of the function, although the cursor is beyond the range settings of the window.

Students' experiences in interpreting graphs can be further expanded by reversing the process. For example:

> Suppose that in the three-month period since joining the club, Johnny spent $126.50 on video game rentals. How many video games did he rent?

In this case, students would TRACE along the graph until the $y$ display at the bottom reads 126.5. The corresponding value of $x = 29$ represents the number of games rented. Students could confirm algebraically that the solution to the equation, $126.50 = 3.5x + 25$, is indeed 29.

Through the example above students can find functional values, given the value of the independent variable, and the converse. Additionally, they learn to interpret the meaning of the slope and the $y$-intercept within the context of a real-world application. Through the use of a graphing calculator, students who might otherwise be denied rich opportunities for thinking mathematically are provided ready access for doing just this.

## Extending the Mathematics

> Videos-R-Us offers free membership and charges its customers \$4.99 to rent a Super Nintendo game for three days. Video Palace, across the street, charges its customers only \$3.50 to rent the same video game for three days, but the store has a one-time-only membership fee of \$25.00. Which store offers the better deal? For the number of video games you rent, which store would you patronize?

Students could be encouraged to solve this problem using any combination of tables, graphs, and equations. Using the graphical technique developed in the vignette, students can enter the equations $y_1 = 4.99x$ and $y_2 = 3.50x + 25$ in the $Y=$ menu of the graphing calculator. Students need to select an appropriate viewing window for the problem situation. If the viewing window is too small, the lines appear to be parallel as in figure 14.5, where the equation is graphed in the standard $10 \times 10$ viewing window.

From this diagram, students might suspect that the better deal will always be the store without the membership fee, since it appears that $y_1 = 4.99x$ will always lie below $y_2 = 3.50x + 25$. Clearly, a discussion of the need for appropriate scaling that reflects the problem situation must be a component of this lesson. Students should be encouraged to experiment with various range and domain settings. One such setting might be $0 < x < 40$ and $0 < y < 150$ (scl = 0).

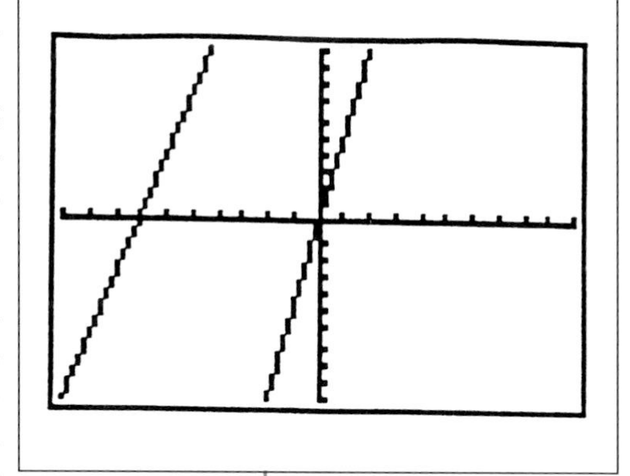

Fig. 14.5

The graphs of the same equations in this new viewing rectangle lead students to an entirely different interpretation of the problem (see fig. 14.6). It is now evident that Videos-R-Us appears to be the better deal to the left of the intersection point, whereas Video Palace appears to be the better deal to the right of the intersection point.

Students might be asked to determine the number of video game rentals needed in order to make it worthwhile to become a member of the Video Palace club. Using the TRACE feature, it is difficult to tell whether the point of intersection is closer to sixteen or to seventeen video rentals. The ZOOM BOX feature can be used to create a magnified viewing window around the point of intersection. Using this feature successively, Video Palace appears to be the better deal after at least seventeen video rentals.

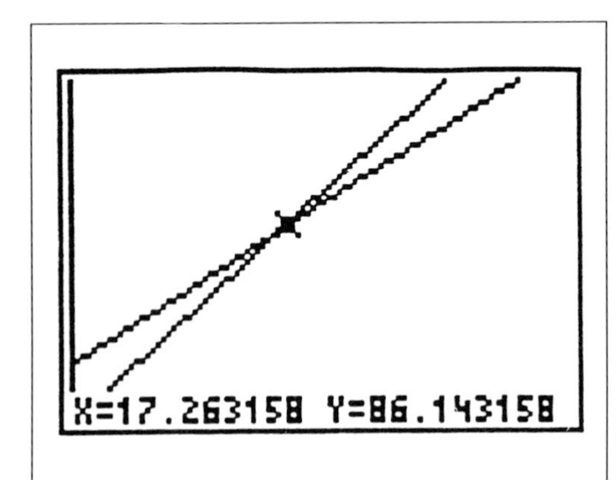


Fig. 14.6

Students should be encouraged to confirm this result algebraically. The algebraic solution of $3.5x + 25 = 4.99x$ yields $x = 17.36$. This is a natural lead into a discussion of the reasonableness of answers. Because of the discrete nature of this real-world application, Video Palace is the better buy when at least eighteen games are rented.

An open-ended classroom discussion might follow, to determine which store really offers the better deal. Some students may not rent sufficient numbers of video games to make the $25 investment worthwhile. Others may reason that over the course of a lifetime the chances are that you will always break even. But the skeptic might say, "What happens if the store goes out of business?"

## Proceeding in Algebra

Mr. Douglas's class of high school resource room students has been using algebra tiles to solve equations such as $ax = b$ and $x + a = b$. (For advice and ideas on how to work with algebra tiles, see Howden [1985] and Rasmussen [1977].) Two students each share a mat with an "=" sign and a set of tiles. To this point, the students themselves have not written equations. They have listened to or have read applications created by their teacher, used tiles to represent the application on the mats, and verbalized their solutions. For example:

**Applications give meaning and context to the tiles. Here each single "one" tile stands for a model airplane.**

> I have a model airplane collection. At a garage sale, I picked up 10 more model airplanes, and now I have 3 times as many. What was the original number of model planes?

Using tiles, the students have learned that "equations" like $x + 4 = 9$ and $x + 4 = 1$ can be solved by "taking 4 ones from each side." They know how to use positive and negative tiles and understand that if you take 4 ones from a side that has a single one, you must add 3 of the negative-color ones. Students have seen that situations of the form $2x = x + 5$ can be solved by "taking a bar from each side." For "equations" of the form $3x = 12$, they learned they could "match up the bars and ones" to solve—that is:

- Separate the bars on the left and then separate the ones on the right into equal sets.
- Match up each bar on the left with one set of ones on the right.

The model plane example above, one of the more recent examples worked, combines several of these ideas.

Throughout today's lesson, Mr. Douglas hopes to have students see how the writing of the algebraic equation comes from their work and the verbal or written instructions. To begin the lesson, Mr. Douglas presents the following problem:

> I was weighing two puppies. When I weighed them together in their pet carrier, they weighed 21 pounds. When I weighed the pet carrier on its

own, it weighed 5 pounds. Assuming the puppies weigh roughly the same amount, what is the weight of a single puppy?

The students use tiles to picture the conditions of the problem and, in response to Mr. Douglas's query, agree that the equation $2b + 5 = 21$ represents what they have shown (fig. 14.7).

**The use of the variable *b* matches the terminology "bar," which has been used from the outset of the work with algebra tiles.**

*Mr. Douglas:* What can you do to solve?
*LaShonda:* Take 5 ones from each side.

As the students carry out this instruction, Mr. Douglas has LaShonda illustrate it with the tiles he has placed on the overhead while he writes, "Take away 5 from each side" on the chalkboard underneath the equation.

Fig. 14.7

*Mr. Douglas:* Who can tell what remains?
*Elliot:* Two bars balances 16 ones.

Mr. Douglas writes "$2b = 16$" underneath the written instruction on the chalkboard.

**The writing of the instructions in words reinforces the prior verbal instructions that the students have used.**

*Mr. Douglas:* Go ahead.
*Elliot:* Match up the bars and the ones.

As this is said, the teacher writes this instruction on the chalkboard under "$2b = 16$" while Elliot matches up the bars and ones on the overhead projector (fig. 14.8).

*Mr. Douglas:* What is the result?

*Jenny:* Each bar balances 8 ones.

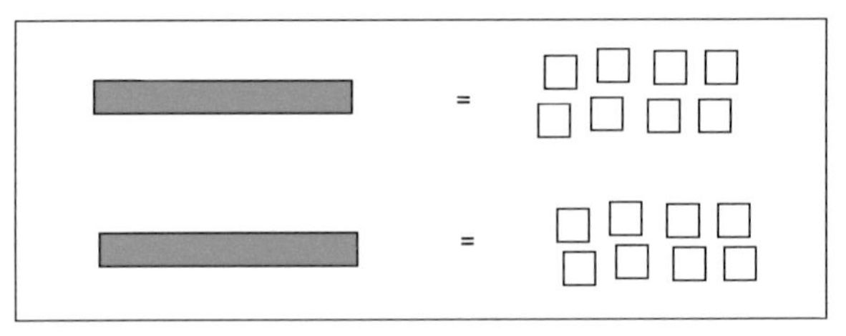

Fig. 14.8

To record the result, Mr. Douglas writes "b = 8" as a last line on the chalkboard.

**It is important that students get in the habit of determining if their answer is reasonable. Often this involves seeing if the number seems to be right; in applications, students should always explore whether the answer fits the context of the problem.**

*Mr. Douglas:* What does this mean for the original problem?

*Jenny:* Each puppy weighs 8 pounds.

*Mr. Douglas:* Does that make sense?

A little discussion ensues about the sizes of puppies, and it is noted that different breeds and different ages lead to different weights.

*Mr. Douglas:* Now let's solve this same problem again, except this time, as I write the equation, I'll write the instructions in a shorthand way. To start, set up 2 bars and 5 ones balancing 21 ones.

As students do this, Mr. Douglas erases the work on the chalkboard and writes "2*b* + 5 = 21" on an overhead transparency.

**To increase student participation and the potential for greater thinking and retention, Mr. Douglas asks students to share their *predictions* about what should be written.**

*Mr. Douglas:* What was done first to solve the problem?

*Mark:* Take 5 ones away from each side.

The teacher waits for the students to complete this with their mats and then asks them to predict how he will record this result.

*Esdra:* Just subtract 5 from both sides.

All agree that Esdra's suggestions make sense, so Mr. Douglas writes "–5" under each side of the equation. (See fig. 14.9.)

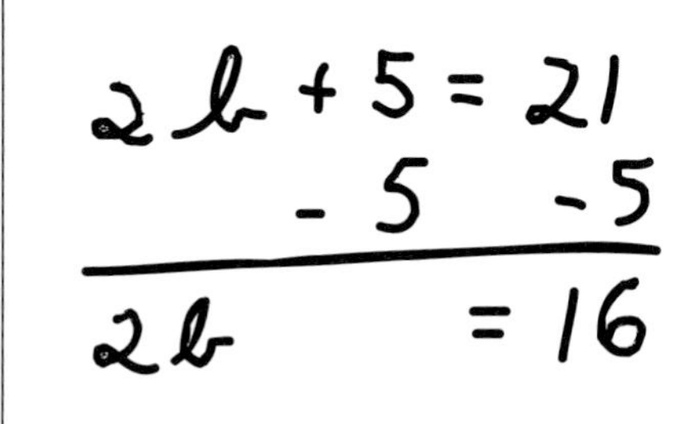

Fig. 14.9

*Mr. Douglas:* What remains on your mat?

*Elliot:* 2 bars and 16 ones.

*Mr. Douglas:* Tell your partner what math shorthand I might write for this.

Mr. Douglas listens to the students and then records what he hears: "2b = 16."

**The symbolism parallels the verbal and written descriptions used by the students to describe results at each step toward their solution. It is important that explicit connections be made between reported and written work done with tiles.**

*Mr. Douglas:* What's next?

*Maria:* Match up the bars and ones.

*Mr. Douglas:* And what's the result?

*Mark:* Each bar balances 8 ones.

*Mr. Douglas:* Tell your partner what I should write for this.

After listening to the students, Mr. Douglas smiles and writes, "$b = 8$" as a last line. "This is what I heard you say," he said.

The teacher now presents a transparency with four equations written in symbolic form. The students are asked to solve the first two equations using the tiles and to record the instructions in words. For the last two equations, they are challenged to record their work symbolically, as in the last part of the lesson. □

**Equations given in this section will have both positive and negative numbers, but only positive coefficients of *b*. Depending on the level of attainment, students could do all four equations in the same way, either with written instructions or with symbolism.**

## Extending the Special Education Thrust

Using algebra tiles for mathematics instruction presents the student with a concrete model for working with algebraic thinking and processes. This type of manipulative strategy is especially effective with special education students. The long-recognized use of multisensory techniques as an effective tool for learning has been clearly demonstrated in many classrooms.

In using this delivery system with special education students, some general instructional strategies and principles might be considered. Consistency in the use of terms, directions, and processes is critical when working with students with learning disabilities. It is necessary to avoid confusion by initially establishing understanding through focused attention on new vocabulary. This can then be reinforced through visual reminders in the classroom such as charts, displays, or posters.

In modifying the equation lesson for auditory learners, as well as for those with auditory perception or memory deficits, the use of an overhead projector rather than the chalkboard has the advantage of eliminating the distraction of chalk noise and provides for greater visual contact as the teacher is speaking. In the multiple steps of solving an equation, most students will need to clearly understand each segment before proceeding. Because the goal is developing *intuitions* and understandings for solving equations, colloquial language used by the students themselves should be encouraged (and modified when necessary for accuracy) to describe ideas and focus students' attention.

Visual learners, like those with attention deficit disorders or hyperactive tendencies, may require extensive demonstration and modeling activities. These students, as well as those with memory deficits, may benefit from several additional examples and work in pairs in which one student moves the tiles while the other records algebraically. The written work can then serve as a model for completing assigned work.

Algebra tiles obviously form a strong basis of understanding for tactile learners. Through connecting activities like that suggested in the preceding paragraph, in which students are required to record after each phase of moves with the tiles, understanding the relationship between the tile moves and the algebraic representation is nurtured.

For students with physical disabilities, the use of a nonslip work surface, such as that provided by felt, or the use of magnetic or enlarged tile models may be necessary. For students with visual impairments, attention to the work space and the size of the tiles would also be important. For this latter group of students, clarity in any verbal message is imperative. For many such students, using different-textured surfaces will be necessary to allow them to distinguish "positive" from "negative" tiles. A redesigned mat with a raised equals sign and containers to either side for holding materials may also be useful.

## Extending the Mathematics

The process used in the classroom episode above allows special education students to model both simple and complex algebraic equations. One extension is to use tiles for solving problems involving the distributive property. Students begin by setting up sets of tiles, such as three sets of 4 bars and 2 ones (fig. 14.10). By rearranging these sets of tiles, students find that they have 12 bars and 6 ones.

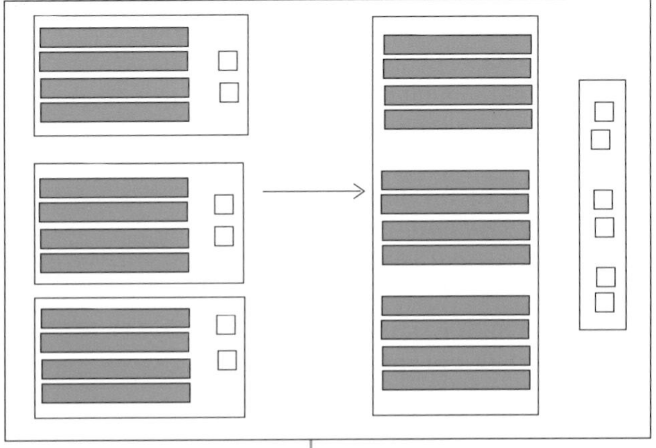

Fig. 14.10

When this idea is applied to equations, the examples that can be addressed can become fairly complex. For instance, without tiles, many students attempting to solve the equation

$$4(2b + 3) = 2(3b + 5) + 8$$

commit a myriad of errors, not the least of which is distributing the 2 on the right-hand side also to the 8. However, through tiling, the sense of the equation and the direction for solution are quite clear.

A further extension of applying tiles to linear equations comes when students are confronted with the equation $2b = 9$. When the student matches up the bars on the left and the ones on the right, there is an "extra" one (fig. 14.11). Students can then see that each bar balances 4 full ones. With some discussion, whether it be with the teacher or with other students, it can easily be determined that half the extra one can be matched with each bar, making each bar balance 4 1/2.

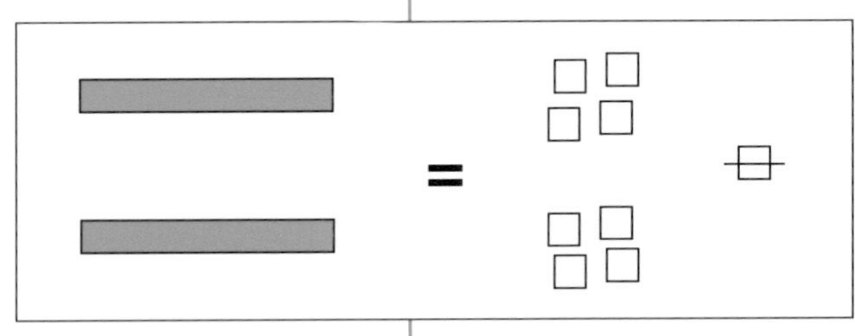

Fig. 14.11

With a little help, students should approach the point where they can use negative as well as positive coefficients for $b$. Often it is easy to change the colors of all the tiles! This is of course equivalent to multiplying both sides by –1. With the use of techniques such as these, students who are familiar with tiles are more likely to know that $-(x - y)$ becomes $-x + y$, a manipulation carried out incorrectly by many algebra students.

## Problem Solving with Algebra

Ms. Boswell's eleventh-grade algebra class was about to preview a unit on graphing rational equations. The students had prior success solving linear and quadratic equations and inequalities and were fully familiar with graphing functions on a calculator.

At the beginning of the period, Ms. Boswell presented the following problem:

Consider rectangles made from pennies, such as a $3 \times 4$ rectangle made with 12 pennies (see fig. 14.12). In this rectangle, there are 10 pennies along the perimeter and 2 pennies completely inside. What (if any) rectangles can you form such that the number of perimeter pennies is equal to the number of pennies completely inside?

Fig. 14.12

The class was given the opportunity to work in small groups in order to explore different strategies to initiate the solution. During this time, one student said, "Ms. Boswell, this problem doesn't have anything to do with algebra; it's just counting."

Ms. Boswell replied, "What do *you* think it has to do with algebra, Andy?"

Andy said, "We've pushed symbols around a page; I guess today we're pushing pennies around a page."

Ms. Boswell laughed. "That's not exactly what I had in mind, but we'll think about this question and get back to it later."

While circulating, Ms. Boswell stopped by a group that had found an answer. She said, loud enough for the whole class to hear, "Terrific, you found a solution! Now see if there are any others!"

After several more minutes, Ms. Boswell elicited responses from the students.

Mark volunteered that his group tried to guess numbers that would satisfy the conditions of the problem. Ms. Boswell asked, "How did you determine what numbers to guess?"

**The use of small groups in a problem-solving situation provides additional support for all students, especially those with special needs. Within the group, students explore, discuss, clarify, and brainstorm ideas related to the solution of the problem.**

**With students having special needs, humorous exchanges should be encouraged. Ms. Boswell does not answer the question posed but waits until the connection with algebra will become clear.**

**Often knowing that an answer exists can help a student solve a problem. Here Ms. Boswell's dual purpose is letting the class know that an answer exists and keeping the group that found an answer working on looking for more answers.**

Carefully sequenced questions help students increase the complexity of their thought process. Consequently, they can see how the components fit together.

Most students need the experience of generating a table of values consistent with the stated problem. It helps them to identify possible solutions to the problem. To support the construction of the table, students might be asked to confer with a partner and later share their thinking.

Flexibility in problem solving is advanced by having students look for and compare alternative approaches. Here both the helpful aspects and not so helpful aspects of using the table are noted.

Mark said they guessed numbers that differed by 3. They started with 6 and 3.

Ms. Boswell inquired, "What did you do next?"

Mark replied, "We made a rectangle with 18 pennies. But it didn't work. So we guessed another pair of numbers."

"Did you find a pair that worked?" asked Ms. Boswell.

Mark replied, "Not yet."

Ms. Boswell asked, "Did any other group approach the problem in the same way? In a different way?"

Mary Ann responded that her group tried to figure out the relationship between the perimeter pennies and those completely inside. Ms. Boswell suggested that the class explore this approach in greater detail. She suggested that they make a table of values in order to organize the data and see if any patterns occurred.

| Length | Width | Perimeter | Inside Pennies |
|---|---|---|---|
| 4 | 2 | 8 | 0 |
| 5 | 2 | 10 | 0 |
| 5 | 3 | 12 | 3 |
| 6 | 3 | 14 | 4 |
| 6 | 4 | 16 | 8 |
| 7 | 4 | 18 | 10 |
| 5 | 5 | 16 | $3 \cdot 3 = 9$ |
| 6 | 5 | 18 | $4 \cdot 3 = 12$ |
| 7 | 5 | 20 | $5 \cdot 3 = 15$ |
| 8 | 5 | 22 | $6 \cdot 3 = 18$ |
| 7 | 6 | 22 | $5 \cdot 4 = 20$ |
| 8 | 6 | 24 | $6 \cdot 4 = 24$ |
| 9 | 6 | 26 | $7 \cdot 4 = 28$ |

"What patterns can you notice?" asked Ms. Boswell.

Several students responded that "the $8 \times 6$ rectangle works." It was also noted that the inside pennies form a rectangle, so you can just multiply to get those. The perimeter could then be found by subtracting the inside from the total number of pennies.

"What is the problem with using this list to help us solve the problem?" asked Ms. Boswell.

Several students responded that the problem is that it doesn't give a relationship between the number of inside pennies and the perimeter and that it is unclear whether there were other solutions besides $6 \times 8$.

Tasha called out, "The length and width of the inside rectangle is 2 less than the length and width you start with. See the $9 \times 6$ at the bottom. Take two from each and you have $7 \times 4$."

After Ms. Boswell saw that the class agreed with Tasha's observation, she asked if there was a way of stating the problem using this new information.

Sensing some confusion, Ms. Boswell wrote on the overhead projector and asked the following questions:

1. If the number of inside pennies equals the number of perimeter pennies, what fraction of the total pennies is the number of inside pennies?

2. Can you express this problem using variables for the length and width?

Ms. Boswell asked students to work with their groups to answer these questions. For the first question, students soon realized that the inside pennies were half the total number. For the second question, groups could tell that the number of pennies in the inside rectangle was $(L - 2)(W - 2)$, but they weren't sure how this helped.

Ms. Boswell got the class's attention. "Let's put all this knowledge together. What is the total number of pennies used for a rectangle of length $L$ and width $W$?"

Elaine said "$LW$" and then wrote it on the board.

Ms. Boswell said, "Now what is the total number of pennies used for the inside rectangle?"

She then had Dan write $(L - 2)(W - 2)$ to the right of what Elaine had written.

"Now from question 1 on the overhead projector, what is the relationship between Elaine's total and Dan's total?" she asked. After the class responded one-half, Ms. Boswell transformed the board to show

$$\frac{LW}{2} = (L-2)(W-2).$$

"I want you now to solve this equation for $L$ and then graph it on your graphing calculator," she continued. "You might want to replace $L$ with $y$ and $W$ with $x$ so that it is easier to type in."

Ms. Boswell circulated to those groups who needed help with the manipulation. The final form of the equation became

$$y = \frac{4(x-2)}{x-4},$$

which the students displayed on their graphing calculators.

Ms. Boswell then asked, "Given the original problem, what can be the domain for $x$ and $y$?"

The students quickly answered that both must be positive and both must be integers because you can't have a negative or fractional penny.

Yesenia added that "$x$ can't be 4 or the denominator is 0."

By graphing on the integer setting under the ZOOM menu, the students could quickly get a list of values: (1, 4/3), (2, 0), (3, -1), (5, 12), (6, 8), (7, 20/3), (8, 6)....

Ms. Boswell said, "From this short list, we can see that a 5 × 12 and a 6 × 8 rectangle are solutions to the original problem. For tomorrow, explain why we don't have to go further on this list. How can I state that the 5 × 12 and the 6 × 8 are the *only* solutions?" ❑

**Having the students get up and be an active part of the discussion empowers students with special needs and keeps them engaged. The students were called on after they had confidently given the correct answers.**

**Although this is a complex manipulation, it arises from the context of a problem situation.**

## Extending the Special Education Thrust

As students enroll in higher-level mathematics courses, it is equally important that they have hands-on activities. For special education students, their senses are often highly sophisticated; thus, approaches should be visual, verbal, written, and tactile. All these approaches were used in this episode. It is a great misconception that students will automatically think abstractly at this level. A mixture of the concrete and the theoretical is of great benefit for all students as they construct knowledge.

For students having special needs that affect learning, it is important to provide ample wait time for processing information. Frequent use of "think-pair-share," in which students summarize their thinking or solution to a partner or small group before sharing with the larger group, is valuable. This approach not only increases wait time but provides important opportunity for students to organize and verbalize their thinking, needed by many of these students if they are to internalize and retain important ideas. Other forms of cooperative learning provide additional support for special learning needs and, by increasing active participation, enhance the potential for learning.

## Extending the Mathematics

With this episode as a precursor, the logical step mathematically is to examine the graph of

$$y = \frac{4(x-2)}{x-4}.$$

A discussion can be made of the vertical asymptote at $x = 4$ and how the graph behaves as $x$ approaches 4 both from the right and from the left. Another discussion of this graph concerns end behavior. The value of $y$ approaches 4 as $x$ gets larger. Why does it not reach 4? What does this mean in the context of the original problem? (It should be noted, however, that this particular equation is not the best choice for a full abstraction of these issues, since there are asymptotes at both $x = 4$ and $y = 4$ and the students might have a hard time differentiating.) Students might be asked to consider, "How would the equation

$$y = \frac{5(x-2)}{x-3}$$

behave? Make a prediction and then check on the calculator."

The situation in the episode above provides important opportunities for encouraging students to suggest generalizations. For example, when Ms. Boswell asked the question, "What is the problem with using this list to help us solve the problem?" she was helping students recognize that to determine a solution, you need to continually evaluate how close you are getting to the goal. In neither confirming nor denying the correctness of student responses, Ms. Boswell empowered the students to operate in a risk-free environment that stimulated student willingness to explore and make conjectures.

When students are given the opportunity to explore, amazing things can happen. An extension of the given problem is to look for the same phenomena in a three-dimensional box or other two-dimensional shapes such as equilateral triangles. In particular, when students with special needs are given time for exploration, they often produce exceptional findings (having seldom had this opportunity before).

## CREATING WINDOWS OF OPPORTUNITY

All teachers are undergoing the challenge of viewing (1) the mathematics content they teach in light of the *Curriculum and Evaluation Standards for School Mathematics* (NCTM 1989) and (2) the way they teach mathematics in light of the *Professional Standards for Teaching Mathematics* (NCTM 1991). Since many of these recommendations came out of research on students with special needs and abilities, teachers of these students have, perhaps, the most to gain by incorporating these new content and pedagogical ideas in their classrooms.

Specifically for algebra, the new agenda for the typical first-year algebra course focuses on expressing patterns with variables; exploring relationships between quantities using tables, graphs, and other displays; and studying certain procedures that often occur in the context of solving rich and real problems. Applications, explorations, connections, and conjectures should permeate all algebraic activities.

Greater potential for real learning in algebra occurs when—

- a more experiential, experimental, even laboratory type of atmosphere prevails;
- all activities connect to applications, technology, or manipulatives, often incorporating all three;
- students are treated with respect as members of a community of learners where they work in partnerships, small groups, or a full class.

The new role for algebra teachers is that of leader and advisor for their students. In ascertaining the students' needs, teachers will be able to select which algebraic tasks to present to their students. Once the tasks are selected, the teacher will work with the students and give them as much independence of inquiry as is possible. The special needs of each individual in the classroom (including the teacher) should be acknowledged and respected so that mathematics can remain the focus in the algebra classroom.

Almost all students have difficulty making sense of algebra. Almost all the ideas in this chapter apply to any classroom of students, whether they have been identified as having special needs or not. But the usefulness, power, and excitement of algebra has been hidden from students with special needs for too long. It is time to let these students enjoy the fresh air that the window of algebraic success brings when open.

## REFERENCES

Flanders, James R. "How Much of the Content in Mathematics Textbooks Is New?" *Arithmetic Teacher* 35 (September 1987): 18–23.

Howden, Hilde. *Algebra Tiles for the Overhead Projector*. New Rochelle, N.Y.: Cuisenaire Company of America, 1985.

Mullis, Ina V. S., John A. Dossey, E. H. Owen, and G. W. Philips. *State of Mathematics Achievement: NAEP's 1990 Assessment of the Nation and the Trial Assessment of the States.* Washington, D.C.: National Center for Education Statistics, United States Department of Education, 1991.

National Council of Teachers of Mathematics. *Curriculum and Evaluation Standards for School Mathematics.* Reston, Va.: The Council, 1989.

———. *Professional Standards for Teaching Mathematics.* Reston, Va.: The Council, 1991.

Rasmussen, Peter. *MathTiles Manual: A Concrete Approach to Arithmetic and Algebra.* Berkeley, Calif.: Key Curriculum Press, 1977.

Usiskin, Zalman. "Conceptions of School Algebra and Uses of Variables." In *The Ideas of Algebra, K–12,* 1988 Yearbook of the National Council of Teachers of Mathematics, edited by Arthur F. Coxford, pp. 8–19. Reston, Va.: The Council, 1988.

———. "Why Elementary Algebra Can, Should, and Must Be an Eighth-Grade Course for Average Students." *Mathematics Teacher* 80 (September 1987): 428–38.

**Francis Gardella** is the supervisor of mathematics, K–12, in the East Brunswick Public Schools in East Brunswick, New Jersey. His teaching experience includes junior high school and college-level mathematics. His background includes a bachelor's degree in mathematics and a doctorate in mathematics education. The coauthor of five mathematics texts, his major interest is in initial teaching experiences in mathematics and the language level necessary for initial understanding of mathematical concepts.

**David Glatzer** is a mathematics supervisor (grades 6–12) for the West Orange Public Schools, West Orange, New Jersey. He recently served on the Board of Directors of the NCTM. He received his master's degree at Montclair State College, Upper Montclair, New Jersey. He is a past president of the Association of Mathematics Teachers of New Jersey.

**Joyce Glatzer** is a mathematics education consultant. She received her master's degree at Montclair State College, Upper Montclair, New Jersey. She is a past president of the Association of Mathematics Teachers of New Jersey. She was formerly mathematics coordinator (grades K–9) for the Summit Public Schools, Summit, New Jersey.

**Daniel Hirschhorn** is an assistant professor of mathematics at Illinois State University. He is an experienced middle school and high school teacher, and he is an author of UCSMP *Algebra* and UCSMP *Geometry.*

**Carolyn Rosenberg** is supervisor of mathematics, science, and computer education at Somerville Public Schools in New Jersey. She is the recipient of the Princeton Prize for Distinguished Secondary School Teaching and the Presidential Award for Excellence in Mathematics Teaching. She serves on numerous state and national committees that strive to improve the teaching and the learning of mathematics.

**Margaret Walsh** is the department chairperson for special education in the East Brunswick Public Schools, East Brunswick, New Jersey. A certified learning disabilities teacher and consultant, her experience includes teaching children with a wide range of disabilities including neurological impairment, mental retardation,and emotional disturbance. She is currently completing doctoral studies in special education, and her research interests are inclusive programming and planning for special-needs students.

**Angela Waters** is a special education teacher at East Brunswick High School, East Brunswick, New Jersey. Her background includes undergraduate work in speech pathology and additional graduate study in special education. She has taught mathematics to special education students for the past six years and has presented a variety of mathematics workshops.

**Joan Stanton** is currently a special education teacher at East Brunswick High School, East Brunswick, New Jersey. Her twenty-year teaching record includes experience with children with a variety of disabilities in elementary through high school. Her background at the undergraduate and graduate levels is in special education. She is also certified in elementary education.

B'
C'
A'

# 15

# Flexible Pathways
## Guiding the Development of Talented Students

*Sue Eddins*
*Peggy House*

ANOTHER new school year. Another group of students, each with unique experiences, abilities, and needs. Among them are those who have been identified as mathematically gifted. Your challenge: to open wider the windows of opportunity for the mathematically talented. The following student profiles provide a context for that challenge.

Meet DAVID, a bright, enthusiastic young man, always ready to volunteer answers, the one student you can count on to carry the class even on a slow day. But David has never really known what it means to be challenged in a mathematics class. At least, he has never experienced that challenge on more than an infrequent, episodic basis.

Whereas other students struggle to understand a lesson, David quickly grasps the concept, translates it into action, and completes the day's assignment with ease. Even though David can see that many of his classmates are struggling with the lesson, he knows that it is inherently simple. Like many capable students, he often seems to know intuitively when a topic is routine and when it requires special insight or cleverness.

Although David is realistic about his past academic successes, he has not yet developed real respect for himself as a learner. He does not yet realize what he is truly capable of doing, and he needs an educational experience that can help him identify and test his intellectual limits. David needs to become engaged with challenging ideas and problems, ones whose difficulty he respects, so that he can begin to investigate and enhance his intellectual self-confidence. To do this, he needs a very safe classroom environment, since he is not used to making mistakes, and it takes a tremendous effort for him to verbalize an answer of which he is not confident.

ANGELA, however, has demonstrated her exceptional talent by the mathematics she has been able to "create" on her own. While learning much standard mathematics, she has also invented along the way some unique and sometimes slightly erroneous ways of solving conventional problems. Because of her creativity, she has amassed a collection of partial procedures and facts that have, until now, served her quite well. Because Angela is a good student who gets the right answer most of the time and because many of her novel methods were unfamiliar to her previous teachers, her procedures and her understanding rarely have been challenged. Angela needs to "regress" a bit in the content she has already studied so that she can build a more solid foundation for future mathematics.

Despite her considerable achievements, Angela's confidence is extremely fragile. She isn't at all sure that she is "smart," and in an environment in which she is surrounded by talented peers, it is all too easy to convince her that she "isn't really very good at math." Angela also needs to learn how to study. Since she rarely has had to seek help from a teacher or a peer, she does not comprehend what it means to struggle with a concept or a problem, and she is not good at identifying when the amount of work devoted to a problem is enough.

A third student, BRIAN, is younger than most of his peers, since he was accelerated in elementary and junior high school. Brian does not need to regress in his study of mathematics. The foundation he has built, despite acceleration, is a very strong one. But Brian does need to improve his ability to communicate with other students, both those less able than himself and those who are his intellectual peers.

Recently, as he has wrestled with tough problems, he has begun to share his struggles and explain his approaches to classmates, and he has been motivated and stimulated by the opportunity to share in their thinking. Brian continues to need problems difficult enough to encourage such communication.

LUKE, like Brian, was accelerated in mathematics from an early age. By the time he entered junior high school, Luke was studying advanced mathematics in a special program that enabled him to complete the standard secondary school curriculum and begin college-level courses in the tenth grade. But Luke's straight-A record belies the depth of his understanding and the weak conceptual foundation of his mathematical knowledge. Faced with a procedure similar to one taught in class, Luke's performance is flawless; but given a novel problem, or a learned concept in an unfamiliar context, he has few personal resources on which to draw, and his tolerance for any problem whose solution is not immediately apparent to him is extremely low.

Luke needs to reflect on the mathematical concepts and principles that underlie the procedures. He needs to build connections among the procedures that he has mastered in isolation and to probe the richness of related mathematical topics. And he needs to develop a repertoire of problem-solving strategies, the self-confidence to use them, and the perseverance to struggle in the face of adversity.

LAURA, too, is an all-A student, not only in mathematics but in her other subjects as well. She was frequently identified as gifted by her previous teachers. But Laura's success has, in large part, been attributable to her rapid and good memory and to her conscientious attention to detail. She conforms easily to classroom norms: completing assignments, staying on task, displaying a willing attitude, cooperating with teachers and other students, doing extra work. These qualities contributed to Laura's success in elementary and middle school mathematics and in elementary algebra.

But Laura's confidence began to waiver when she encountered mathematics that required more than computation. Laura is at a critical point in her academic development. On the one hand, her parents, teachers, and counselors are encouraging her to excel. On the other, she is experiencing anxiety about her own abilities, fear that she may blemish her perfect academic record, and peer pressure to replace academic success with other adolescent pursuits.

Laura needs emotional support as well as academic challenge. She needs to build a base of mathematical understanding, realistic expectations, and self-reliance. She needs encouragement to continue her commitment to academic excellence so that opportunities remain open to her.

## LOST OPPORTUNITIES

These students typify the diversity one finds among those who are identified as mathematically gifted. By most conventional measures, all have been successful, and many will argue that the educational system must have served them well or they would not have achieved as they did. But what is an adequate curriculum, even a challenging one, for some is deprivation for others. For all the students characterized above, their mathematics curriculum has deprived them in one or more of the following ways:

- Failed to empower the students by setting expectations that are too low and by accepting, even rewarding, performance that is below their capabilities
- Failed to nurture self-reliance and independence of thought, often by substituting teacher approval for originality and self-assessment
- Allowed computational or procedural facility to substitute for conceptual understanding
- Substituted radical acceleration through the traditional curriculum for deeper, broader mathematical exploration
- Emphasized abstract, symbolic manipulations to the neglect of hands-on experiences and concrete applications
- Overlooked or undervalued the original approaches and divergent thinking of gifted students
- Failed to develop the dispositions and resourcefulness needed to struggle with challenging, unfamiliar problems
- Failed to reveal the connectedness of mathematics or to help students grasp its underlying structure
- Failed to nurture and engender an appreciation for the value of clear and deep communication of mathematical ideas

Currently there is a movement toward heterogeneous classes that is founded on the belief that the inclusion of gifted students in the regular classroom can serve to raise the expectations and performance of the entire group. And, although educators have long recognized that much of what is good for gifted students also is good for their less-talented peers, the fact remains that gifted students have special needs that require both an enriched curriculum and a challenging delivery system.

## FLEXIBLE PATHWAYS

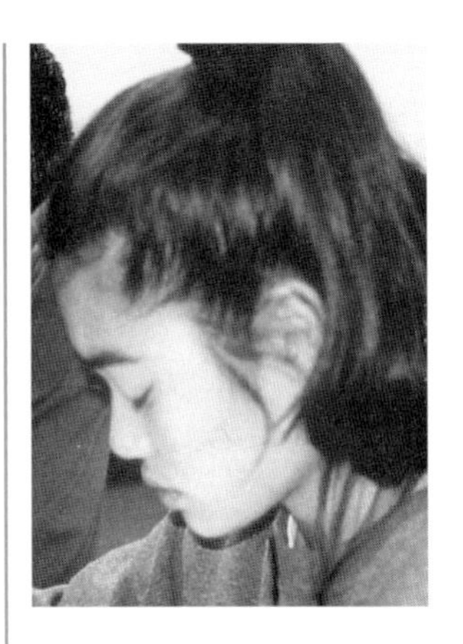

The ways in which we respond to mathematically gifted students may be as varied as the students themselves and the educational settings in which we find them. Provisions can range from differentiated assignments in a heterogeneous classroom to special schools for the gifted (see House [1987] for a discussion of organizational alternatives).

But whatever the arrangement, our responsibility as educators is to offer flexible pathways along which gifted students can encounter rich ideas through challenging, nonstandard learning experiences that cause them to assess and stretch their own abilities. As we work with mathematically talented students, we try to create an environment that encourages them to communicate their thoughts, that minimizes boredom, and that ensures early significant success for all students.

The following examples illustrate several flexible pathways through geometry that we have used to challenge gifted students at the Illinois Mathematics and Science Academy and at the Seaborg Summer Academy. (The Illinois Mathematics and Science Academy [IMSA] is currently the only *three*-year public residential academy for high school students with high ability in mathematics and science. IMSA offers a comprehensive high school program for Illinois students. The Seaborg Summer Academy is a residential summer program at Northern Michigan University for gifted secondary school mathematics and science students.) The activities represent an attempt to address the need for flexible pathways and challenging, open-ended investigations in mathematics. The first activities from a unit on geometric transformations and matrices (Eddins, Maxwell, and Stanislaus 1994a, 1994b) were written to address the need for creating a safe classroom environment. The topic was selected in part because it was believed that the material would be generally unfamiliar to students regardless of their background. The introductory activity for the unit, adaptable to a variety of situations and levels, is described below.

### Getting Started

As an introductory activity, students work in groups of three. The two individuals designated the *teacher* and the *student* in the group sit back to back. The group's *observer* sits off to the side. During the entire activity, the only person permitted to speak is the "teacher."

The "teacher" is given a master graph showing a figure in the lower left corner of the paper and a congruent figure, whose orientation has been changed, in the upper right (fig. 15.1).

The "student's" sheet has the figure in the lower left only (fig. 15.2).

Both "student" and "teacher" have a cardboard shape that fits the given figure. The "teacher's" task is to give a series of instructions that will enable the "student" to move the shape from the original figure onto its image. The "observer" writes down what is said in order to mediate disputes if the desired result is not accomplished. Students can check their own results. They discuss what was said, where misunderstandings occurred, and how to communicate more clearly. The activity is repeated two more times with different figures so that each student can assume each of the three roles.

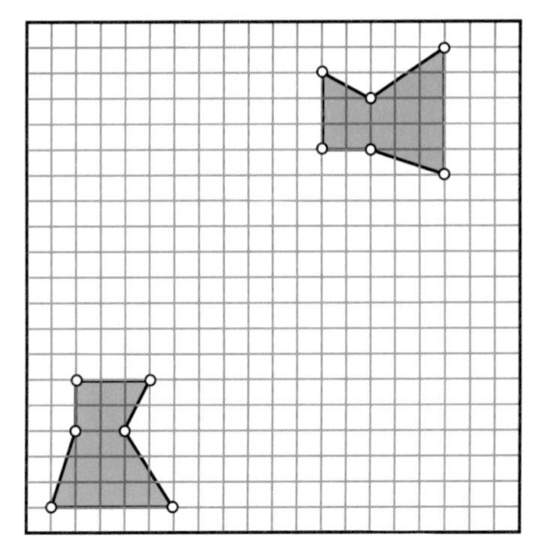

Fig. 15.1. Teacher's page

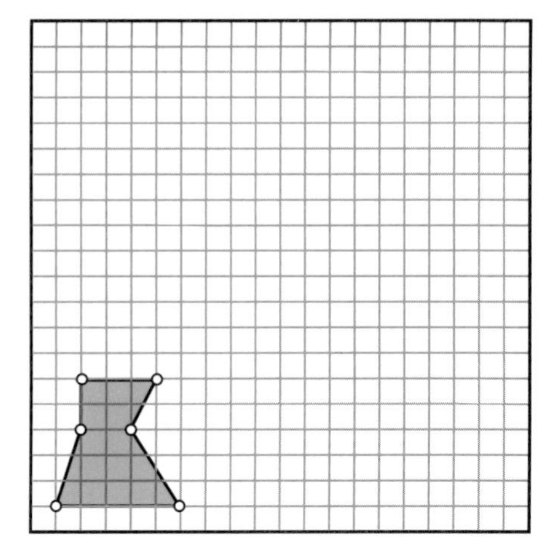

Fig. 15.2. Student's page

Following the activity, discussion focuses on questions like the following:

- What did you discover?
- What actions did "teachers" want "students" to perform?
- What verbs were used to describe those actions?
- What information was essential for each of the actions to be unambiguous?

In this way students are able, for themselves, to develop the standard mathematical vocabulary for rigid motions in the plane. Furthermore, they are able to identify for themselves the principal components necessary to describe these motions unambiguously. Discussion then focuses on the constants and variables under this type of transformation.

## Substantial Benefits from a Simple Activity

Clearly, this activity can be used with students of many different learning styles and abilities, although we might expect gifted students to analyze and synthesize more quickly and at a somewhat higher level. Its role at the beginning of the unit is to provide an interesting activity that permits the teacher to assess the students' abilities to formulate a set of instructions, to understand and act on that set of instructions, to learn from their own mistakes, and to synthesize the results into meaningful mathematics.

The activity also sets the tone that it is the *students themselves*, not the teacher, who will often judge the correctness of results and that an incorrect result should lead to learning rather than to concern. This furnishes an opportunity for student interaction, and it introduces content that is usually new and intriguing enough to whet their appetites for new discoveries.

## Extending the Inquiry

After experiences with transformations on noncoordinatized paper, examples are introduced for which coordinate axes have been added. By introducing column vectors to represent the coordinates of each vertex, the students learn to describe an $n$-sided polygon that has been drawn on coordinate graph paper by encoding its vertices in a $2 \times n$ matrix (see Coxford [1991, pp. 9–11]; Froelich [1991, pp. 19–20]). Thus, for example, the matrix for $\triangle ABC$ with $A(-4, 3)$, $B(0, -1)$, and $C(2, 5)$ will be as follows:

$$\begin{bmatrix} -4 & 0 & 2 \\ 3 & -1 & 5 \end{bmatrix}$$

We call this the *vertex matrix* for $\triangle ABC$. The students can now explore the operations of matrix addition, scalar multiplication, and matrix multiplication through application problems that integrate geometric transformations with matrices. A sample investigation is presented in figure 15.3.

Working sometimes in small groups, sometimes individually, and sometimes as a whole class, the students explore this special set of two-by-two matrices and present their conjectures. The open-ended investigations permit teachers to accommodate a wide range of needs.

Angela has an opportunity to review coordinate graphing of points and lines and to practice writing equations of lines while Laura is forced to grapple with the deeper meaning of concepts like *coordinates* and *congruent*. Brian learns to appreciate the insights of the other students and begins to engage in lively dialogue, even arguments, with David and

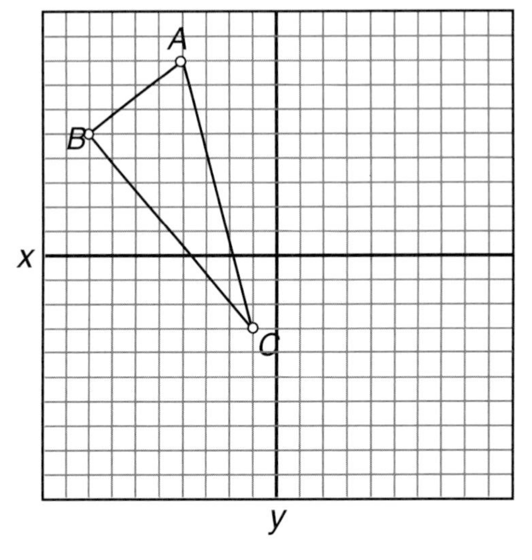


Complete the ordered pairs for $\triangle ABC$ and write its vertex matrix. Translate $\triangle ABC$ 10 units to the right and label the new triangle $\triangle A'B'C'$. Complete the ordered pairs for $\triangle A'B'C'$ and write its vertex matrix.

Write a matrix that you can add to the vertex matrix of $\triangle ABC$ to get the vertex matrix of $\triangle A'B'C'$.

Construct several other polygons and write their vertex matrices. Experiment with adding different matrices to each vertex matrix. Is the result always a translation? What can you conclude about the relationship between the original figure and the image?

In this figure $\triangle XYZ$ has been reflected over the $x$-axis. Describe the similarities and differences between the vertex matrix for $\triangle XYZ$ and the vertex matrix for $\triangle X'Y'Z'$.

This transformation cannot be obtained by addition, but it is the result of multiplication by a $2 \times 2$ matrix. Can you find a matrix $r_x$ for which the matrix product $r_x \cdot \triangle XYZ = \triangle X'Y'Z'$? Will the same matrix produce a reflection over the $x$-axis for other polygons as well?

Can you find a matrix $r_y$ that will produce a reflection over the $y$-axis? A matrix that will reflect a figure over the line $y = x$? A matrix that will produce a rotation of the original figure?

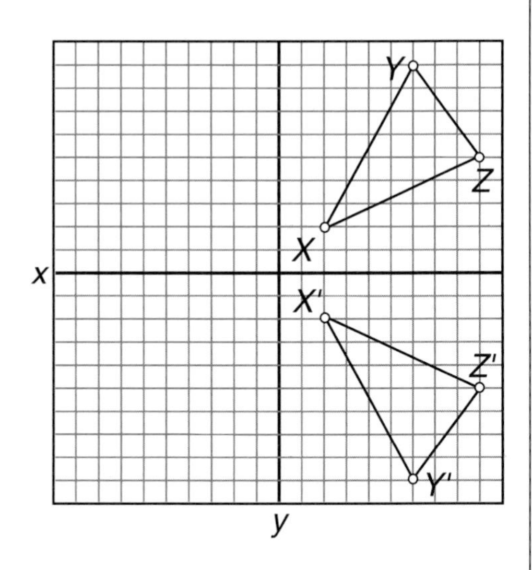


It looks as if the multiplication of vertex matrices by two-by-two matrices with entries of 0, 1, and −1 has something to do with geometric transformations in the plane. How many such two-by-two matrices are there, and which ones seem interesting? Explain your conclusions.

Fig. 15.3

Luke. David writes a program for the graphing calculator that gives him the vertex matrix of an image triangle and then pictures both preimage and image for a given two-by-two transformation matrix. He later expands this to a computer program that cycles through all possible matrices in this set.

## Challenging the Familiar

The activities with coordinates and transformations lead naturally to lattice problems that can be modeled on the geoboard. After becoming familiar with the geoboard and developing strategies for finding areas and perimeters of geoboard polygons, the students are challenged to discover a relationship between the area of a polygon and the number of pegs (lattice points) in and on the figure. Their investigation of various examples leads them to Pick's rule:

$$A = \frac{1}{2}(B - 2) + I,$$

where $B$ is the number of pegs on the boundary and $I$ is the number of pegs in the interior of the polygon.

The students are generally amazed by the simplicity of the relationship, and Angela and Brian, curious about polygons with "holes" in their interiors, discover that they can generalize the formula even further. But David, who admits that the relationship is "neat," remains relatively unmoved until the conventional square geoboards are replaced by isometric ones and the students are challenged to explore questions such as the following:

- Compare the square and isometric geoboards. For example, on the square geoboard you discovered that it is impossible to make an equilateral triangle with vertices at the pegs. On the isometric board, the equilateral triangle is easy, but now it is impossible to make a square. Why is that? What other differences do you observe?
- Since we cannot make unit squares on the isometric board, how should we measure area? What other units of area might we use? Can we do this? Does *area* mean the same thing on the two different boards?
- Can you find a formula for computing the area of a triangle on the isometric board similar to the familiar $A = (1/2)bh$ that we usually use?
- Does Pick's rule hold for the isometric geoboard? How about the Pythagorean theorem?

The introduction of the unfamiliar coordinate system poses challenges for all the students. After a short debate, they agree to use an equilateral triangle as the unit of area, and they then quickly discover that for any equilateral triangle with sides of length $s$ lying along the axes of the board, the area is given by $A = s^2$. Luke points out the similarity between this relationship and the area of a square on the other geoboard. Angela notes that $A = s^2$ can be generalized to isosceles triangles with their two equal sides ($s$) lying along axes, and Laura generalizes even further to describe any triangle with two sides along the axes, where, she finds, $A = s_1s_2$ when $s_1$ and $s_2$ are the lengths of the two sides on the axes. (See fig. 15.4.)

But further generalizations elude the students until David theorizes that "altitude" in this system must mean something other than "intersects at a right angle"; he suggests that *altitude* be redefined to mean the distance *along an axis* from one vertex to the opposite side, and he proposes they name it $h^*$. With growing excitement, he explains his theory to the class, who quickly begin to create expressions for area and to verify that Pick's rule does, indeed, seem to work—almost! All their relationships appear to be "too large" by a factor of 2:

| Square Geoboard | Isometric Geoboard |
|---|---|
| $A_\Delta = \frac{1}{2}bh$ | $A_\Delta = bh^*$ |
| $A = \frac{1}{2}(B-2)+I$ | $A = (B-2)+2I$ |

It is Luke's insight that helps them resolve the dilemma. He notes that two unit triangles make a rhombus and relates the rhombus to a sheared square. He demonstrates how a rhombus could be taken as the unit of area, with the resulting isometric formulas "all being divided by 2 because when the unit is twice as big, we only need half as many."

Other discussions follow. Angela, Laura, and Brian become excited when they discover that Pythagoras's theorem is just as true for equilateral triangles constructed on the sides of a right triangle as it is for squares, and before long they are generalizing the Pythagorean relationship further to other regular polygons, nonregular polygons, semicircles, and irregular shapes.

Their constructions of certain irregular shapes later leads them into studies of symmetry and back again to transformations. But this time they explore the transformations involved in tessellating the plane, and before long they are asking one another questions about how they might go about tessellating the surface of a polyhedron and whether polyhedra themselves can "tessellate" space.

Laura decides to apply her discoveries about tessellations to a project that has been assigned in her art class; Angela prepares a paper for biology on the symmetries found in living organisms. Brian, fascinated by the generality of the Pythagorean theorem, also embarks on an investigation of his own, and soon he is asking David to help him program his calculator to generate Pythagorean triples so that he can study their properties.

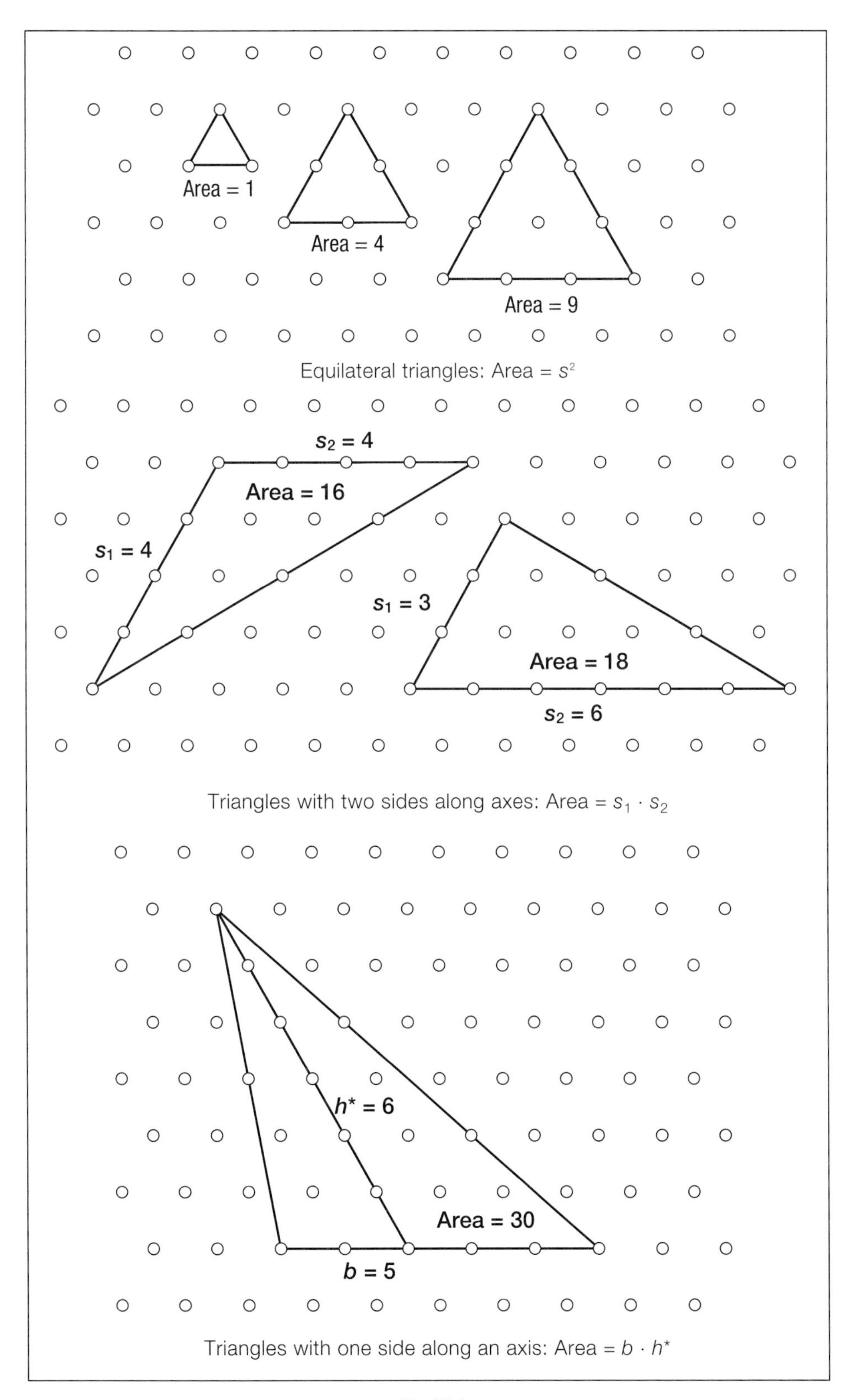

Area = 1
Area = 4
Area = 9
Equilateral triangles: Area = $s^2$
$s_2 = 4$
Area = 16
$s_1 = 4$
$s_1 = 3$
Area = 18
$s_2 = 6$
Triangles with two sides along axes: Area = $s_1 \cdot s_2$
$h^* = 6$
Area = 30
$b = 5$
Triangles with one side along an axis: Area = $b \cdot h^*$


Fig. 15.4

Meanwhile, David continues to think about his realization that *altitude* could take on a different meaning in a new context, and he inquires whether there are other systems where familiar objects take on new dimensions when the context is changed. This provides the perfect opportunity to introduce David to "taxicab geometry" (see Krause [1975]), in which the notion of *distance between two points* is changed from the conventional Cartesian definition of

$$d = \sqrt{(x_1 - x_2)^2 + (y_1 - y_2)^2}$$

to a "taxicab metric" defined as the sum of the horizontal and vertical components:

$$d_T = |x_1 - x_2| + |y_1 - y_2|$$

Before long, David is investigating the shapes and properties of "taxicab circles" and the various "taxicab conics" and describing his discoveries to the class. Meanwhile Luke, amazed to learn that geometry involves so much more than axioms and proofs, begins to search for books about the history of mathematics with special emphasis on Euclid and the Pythagoreans. David and Luke decide that mathematics is as much created as it is discovered, and they set out on a quest to develop an original mathematical system of their own.

Clearly, this is going to be a challenging, exciting year for everyone!

## LOOKING FORWARD AND BACK

Students' responses to these flexible pathways, *modulated by open-ended investigations*, indicate that many of our objectives for talented students are being achieved:

LAURA: Early in the year, working in groups was hard because we didn't know how to "think in public." Now we've learned to teach one another.

LUKE: It's a lot easier for me now because I've learned how to think the problems through, and that's usually as important as solving the problems themselves.

DAVID: When you look for more than one way to do a problem, it makes mathematics more interesting. Now our teachers aren't the only ones who are excited about mathematics.

ANGELA: Most of the time I can figure out for myself whether my solution is correct or not, and that helps me to tackle the next problem.

BRIAN: Sometimes you just stare at the paper for a while. If you get too stuck, the teacher can give you a hint, but then you're on your own again. Usually when you're almost ready to give up, someone will see what to do next. It can be frustrating—but it's great!

Although we don't always achieve everything we hope for, our goals in working with very talented students are simple:

- We want them to build a strong knowledge base, deep in conceptual understanding as well as in procedural skills.
- We want each one to grow in self-confidence and to develop his or her own sense of mathematical power.
- We want to motivate them to think; to look for connections; to conjecture, test, and justify; to verify the validity of their own thinking; and to offer their ideas for public examination.
- We want to excite them about the wonders of our discipline and to encourage them to invest their considerable talents in mathematics.

Mathematically talented students are too valuable a resource for us to base their education on the assumption that they will survive, prosper, and "get it" on their own. Students like David, Angela, Brian, Luke, and Laura do, most certainly, have special needs, and the environment and educational experiences that we offer them must address those needs. Unstimulated, these highly capable students may be lulled into thinking that algorithmic processes constitute the entirety of mathematics. Under such conditions it can become easy for them to avoid our courses in favor of "more interesting" subjects at which they also often excel.

Opportunities for flexible pathways through engaging problems and open-ended investigations *can* provide the key to help gifted students in many different settings maintain their interest in mathematics while stimulating their intellectual development. And such pathways will, most assuredly, stimulate and challenge their teachers.

## REFERENCES


Coxford, Arthur F., Jr. *Geometry from Multiple Perspectives.* Reston, Va.: National Council of Teachers of Mathematics, 1991.

Eddins, Susan K., Evelyn Osman Maxwell, and Floramma Stanislaus. "Activities: Geometric Transformations—Part 1." *Mathematics Teacher* 87 (March 1994a): 177–81.

———. "Activities: Geometric Transformations—Part 2." *Mathematics Teacher* 87 (April 1994b): 258–61, 268–70.

Froelich, Gary W. *Connecting Mathematics.* Reston, Va.: National Council of Teachers of Mathematics, 1991.

House, Peggy A., ed. *Providing Opportunities for the Mathematically Gifted, K–12.* Reston, Va.: National Council of Teachers of Mathematics, 1987.

Krause, Eugene F. *Taxicab Geometry.* Menlo Park, Calif.: Addison-Wesley, 1975.



**Sue Eddins** is on the mathematics faculty of the Illinois Mathematics and Science Academy (IMSA) in Aurora, Illinois. She has taught a wide range of mathematics courses for over twenty years. She received the Presidential Award for Excellence in Mathematics Teaching, and she is coauthor of UCSMP *Algebra*. She currently serves as team leader for curriculum and professional development for the math team at IMSA and chairs the Editorial Panel for NCTM's *Student Math Notes*.

**Peggy House** is a professor of mathematics education at the University of Minnesota in Minneapolis, Minnesota. She was the lead instructor in the Seaborg Summer Academy, a summer residential program for gifted high school mathematics and science students at Northern Michigan University in Marquette, Michigan. She also was editor of the NCTM publication *Providing Opportunities for the Mathematically Gifted, K–12.*

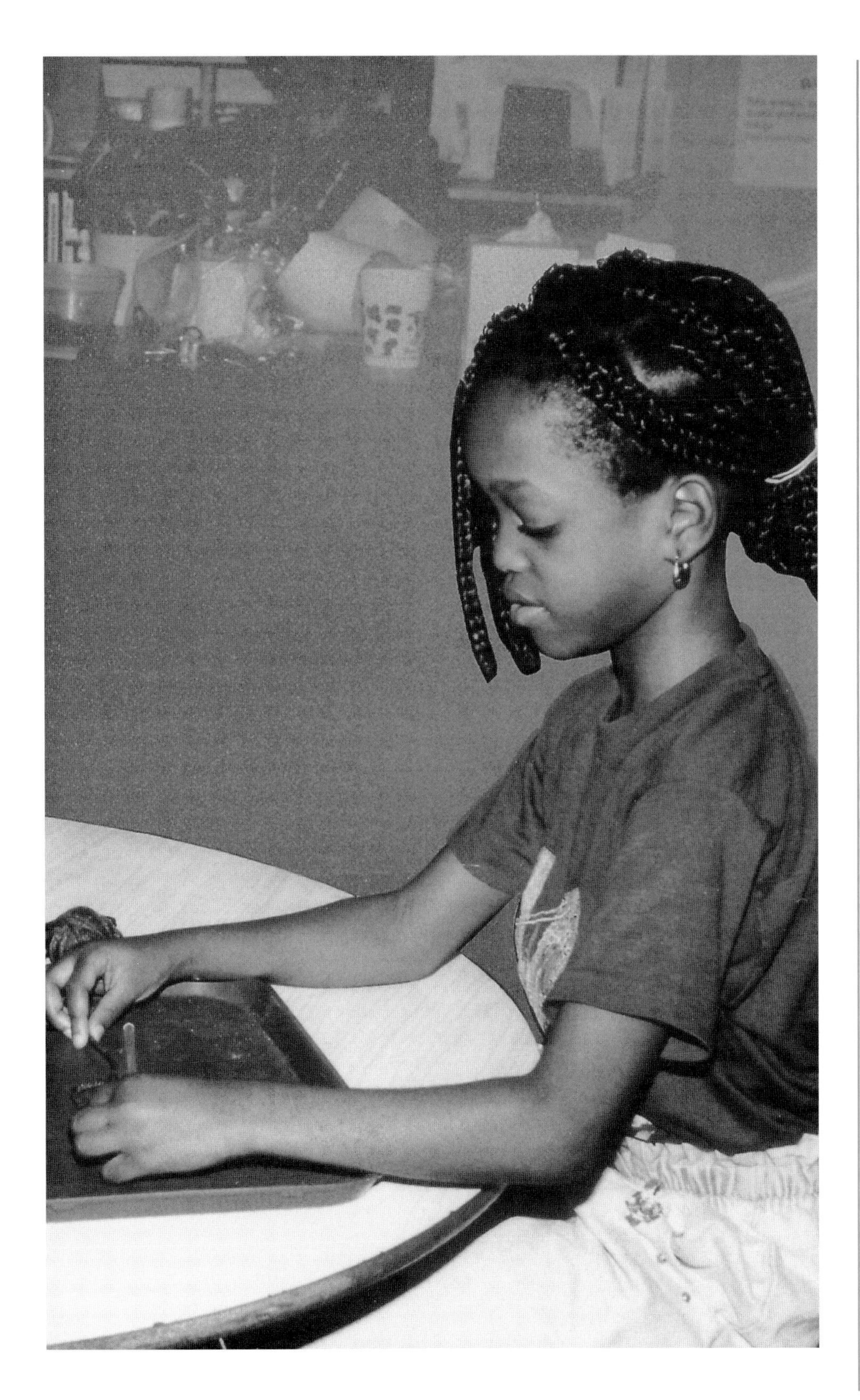

# Part 3

## Promising Practices

| Weight | Length |
| --- | --- |
| 0 | 4 |
| 50 | 6.4 |
| 150 | 11.5 |
| 170 | 12.5 |
| 190 | |

16

# A Case of Applied Mathematics

*Linda P. Ware*

*in collaboration with*

*Jack Krebs*

IF THIS story could have been begun four years ago when the study of mathematics for special education and "at risk" students at Oskaloosa High School in Oskaloosa, Kansas, was characterized by prescriptive routines in general mathematics courses, then the degree of change and accomplishment reflected in this chapter would be even more dramatic. At that time, prevailing practices, policies, and beliefs dictated that these students should be placed in a series of "general math" classes. *Tracking* was rarely said aloud, but this practice was essentially an accepted policy, deemed in the "students' best interests."

Today these students have the option of enrolling in a two-year applied mathematics (AP Math) sequence that was developed to eliminate the tracking of marginally successful students into general math classes. The discussion that follows describes two lessons that represent a promising practice for these students at Oskaloosa High School who, through the AP Math sequence, are routinely engaged in the investigation of patterns, data analysis, and modeling as emphasized by the National Council of Teachers of Mathematics (NCTM) in its *Curriculum and Evaluation Standards for School Mathematics* (1989) and its *Core Curriculum* (1992).

The first lesson is aligned with the Algebra Standard (NCTM 1989, p. 150), which recommends that students be given opportunities to (1) represent situations that involve variable quantities with expressions, equations, inequalities, and matrices; (2) use tables and graphs as tools to interpret expressions, equations, and inequalities; (3) operate on expressions and matrices and solve equations and inequalities; and (4) appreciate the power of mathematical abstraction and symbolism.

AP Math students are encouraged to get their "hands on" mathematical activities and their "minds around" mathematical concepts. However, for the most part, these students have not been successful in school and in mathematics,

and many have been, or are currently receiving services, in special education. Now thrust into wholly different mathematics experiences, these students tend to approach schoolwork in general with apprehension and reluctance. The irony is that at this stage of their academic careers, active involvement with the content is vital to the development of mathematical power.

Over the past weeks, the students had carried out activities in which (1) they explored linear equations where a constant rate of change $m$ and a starting value $b$ were common aspects of real-world situations; (2) they graphed pairs of numbers to represent related events in the world (e.g., Fahrenheit and Celsius temperatures, sides and perimeters of squares, simple and compound interest); and (3) they demonstrated that data could be represented in a common pattern depicted by a straight line. These activities laid the foundation for dealing meaningfully with the content of the lesson that follows.

## A Linear Equation $y = mx + b$ for a Spring

The students sauntered into the classroom in pairs, excited with the chatter that echoed in the hallways: the score from Friday night's football game, the unyielding dress code, the approaching deadline for ordering class rings. Most paused to greet their teacher as he wrote the formula below on the chalkboard:

$$m = \text{slope} = \frac{\text{rise}}{\text{run}} = \frac{\text{change in } y}{\text{change in } x} = \frac{\text{second } y - \text{first } y}{\text{second } x - \text{first } x} = \frac{Y_2 - Y_1}{Y_2 - Y_1}$$

When the bell sounded, the students were issued an activity sheet entitled, "A Linear Equation $y = mx + b$ for a Spring," and Mr. Krebs said, "Just try to answer as many questions as you can for now."

The activity sheet was developed by the teacher to set the stage for the day's instruction—to help students focus on the material, to reflect on what they had learned in previous instruction, and to collect their thoughts before they launched into the lab activities. As students worked together and wrote their responses, Mr. Krebs observed them to informally assess their thinking before formally beginning the day's session.

As the students interacted in pairs and recorded their answers, their attention was drawn to the assortment of small weights and springs displayed on the demonstration table at the front of the classroom. The materials were on loan from the physics class, and their display piqued the students' curiosity and provoked speculation relative to the day's lesson. When Mr. Krebs noted that most students were ready, he began class with a quick oral review and discussion of their written responses.

"I wrote that the purpose of this lab is that we're gonna find the 'stretchiness' of the spring," Laurie offered.

"I'm not sure that *stretchiness* is a word, but it works," Mr. Krebs laughed.

"Weight in grams," Randy announced before the next question was posed.

"Length in centimeters," Kristin followed.

"And how long ..." Mr. Krebs began, interrupted by Mike's bold reply, "Four centimeters."

"Good. Now, let's speculate here for a moment before we get started." Mr. Krebs moved to one side and selected a small spring from the display. "What do you think will happen when we put more weight on the spring?"

This question was not on the activity sheet, and after a long pause, Tony offered, somewhat in jest, "It depends how much weight." In exaggerated gestures with his hands, he imitated an uncoiled spring and provoked the laughter of several students.

"Well, that's true," Mr. Krebs agreed, "but think about what Laurie said. Will the spring become less stretchy—because it has become longer—or more stretchy—because it's stretched out more?"

Students quickly offered support for both ideas, and Megan added, "I think it'll stay the same."

"Hm, that's three different answers," Mr. Krebs replied and turned to the chalkboard to draw two columns. "Let's experiment. Take out a sheet of paper, and we'll use a graph to help us find out."

Two students volunteered and proceeded with the demonstration. The rest of the class called out different weights, which Tony placed on the spring, as Diane generated table 16.1.

"Remember, try to look for a pattern," Mr. Krebs reminded them. His clue was intended to facilitate the students' speculation about the relationship between the variables.

"Then we should add the weights, right—and not skip around? It needs to be 50 then 100, then 150 ... and like that," Jeremy insisted.

"That's right. Let's just start over," Megan suggested.

"No, no—that would be a waste of time," JP said. "Just fix it from 170 on—you know, add 20 at a time." Most of the class looked on somewhat confused and unsure about JP's logic. On impulse, he approached the chart and illustrated his thinking. "See, it's easy!" he exclaimed.

Table 16.1

| Weight | Length |
|---|---|
| 0 | 4 |
| 50 | 6.4 |
| 150 | 11.5 |
| 170 | 12.5 |

After reflecting on JP's solution, Diane then posed, "So, 100 might be OK, too, wouldn't it?"

At that point, a debate ensued concerning the "more right" answer and the "one right way" argument. Mindful of the importance of spontaneous discourse and reluctant to interfere, Mr. Krebs nevertheless deferred to the constraints of time and interrupted the students. "These are all good answers," he said. "But remember, I don't expect everyone to turn in the same graph. We have only twenty-five minutes of class left, so let's move into our groups."

Most of the AP Math students in this class lacked experience with dialogical instructional methods, hands-on learning activities, cooperative problem solving, and self-regulated learning expectations. The bulk of their previous school experience had been dominated by traditional remedial instruction with an emphasis on basic skills, drill and practice, and monological instruction.

For many, long-established patterns of learned helplessness were apparent and interfered with their development of mathematical power. As a consequence, students in AP Math, particularly those identified as special education students, required additional effort on the part of the teacher (1) *to defuse the learner defensiveness* that was provoked when students realized that mathematical problems often have more than one right answer, and (2) at the other extreme, *to ignite learner curiosity*, which was essential to combat learned helplessness.

Following the demonstration, the students worked in pairs or triads with their own lab materials, adding weights and recording the "stretchiness" of the spring on a graph. The teacher moved between the groups, posing questions such as: "Why did you need to measure the length of the spring before it has any weight on it?" "Show me where you started to measure the length of the spring." "What do you suppose is the heaviest weight you could add?" "Do you recognize any patterns yet?" These questions were intended to extend the students' capacity for reasoning and to encourage them to communicate their thinking aloud.

The experiment was intended to provide experience with the process of gathering data, using graphs to find a pattern, and reaching a numerical conclusion in support of their intuitions about the "stretchiness" of a spring. As Mr. Krebs stepped back to observe and reflect on what they were doing, he hoped that his students would find that for every spring the data points result in a straight-line graph and that the rate of change for each spring is a certain number, such as 0.6 cm/gram of weight attached. By comparing data with the other groups, his students, he hoped, would see that the less stretchy springs have smaller numbers for the rate. "We'll see what surfaces in the discussion and decide where to go from there," he thought.

The idea for the preceding lesson was drawn from materials Mr. Krebs obtained from the Center for Occupational Research and Development (Pedrotti 1988), although a number of modifications have been necessary to accommodate the recognized needs of the students. In particular, the pace of instruction was revised, and, overall, fewer assumptions were made about the general knowledge and skills possessed by the students. The lesson that follows, in which students rely on both inductive and deductive reasoning skills to construct three-dimensional geometric models, was drawn from the same materials.

## Building Three-Dimensional Models

The two activities described below, "Building a Cylinder" and "Building a Pyramid," were part of a unit on measurement and applications in which students used real occupational tools (calipers, T-squares, micrometers, levels, and scales) and applied beginning-level drafting skills to construct cardboard three-dimensional geometric models.

Using calipers, students measured various sizes of pipe, the interior and exterior diameters of an assortment of tin cans, and the diameter of pencils, nuts, and bolts. Prior to the lesson, none of the students had used calipers, nor could they speculate about the possible purpose of the tool. Coaxed by their teacher to attempt a guess, they considered likely and unlikely uses and finally agreed that calipers would most likely be used by carpenters, plumbers, and mechanics.

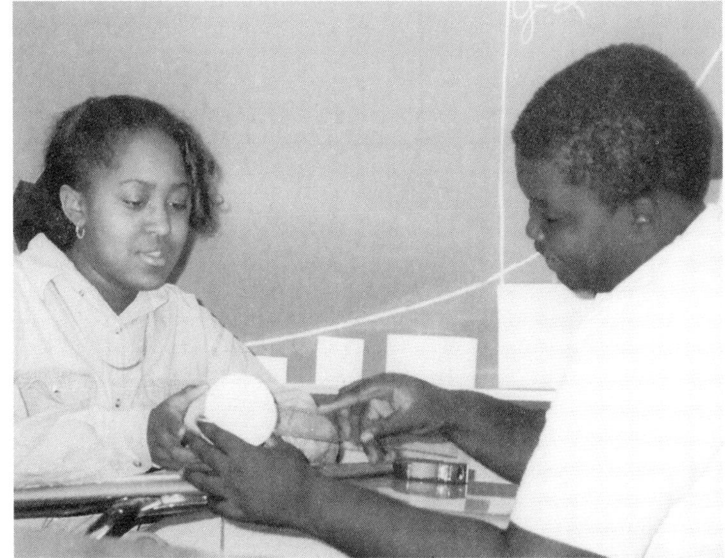

Following an introduction to the tools and a brief discussion of the difference between inside and outside diameter, students worked in small groups to construct a cardboard cylinder. The problem was posed to the students as an example of "working to specs," wherein the specifications were presented and the students were expected to build from this information. This was contrasted to their previous work in this unit, which consisted of measuring items that already existed.

Throughout the activity students were reminded that both *their* approach to creating the model and their production of a "correct to specs" model were important. On completion of the lab, the cylinders were evaluated by the students using categories within the range of "right" to "slightly off right but close." The students collectively determined a holistic score in which points were assigned for following directions, accuracy, and overall product appearance.

Although several students were "close" with their measurements, others failed to carefully follow the directions, which stressed the need for accuracy. In an effort to stress the relationship between accurate measurement and a quality product, Mr. Krebs began the next activity with a demonstration.

The directions for "Building a Pyramid" (see fig. 16.1) were less text-based and more visually rendered than those in the prior activity, which for many students enhanced the challenge of the project. Also, each student was issued an odd-shaped scrap of posterboard to insure that students had to form their own right angles for the base of the square pyramid.

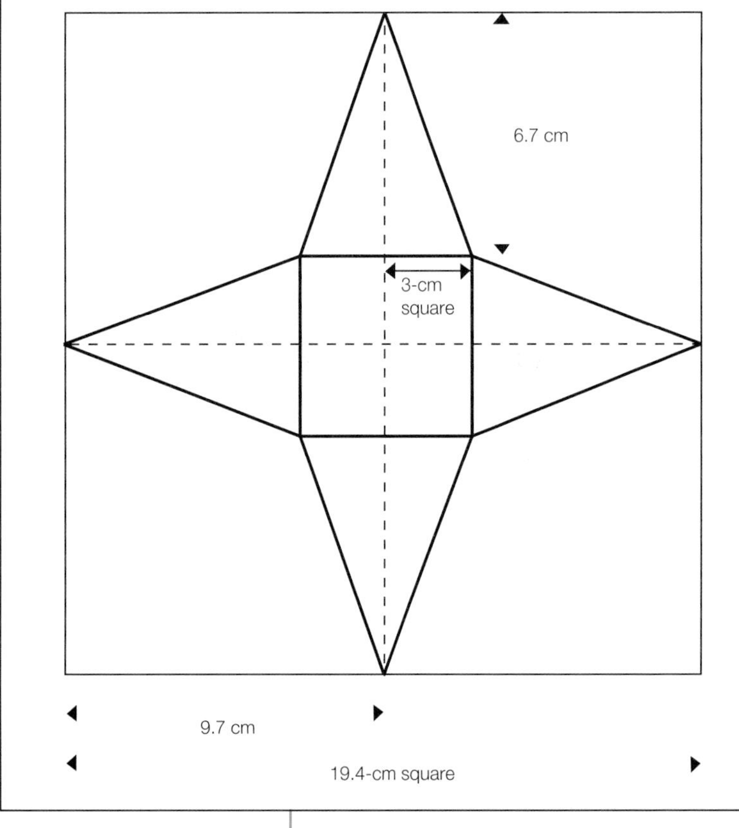


Fig. 16.1

Mr. Krebs stressed the accurate and careful use of a T-square to insure the construction of a "square" base. He likened this initial step to building a house and the need for a solid, level foundation as the necessary starting point. This prompted several students to relate their experiences with the construction of houses, barns, or stables and episodes in which they, too, had been foiled by an inadequate foundation.

After the demonstration, most students began drawing lines, positioning their T-squares, carefully measuring right angles on their scraps of posterboard, and, in general, following the steps outlined by their teacher. Kristin, however, envisioned an alternative approach to building the pyramid. Once she was confident that her solution would work, she challenged a classmate who had been carefully following the directions modeled by the teacher.

"What are you doing that for?" Kristin asked, as Danny positioned his T-square.

"You have to start with a square first," Danny answered with confidence, leaning close into the T-square as he completed the square.

"No, not necessarily. Watch." Roughly in the center of his scrap of posterboard, Kristin drew two perpendicular lines each measuring 19.4 cm in length (taken from the directions on his activity sheet). Next she measured 3 cm from the midpoint on all four sides. These points then became lines that formed Kristin's interior square.

"That's the only square you really have to worry about," Kristin maintained. Raising one eyebrow in response, Danny continued to draw the base angle needed to complete his pattern. He kept a close eye on Kristin as he constructed his model, but throughout his observation, he refrained from comment. Once both students completed the activity, Danny realized that the pyramids were in fact identical, and, somewhat sheepishly, he accepted this proof with a grin and a nod.

In a follow-up discussion, Mr. Krebs asked if there were different correct ways to "Build a Pyramid" or whether everyone needed to follow the exact same steps. Danny considered the question for a moment as if he were still perplexed by Kristin's outcome, and unwilling to attempt to "explain it with words," he said, "I get it—I get it, she just started on a different foundation—that's all."

Mr. Kreb's question was intended to prod students to be more flexible in considering and analyzing different solution approaches. Sensing Danny's reluctance to discuss further now, he decided to wait until tomorrow to revisit the model-building results and to analyze *why* Kristin's solution also "worked." Some days the students were more willing to be reflective and communicate their understanding, but Mr. Krebs knew this could never be assumed or forced with the students. What was apparent nonetheless was that they understood the important concepts presented in the activity.

Providing students with concrete learning activities in cooperative group settings as a basis for discussing, comparing, justifying, reformulating, and refining their thinking about, and personal explanations of, mathematical ideas is a major objective of the AP Math course. This type of activity, grounded in significant mathematics but modified by the teacher to accommodate individual interests and backgrounds, is seen as foundational, empowering *all* students, including those with special learning needs, with mathematical reasoning.

The irony is that the lines that have tended in the past to separate individual learners by strengths and weaknesses can easily be diminished in activities such as those described in this chapter. Through the use of multiple interpretations, approaches, and solutions to classwork, the students themselves present a "range" of responses that furnish a basis for future planning and modification of learning activities.

For example, both Danny and Kristin met the objective of the unit and both experienced success in the process, although some might argue that Kristin was successful at a higher cognitive level of reasoning. However, given the context, it could also be argued that Danny succeeded twice over in that he essentially learned two approaches, his and Kristin's solution of the problem.

Recall that even though Danny was initially perplexed by Kristin's theory, he grappled with his doubt until he *saw* that the two physical models—his and Kristin's— were the same. The physical proof enabled him to then "see" the meaning behind Kristin's argument. Had all the students been expected to conform to one interpretation or to follow one single approach, neither student would have profited to the same degree of success described. When students engage in concrete, cooperative learning experiences, they realize firsthand the idiosyncratic nature of learning, both their own and that of their peers.

## NURTURING MATHEMATICAL POWER

The goal of AP Math has been to offer a dynamic curriculum and a safe environment for experimentation, failure, and success—essential to the goal of nurturing mathematical power. An overriding focus throughout the course has been on providing learning opportunities like those described in this chapter that enable all students to *make meaning* rather than to *accept* meanings made by others.

NCTM's *Core Curriculum* (Meiring et al. 1992) has explicitly addressed the inclusion of the educationally "at risk" population in its framework for change, although it reiterates that the transition to this change may not be "an uninterrupted success story" (p. 113). Activities characterized as constructivist (e.g., Goldin 1990) can pose a range of problems not only for "at risk" or special education students but for all students with limited backgrounds whose mathematics experiences have consisted of little more than paper-and-pencil exercises.

Consistent with a constructivist philosophy, the activities described in this chapter were designed to promote learner independence and self-monitoring and *de-emphasize* learner dependence and teacher monitoring. Although these are worthy goals, none are being easily or quickly attained, given that the students enrolled in AP Math have grown accustomed to a more dependent rather than independent learning and teaching style. If students like Kristin and Danny had had the opportunity to experience success in mathematics earlier in their educational careers, if they had benefited earlier from *constructivist*-oriented programs such as those described by Goldin (1990), their story now would not be marked by a slow and difficult undoing of previously learned beliefs and behaviors.

Many students enrolled in AP Math, Kristin and Danny included, began the school year with little or no interest in mathematics and, more important, generally lacked the confidence to view themselves as capable learners. The actual degree of change this curriculum and approach to teaching and learning mathematics has produced would be difficult to measure in the absence of documenting these two important learner characteristics and attitudes.

Anecdotes from the two lessons illustrate how progress is being made toward actively involving the students in AP Math as a community of learners, beginning to interact and "think mathematically." The determination of Kristin and the open-minded speculation of Danny affirm the promise of these first steps toward nurturing mathematical power for *all* students at Oskaloosa High School.

## REFERENCES

Goldin, Gerald A. "Epistemology, Constructivism, and Discovery Learning in Mathematics." In *Constructivist Vews on the Teaching and Learning of Mathematics*, edited by Robert B. Davis, Carolyn A. Maher, and Nel Noddings, pp. 31–47. Reston, Va.: National Council of Teachers of Mathematics, 1990.

Meiring, Steven P., Rheta N. Rubenstein, James E. Shultz, Jan de Lange, and Donald L. Chambers. *A Core Curriculum: Making Mathematics Count for Everyone.* Reston, Va.: National Council of Teachers of Mathematics, 1992.

National Council of Teachers of Mathematics. *Curriculum and Evaluation Standards for School Mathematics.* Reston, Va.: The Council, 1989.

Pedrotti, L. *Applied Mathematics Materials.* Waco, Tex.: Center for Occupational Research and Development, 1988.

**Jack Krebs** teaches mathematics at Oskaloosa High School, Oskaloosa, Kansas, and is the state staff development specialist for applied mathematics. He also serves as the district school improvement director.

**Linda Ware** is the director of the Northeast Kansas Math/Science Consortium. An experienced teacher, she is a doctoral candidate in the Department of Special Education at the University of Kansas. Her interests include teacher-initiated change, technology integration, curriculum revision, and school reorganization.

17

# Legs + Heads + 1 000 000 Floor Tiles = "Thinking Mathematics"

## For Students with Learning Disabilities

*Barbara W. Grover*
*Susan Hojnacki*
*Debra Paulson*
*Carol Matern*

PERHAPS the most radical idea underlying the philosophy of the *Curriculum and Evaluation Standards for School Mathematics* (National Council of Teachers of Mathematics 1989) is that the curriculum being promoted is "intended to be available to *all* [emphasis in original] students.... [and] that ... all students [be given] the opportunity to appreciate the full power and beauty of mathematics and acquire the mathematical knowledge and intellectual tools necessary for its use in their lives" (p. 69). Although skill in mathematics is necessary for careers in nearly every aspect of our society (e.g., business, finance, health, the trades), very few students continue to study mathematics once it is no longer required.

As traditionally taught, mathematics has been acting as a filter, keeping students out of many careers. In *Everybody Counts* (National Research Council 1989), the goal of the reform movement was aptly characterized as follows: "Mathematics must become a pump rather than a filter in the pipeline of American education" (p. 7). All too prevalent has been the notion that mathematics is a subject to be understood only by the elite, those who were born with a particular ability.

New understandings about how students learn and what educational activities and environments facilitate learning have begun to alter this view. Pursuing the philosophy of the *Curriculum and Evaluation Standards for School Mathematics* and making this goal of educational reform a reality is no simple task. Instruction in mathematics must change, and change is hard to effect. Altering instructional approaches is at the heart of the proposed change, and changing the teacher's educational philosophy is central to altering the instructional approach.

This chapter describes how one reform-spirited program, Thinking Mathematics, guided two teachers in altering their instructional approaches, thus opening windows of opportunity for their students with learning disabilities. We begin and end with descriptions of lessons developed by these teachers that are examples of the high-level reasoning, thinking, and problem-solving activities that have occurred in Thinking Mathematics classrooms with special-needs students. In between, we offer a brief overview of the Thinking Mathematics project, focusing on three of its underlying principles, and on the relationship of these three principles to the *Curriculum and Evaluation Standards for School Mathematics* and the described lessons.

## Dogs Have Heads, Too!

Carol Matern and three of her colleagues team-taught a group of sixty-four students (thirty-eight boys and twenty-six girls) aged nine to twelve. Thirteen of the students were diagnosed severely learning disabled and had previously been assigned to a self-contained special education classroom. Another ten students were case conferenced into the LD Resource Room, and nine other students qualified for Chapter 1 service in reading. Approximately one-third of the students were minorities.

Christina was working in her cooperative group of four students (two boys, two LD girls) on the following problem:

> I took a picture and in the picture there were some children and some puppies. There were 7 heads and 22 legs. How many children were there?

The students had been working in mixed-ability groups only for several weeks, and this particular problem was chosen for the following reasons:

- It was good for group work and discussion but could also be used to assess individual student thinking.
- It could be worked out quite nicely using manipulatives.
- The problem was easily recordable using "math talk" and pictorial representation.
- It could help students from traditional mathematics classrooms ease into ours; it wasn't expected to be particularly demanding.

In the course of facilitating three cooperative groups, I stopped at Christina's table and asked if her group had solved the problem. They all nodded and said that Justin had solved it. I noticed that they all had Justin's same answer of 3 R 1 written in their math recording books, but no other information was recorded. So I asked Justin how he arrived at such an answer. He said he figured it in his head, but he was unable to tell any more. I asked if they all agreed, and everyone but Christina said that they agreed with Justin's response.

I asked how many legs each child had, and they said, "Two." I asked how many legs three children have, and they counted by twos and said, "Six." Then they issued a collective groan when they realized that their original answer was not correct, except for Christina, who was very agitated. The group seemed not to be interested in Christina's idea, since she is in the learning disabilities program. I suggested that we hear what she had to say. She said, "First we put these Popsicle sticks out like this" (see fig. 17.1).

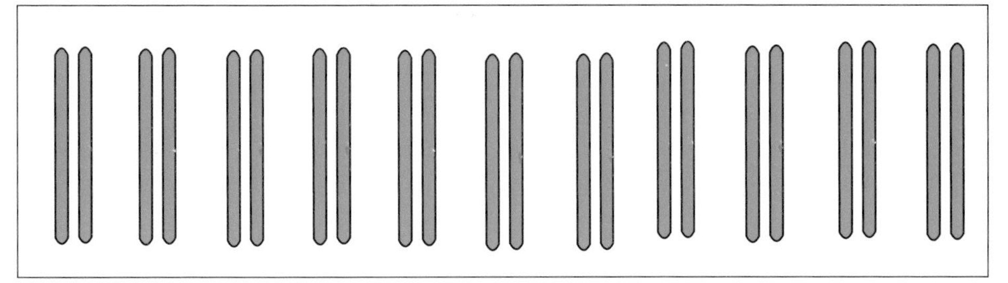

Fig. 17.1. Christina's arrangement of twenty-two Popsicle sticks

I asked, "Why?" Christina explained that they represented the twenty-two legs, two for each person. Then she said, "You take away seven pairs for the seven heads. That leaves four pairs of legs, or two dogs, and the answer is seven children!"

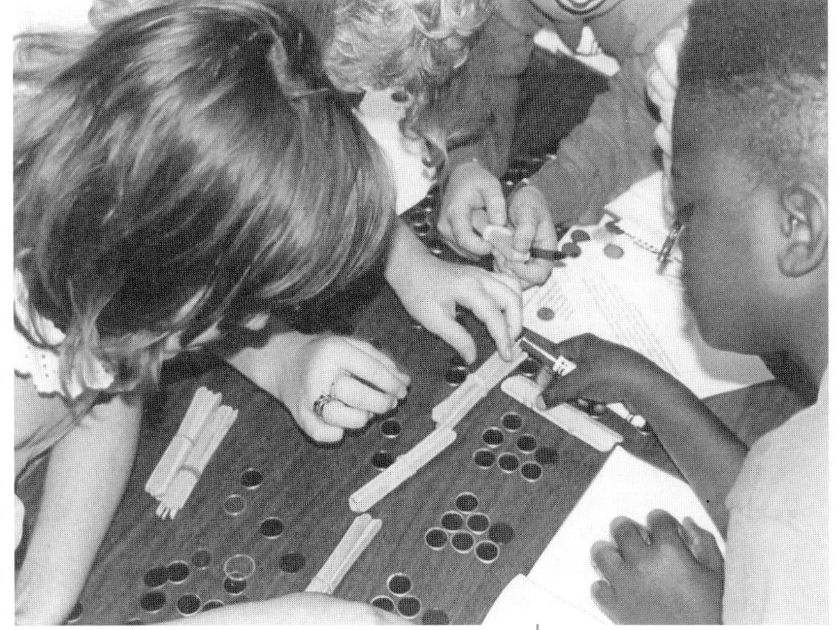

An entirely new attitude could be felt within the group. They had all listened intently as Christina and I discussed her thinking. They followed along and began to see and understand, and then they realized that Christina, although different, was smart and capable of being a contributing member of their group.

They all began recording in their math recording books pictorial representations of sticks in sets of two. So I waited, and I realized that they were all content with Christina's reasoning and logic. Then I asked, "Do dogs have heads?" Christina said, "Yes, but they didn't say anything about dog heads."

Evan, by this time, seemed a bit confused and said, "It is much too obvious. It gives away the answer because it says seven." Evan continued, "It doesn't say anything about dog heads. You kinda assume it means children's heads."

Mindy then suggested that there were fewer kids and more dogs, "like six children and three dogs."

I said, "Remember, how many heads are in the picture?" Then she said, "Maybe two kids and the rest were dogs."

While this conversation was taking place, Evan and Mindy had got out red and purple bingo chips. Evan duplicated Christina's Popsicle-stick representation, and Mindy was playing with her manipulatives. I kept wanting to suggest that they start with seven and place the legs instead of starting with the twenty-two legs and placing heads, but I stopped myself. I knew they could solve this problem, and I really wanted them to think.

Evan had placed twenty-two purple bingo chips in groups of two, and he had placed seven red chips over seven groups of two and had four groups of two "legs" left over. So I said, "What are these purple bingo chips that are left?" Evan responded, "They are the dogs." I asked, "Have you ever seen dogs with legs and no heads?" Evan was obviously thinking and said, "At first I put heads on children, not dogs. Now I know I'll have to change it."

So he and Christina began moving around their red bingo chips (heads) so that they both arrived at an answer of three children and four dogs.

I asked which answer they liked best—seven children or three children and four dogs—and they agreed that three children and four dogs was a better answer because, they said, "Dogs have to have heads, too."

At this point Mindy (also LD) began using her bingo chips in an orderly fashion, referring to Christina's model. As she concentrated, she said, "Oh, I get it." Justin still seemed confused, so I suggested that Christina explain her representation to him, which she readily agreed to do. Once the students had explained their representations, they drew pictures in their math recording books of their manipulative layout.

By completing this problem, the group learned that working together was an effective way to solve problems. Students learned to listen to each other, to think about their thinking, and to extend their thinking to reach logical conclusions. Evan was new to Thinking Mathematics, but he was able to follow Christina's lead and contribute to the process and engage in real mathematics learning. Thinking Mathematics creates an environment for discussing and problem solving with room for trial and error. The emphasis is not on the answer only, as Evan and Justin, who came from traditional mathematics classrooms, had expected.

This "heads and legs" problem paved the way for mathematical exploration for this cooperative group of children. Through the use of manipulatives and "math talk," this cooperative group went from (1) a quick incorrect answer with no thinking to (2) an answer that at first seemed logical and could be explained to (3) the correct answer. All the students were engaged and contributed to the solution. Mindy and Christina performed very well and became partners in their learning group. The Thinking Mathematics principles (discussed in detail below) applied to this mathematics lesson enabled all students to contribute, participate, and learn.

### A New Role for Teachers

Before an educational system can provide mathematics programs that enable and empower students in the way this lesson does, it must furnish programs that enable and empower the teachers. Cohen and Ball (1990) framed the issue of concern as follows: "How can teachers teach a mathematics that they never learned, in ways that they never experienced? That is the dilemma that such reforms pose" (p. 353). The *Professional Standards for Teaching Mathematics* (National Council of Teachers of Mathematics 1991) recognizes the importance of this issue; the Teaching Standards were developed on the basis of two assumptions (p. 2):

- Teachers are key figures in changing the ways in which mathematics is taught and learned in schools.
- Such changes require that teachers have long-term support and adequate resources.

In order to enable and empower students, teachers need to take on a new role in the classroom that is not particularly familiar to them. They must learn how to help students learn how to learn, rather than try to be the sole provider of knowledge to students. Such a new role requires both new instructional techniques and a deeper knowledge of the subject matter as well as assistance and support for teachers who attempt to take on this new role (Darling-Hammond 1990).

## "THINKING MATHEMATICS"

One program that is helping to empower teachers so they can empower students is a National Science Foundation–funded project called Thinking Mathematics. This project, a collaborative effort of the University of Pittsburgh's Learning Research and Development Center (LRDC) and the American Federation of Teachers (AFT), has developed and is currently disseminating a research-based instructional approach that integrates the four overarching thrusts of the *Curriculum and Evaluation Standards for School Mathematics* (NCTM 1989)—problem solving, communication, reasoning, and mathematical connections—and the four themes of the *Professional Standards for Teaching Mathematics* (NCTM 1991)—mathematical tasks, discourse, learning environment, and analysis.

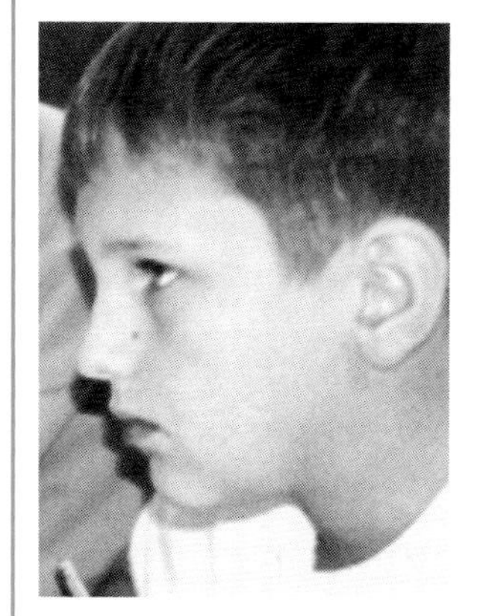

Over a three-year period, a team of AFT teachers and LRDC researchers jointly arrived at the following ten research-based and clinically tested principles that should guide mathematics instruction:

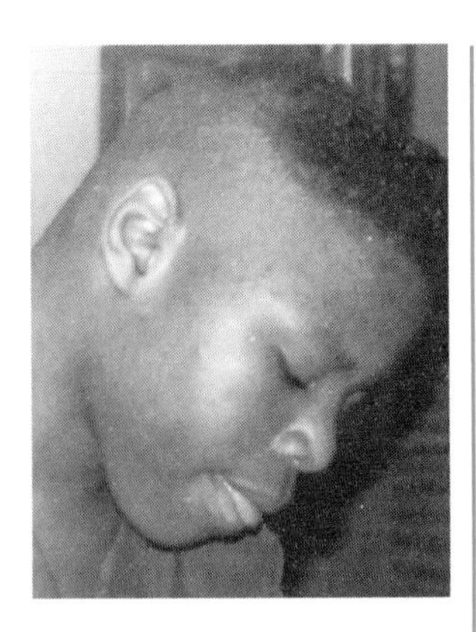

1. Build from students' intuitive knowledge.
2. Establish a strong number sense.
3. Base instruction on situational story problems.
4. Use manipulatives to represent problem situations.
5. Require students to describe and justify their mathematical thinking.
6. Accept multiple correct solutions.
7. Balance conceptual and procedural learning.
8. Use a variety of teaching strategies.
9. Use ongoing assessment to guide instruction.
10. Adjust the curricular time line.

Three volumes of materials—*Analysis of Arithmetic for Mathematics Teaching* (Leinhardt, Putnam, and Hattrup 1992); *Thinking Mathematics*, vol. 1, *Foundations* (Bodenhausen et al. 1993); and *Thinking Mathematics*, vol. 2, *Extensions* (Gill and Grover 1993)—were developed to support the Thinking Mathematics approach.

The first is a book of commissioned chapters that served as core research articles in the developmental stage. Two volumes were designed for in-service training and were pilot-tested during 1990–92 by the AFT teachers at eleven sites across the United States.

Two of this chapter's authors, Carol Matern and Debra Paulson, are special education teachers who are using the Thinking Mathematics approach. On the basis of their classroom experience with students having learning disabilities, they have concluded that Thinking Mathematics could be the restructuring vehicle that will allow children diagnosed learning disabled to learn mathematics in a mainstreamed classroom environment as well as in a self-contained situation. Since these teachers were empowered by their training and the support system of the Thinking Mathematics project, they in turn were able to establish an atmosphere in their classes that empowered their students. (For details about this collaborative project, see Bickel and Hattrup [1991/1992]; Grover, Gill, and Kaduce [1991]; and Leinhardt and Grover [1990].)

## Three Important Principles

Of these ten principles, we focus on only three in this chapter, using the two vignettes as illustrations of the three. The combination of using manipulatives to represent problem situations (principle 4), requiring students to describe and justify their mathematical thinking (principle 5), and accepting multiple correct solutions (principle 6) was a very potent instructional approach for the learning disabled students in Matern's and Paulson's classrooms. We highlight here the recommendations of the National Council of Teachers of Mathematics in its *Curriculum and Evaluation Standards for School Mathematics* (1989) concerning these three principles and also the research rationale that helps explain their effectiveness.

### *Use Manipulatives*

That manipulatives "engage students intellectually and physically" (NCTM 1989, p. 87) is true even for middle-grades students, who are "especially

responsive to hands-on activities in tactile, auditory, and visual instructional modes" (NCTM 1989, p. 87). Better student understanding and performance results when abstract concepts are tied to such models as manipulatives and to pictorial representations (Moyer et al. 1984; Riley and Greeno 1988).

Improved performance results, in part, because the use of carefully chosen manipulatives helps students develop a concrete language for talking about abstract mathematical concepts (Bright 1986) and thus helps students to gain confidence in their own abilities to make sense of mathematical situations (Holden 1987). But it is very important that the manipulatives be connected to the symbolic representations at each step in problem solving; students cannot often make this connection on their own (Fuson 1992; Resnick and Omanson 1987). This difficulty in making connections is especially true for students with learning disabilities (Thornton and Wilmot 1986). Note that in Matern's lesson students used the manipulatives to facilitate their reasoning and were recording pictorial representations in their math recording books.

### *Require Justification*

Initially students find it difficult to describe and to justify their mathematical thinking; they aren't used to "thinking" in mathematics class! This particular kind of student talk is, however, essential for fostering communication both with and about mathematics in the classroom (NCTM 1989, p. 66). The teacher must ensure the development of an atmosphere of mutual respect in which students are enabled to explore ideas together and listen to one another.

As a consequence of learning mathematics in such a way, students will come to trust their own judgments more and to rely less on outside authorities (NCTM 1989, p. 69). Justifying thinking is a tremendously empowering activity for many students, but its benefits are seen quite clearly with at-risk

students (Resnick, Bill, and Lesgold 1992). Again, we can see from the example how Matern acted as a guide in the discussion ("I kept wanting to suggest ... but I stopped myself."). Students articulated their own thinking and convinced each other of the accuracy of, or the flaws in, their solution processes.

### *Encourage Multiple Solutions*

Accepting multiple correct solutions goes hand in hand with requiring students to justify their mathematical thinking and with fostering mathematical communication and mutual respect (NCTM 1989, p. 69). The old notion that only one problem-solving procedure (i.e., the teacher's) can be correct is belied by the finding that logical and mathematically correct procedures can be designed and understood by students (Lampert 1986; Resnick 1986). The teacher's acceptance of students' correct solutions (even if they are inefficient ones at first) leads each student to experience success. As students become aware of other solutions, they can be led toward more efficient procedures while retaining their understanding (Lampert 1986, 1992). Students who do not become comfortable with the most efficient solutions can still operate successfully if somewhat inefficiently.

Fostering this kind of instructional climate is consistent with the recommendations in the *Professional Standards for Teaching Mathematics*. In particular, it facilitates these five major shifts in classroom environments recommended by the *Professional Teaching Standards* (p. 3) for the empowerment of students:

1. Toward classrooms as mathematical communities
2. Toward logic and mathematical evidence as verification
3. Toward mathematical reasoning
4. Toward conjecturing, inventing, and problem solving
5. Toward connecting mathematics, its ideas, and its applications

These shifts were illustrated in the previous vignette and are also illustrated in the second vignette that involves older students in a nonmainstreamed situation.

## A Floor Tile—One in a Million!

Debra Paulson's mathematics class consisted of eight eighth-grade learning disabled students, aged fourteen to sixteen (six males and two females) in a self-contained setting. Two were African American; all were from urban, low-income neighborhoods. All had been retained at least once, and none had experienced success with traditional instruction. Most did not know the basic facts of multiplication and could not accurately subtract multidigit numbers in problems requiring regrouping.

Because secondary school students with learning disabilities have difficulty comparing, computing, or even understanding decimals, I built on their knowledge of money and used the base-ten blocks to connect decimal symbols with quantities and especially to demonstrate how these quantities decrease in size. A thousands cube became our "one." Other labels were assigned: the flat became a tenth, a long became a hundredth, and the ones cube was labeled a thousandth.

I often asked the students how we could show a smaller unit. They conjectured that the thousandths cube could be cut into ten equal pieces, but I never modeled this procedure. One day, I asked what was the smallest decimal place. Unanimously they answered that it was the thousandths. My dilemma became how to show and label smaller decimal places. I shared this problem with my eight students and so began our impromptu lesson.

I posed questions that encouraged brainstorming: "How can we show the next smallest decimal place using this cube?" "How can we show the *next* smallest?" "What happens to the size of the cube?" After a consultation with the industrial arts teacher to check on the feasibility of the idea, the group came to the consensus that it was impractical to divide the thousandth cube any further. Several students remarked that we needed to begin with a bigger ones unit. Suggestions ranged from using the entire school, the football field, or a five-foot-by-five-foot model that we could build to using our classroom. I posed another question: "What in our classroom is one-millionth of the room?" A lively debate ensued, resulting in a decision to investigate the following question: Would one million floor tiles fill our classroom?

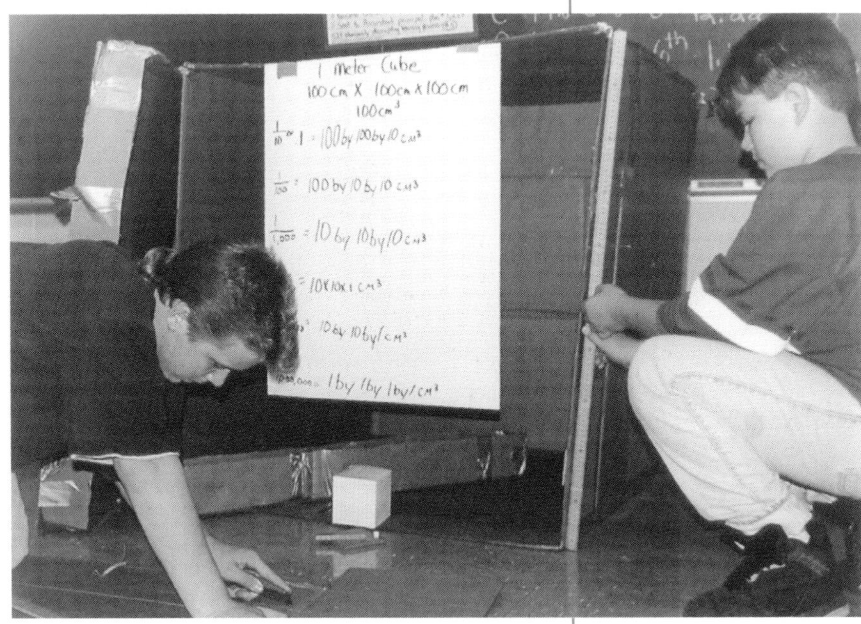


The students decided on the tasks—counting the number of tiles on the floor and counting the number of tiles that would go from the floor to the ceiling, formed two teams, and divided the labor. Team 1 counted the number of tiles on the floor. All helped with counting the number of tiles going from wall to wall one way (to measure its length) and then the other (to measure the width). Using the calculator, they multiplied the two amounts, getting 1260 tiles. I thought it interesting and practical that they measured the floor using actual tiles and not in feet or inches.

Team 2 had the task of finding the number of tiles that would stack from floor to ceiling. Their task was more difficult because it was so abstract. They came to the conclusion that they were missing some important information. How thick was the floor tile? How tall was the ceiling?

Using the measuring tape, they determined the ceiling to be ten feet from the floor. Getting a precise measure of the floor tile's thickness was not so easy. I contacted the Color Tile Supermarket in Anderson, Indiana, and arranged to borrow one-eighth-inch tiles. I decided that the more tiles the students had to work with, the more concrete the concept would become. Unfortunately, only twenty boxes, with twenty-five tiles in each box, would fit in my car. Armed with rulers, the team first determined the number of tiles that equals one foot. They were amazed to find out that ninety-six stacked tiles equal twelve inches, or one foot. I asked them, "How many tiles equal an inch? Can you determine the number without measuring or counting? Using calculators and drawing pictures, they decided to divide 96 by 12, since there are twelve inches in one foot. The result was eight tiles to an inch. One student then connected this information with the eighth-inch thickness and exclaimed, "Of course, eight eighths would equal one!"

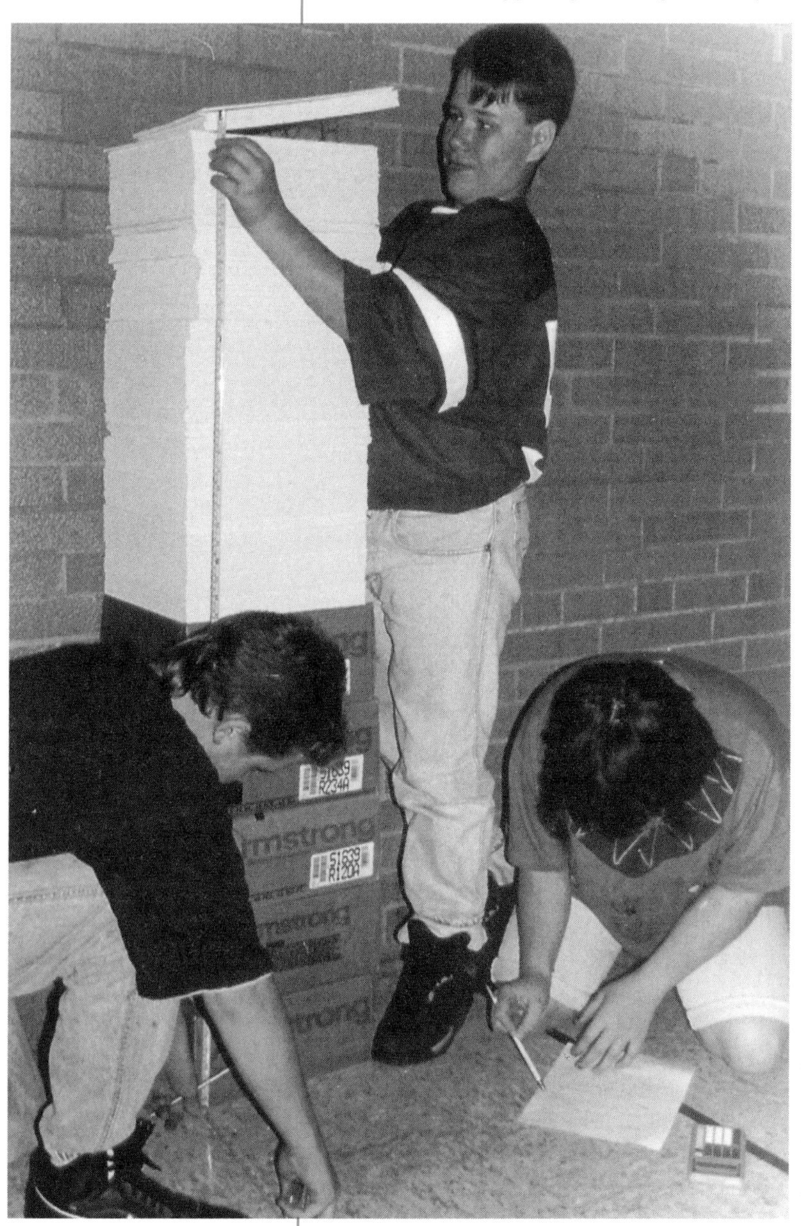

Looking at the stack of 96 tiles, the team discussed how they would determine how many tiles would stack to the ceiling. One student wanted to stack all the tiles—500 of them—and see how high they would go. The rest of the group argued that it would be much faster, and less work, to find out by using their calculators. If the ceiling is ten feet high, then all they had to do was multiply 96 times 10. So 960 tiles would reach from the floor to the ceiling. The magnitude of this project began to crystallize for all the students as they observed that the twenty boxes of stacked tiles would reach only a little over halfway to the ceiling. I asked them how many boxes would be needed to reach ten feet. This question created much discussion. The students estimated that fewer than forty boxes would be required. To get the actual number, they turned to their calculators. Several students divided 960 by 25 and calculated the answer to be 38.4. Two students decided to find out the number of boxes by multiplying 25 until they got to 960: 39 times 25 equals 975, and 38 times 25 equals 950. At that point I asked, "What are you trying to find out? How many boxes would be needed? What if we can get only full boxes?" The students who had multiplied were the first to decide that thirty-nine boxes would be needed, with some tiles left over. They convinced those who had divided that their answer was correct.

I suggested that the group write the total 960 on several pieces of paper and place one piece of paper on each tile in a row on the floor. All eight students could then get a good idea of how it would look if we stacked the tiles to the ceiling on each floor tile. I believe that it helped to see a quantity of tiles stacked, even though it wasn't ten feet of tiles. Using their calculators, the students multiplied the 1 260 tiles on the floor by 960. The students had another surprise when they concluded that it would actually take 1 209 600 tiles to fill our room. I referred the class to the original question, which I had written on the chalkboard: Would one million floor tiles fill our classroom? The answer was no, more than one million tiles would be needed. The class was amazed. No one, including me, had thought that our room would hold one million tiles.

I asked the students if we could consider one floor tile to be one millionth of our classroom, and if so, why. The students all agreed that 1 209 600 could be rounded to 1 000 000. The justifications varied from "I didn't want to do all this work for nothing" or "The number 209 600 is big, but when compared to 1 000 000 it is not even one-fourth" to "There is no way all those tiles would fit in our room. They are too heavy. They would make our room fall into the first floor. So we might as well say 'one million.'" With guided instruction and the use of graph paper and calculators, we divided the room into approximate hundred thousandths, thousandths, hundredths, and tenths. Two students later constructed a one-meter cube out of cardboard, along with models of one tenth (1 m × 1 m × 10 cm), one hundredth (1 m × 10 cm × 10 cm), one thousandth (a base-ten 10-cm cube), one ten thousandth (a base-ten flat, 10 cm × 10 cm × 1 cm), one hundred thousandth (a base-ten long, 10 cm × 1 cm × 1 cm), and one millionth (the 1-cm cube).

The mathematics class then resumed the task of computing with decimals. We continued to use manipulatives to link concrete situations and symbolic representations of them. The students made estimates of computations using all four arithmetic operations, carried out some paper-and-pencil calculations, and became very proficient at reading numbers with decimal places to a millionth.

This unplanned lesson evolved into a week-long project that was characterized by using problem-solving strategies and manipulatives and models and by frequent discourse in which students shared their thinking. The eight students kept on task throughout the project because they had ownership in creating and solving the problem and because they were physically involved. They worked cooperatively and listened to one another. Both these behaviors of remaining on task and working together are difficult for many middle school students with learning disabilities. The complexity of the floor-tiles problem and the strategies used to solve it reflected the students' use of reasoning and problem-solving skills far beyond anything observed by Paulson in her many years of teaching.

## "THINKING MATHEMATICS" PRINCIPLES IN ACTION

The previous two vignettes highlight the integration of three Thinking Mathematics principles (i.e., use manipulatives, require justification, and encourage multiple solutions) into instructional practice. We have already looked at the principles in action in the "legs and heads" vignette. In the "million floor tiles" vignette, breakthroughs in the students' conceptual understanding of decimals occurred through the ingenious use of floor tiles as manipulatives. Students assumed responsibility for coming up with possible solutions and testing them out.

The tasks illustrated were open ended and facilitated reasoning, discourse, and problem solving. In addition, valued mathematical concepts and skills were embedded in the tasks. Students were encouraged to use manipulatives, thus devising concrete representations that they could use to reason and connect the concrete with the symbolic. Through the use of cooperative groups, the encouragement of multiple approaches to solving the problems, and questioning techniques, the teachers supported a learning environment that fostered cooperation, respect, and self-reliance among students.

According to this model of teaching, teachers guide the students' thinking rather than act as arbiters of right and wrong. Students are challenged to be critical thinkers and evaluators. Each student experiences some degree of success. This atmosphere affords opportunities for the intellectual authority to shift from the teacher to the students, which is the essence of the current reform in mathematics instruction.

Changing students' paradigms was no easy matter. Students preferred to rely on algorithmic procedures that often were used incorrectly, with no understanding of the concepts involved. Matern's lesson took place early in the school year and exemplifies the students' resistance to change. It took a bit of prodding by the teacher to accustom the students to mathematical exploration. Much more student progress is demonstrated in Paulson's lesson, which took place in the spring, following months of instructional practice based on the Thinking Mathematics principles and on the recommendations contained in the NCTM's *Curriculum and Evaluation Standards for School Mathematics*.

## QUALITY LEARNING EXPERIENCES IN MATHEMATICS FOR ALL STUDENTS

The Thinking Mathematics program is not specifically tailored to students with learning disabilities. Its goal, like that of the NCTM's *Curriculum and Evaluation Standards for School Mathematics*, is to maximize the learning potential of all students in mathematics by integrating the best of mathematics education research with the professional expertise of classroom teachers. As we discovered, however, the same principles used successfully with students of normal ability can also unlock the mathematical potential of those who are learning disabled. In Paulson's self-contained classroom setting, students were empowered to ask their own questions, come up with their own solutions, and transfer their new knowledge to previously difficult concepts. In Matern's mainstreamed classroom setting, the regular-education students were able to learn from Christina as she described her alternative solution strategy.

It is also clear that Paulson and Matern were able to maintain a learning environment that enabled and empowered their students because they themselves had been empowered by their knowledge of the research and their willingness to become risk takers and problem solvers in their instructional practice. Nurturing mathematical thinking and creating windows of opportunity for teachers as well as students is essential to the success of reform.

### REFERENCES

Bickel, William E., and Rosemary Hattrup. "A Case Study of Institutional Collaboration to Enhance Knowledge Use: Restructuring Practitioner-Researcher Dialogue in Education." *Knowledge and Policy: The International Journal of Knowledge Transfer and Utilization* 4 (Winter 1991/1992): 56–78.

Bodenhausen, Judith, Nancy Denhart, Alice Gill, Margaret Kaduce, Marcy Miller, Barbara Grover, Lauren Resnick, Gaea Leinhardt, Victoria Bill, Marilyn Rauth, and Lovely Billups. *Thinking Mathematics*. Vol. 1, *Foundations*. Washington, D.C.: American Federation of Teachers, 1993.

Bright, George W. "One Point of View: Using Manipulatives." *Arithmetic Teacher* 33 (June 1986): 4–5.

Cohen, David K., and Deborah L. Ball. "Policy and Practice: An Overview." *Educational Evaluation and Policy Analysis* 12 (March 1990): 347–53.

Darling-Hammond, Linda. "Instructional Policy into Practice: The Power of the Bottom over the Top." *Educational Evaluation and Policy Analysis* 12 (Fall 1990): 233–41.

Fuson, Karen C. "Research on Learning and Teaching Addition and Subtraction of Whole Numbers." In *Analysis of Arithmetic for Mathematics*, edited by Gaea Leinhardt, Ralph Putnam, and Rosemary Hattrup, pp. 53–187. Hillsdale, N.J.: Lawrence Erlbaum Associates, 1992.

Gill, Alice, and Barbara W. Grover, eds. *Thinking Mathematics*. Vol. 2, *Extensions*. Washington, D.C.: American Federation of Teachers, 1993.

Grover, Barbara W., Alice Gill, and Margaret Kaduce. "Teacher and Researcher Collaborations: Implementing the NCTM Standards." Paper presented at the research presession of the annual meeting of the National Council of Teachers of Mathematics, New Orleans, April 1991.

Holden, Linda. "Even Middle Graders Can Learn with Manipulatives." *Learning 16* (October 1987): 53–55.

Lampert, Magdelene. "Knowing, Doing, and Teaching Multiplication." *Cognition and Instruction* 3 (1986): 305–42.

———. "Teaching and Learning Long Division for Understanding in School." In *Analysis of Arithmetic for Mathematics*, edited by Gaea Leinhardt, Ralph Putnam, and Rosemary Hattrup, pp. 221–82. Hillsdale, N.J.: Lawrence Erlbaum Associates, 1992.

Leinhardt, Gaea, and Barbara W. Grover. "Interpreting Research for Practice: A Case of Collaboration." Paper presented at the annual meeting of the American Educational Research Association, Boston, April 1990.

Leinhardt, Gaea, Ralph Putnam, and Rosemary Hattrup, eds. *Analysis of Arithmetic for Mathematics Teaching*. Hillsdale, N.J.: Lawrence Erlbaum Associates, 1992.

Moyer, John C., Larry Sowder, Judith Threadgill-Sowder, and Margaret B. Moyer. "Story Problem Formats: Drawn versus Verbal versus Telegraphic." *Journal for Research in Mathematics Education* 15 (November 1984): 342–51.

National Council of Teachers of Mathematics. *Curriculum and Evaluation Standards for School Mathematics*. Reston, Va.: The Council, 1989.

———. *Professional Standards for Teaching Mathematics*. Reston, Va.: The Council, 1991.

National Research Council. Mathematical Sciences Education Board. *Everybody Counts: A Report to the Nation on the Future of Mathematics Education*. Washington, D.C.: National Academy Press, 1989.

Resnick, Lauren B. "The Development of Mathematical Intuition." In *Perspectives on Intellectual Development: The Minnesota Symposia on Child Psychology*, vol. 19, edited by Marion Perlmutter, pp. 159–94. Hillsdale, N.J.: Lawrence Erlbaum Associates, 1986.

Resnick, Lauren B., Victoria Bill, and Sharon Lesgold. "Developing Thinking Abilities in Arithmetic Class." In *Neo-Piagetian Theories of Cognitive Development: Implications and Applications for Education*, edited by Andreas Demetriou, Michael Shayer, and Anastasia Efkides, pp. 210–30. London: Routledge, 1992.

Resnick, Lauren B., and Susan Omanson. "Learning to Understand Arithmetic." In *Advances in Instructional Psychology*, vol. 3, edited by Robert Glaser, pp. 41–95. Hillsdale, N.J.: Lawrence Erlbaum Associates, 1987.

Riley, Mary, and James Greeno. "Developmental Analysis of Understanding Language about Quantities and of Solving Problems." *Cognition and Instruction* 5 (1988): 49–101.

Thornton, Carol A., and Barbara Wilmot. "Special Learners." *Arithmetic Teacher* 33 (February 1986): 38–41.

---

*Note:* For more information about the Thinking Mathematics program, contact Alice Gill, Assistant Director, Educational Issues Department, American Federation of Teachers, 555 New Jersey Avenue, N.W., Washington, DC 20001.

**Barbara Grover** is a research associate at the Learning Research and Development Center at the University of Pittsburgh. She is interested in the professional development of veteran mathematics teachers and the ways in which teachers' knowledge and beliefs influence their instructional practice. She was the coordinator for the Thinking Mathematics project and is currently associated with QUASAR, a national project designed to study the implementation of innovative mathematics programs in middle schools.

**Susan Hojnacki** is a research specialist at the University of Pittsburgh with a background in cognitive psychology. She was the associate coordinator for the Thinking Mathematics project. Having been deeply impressed by the teachers and students in that project, she is now preparing to join them in the classroom by entering a teacher education program.

**Debra Paulson** teaches seventh- and eighth-grade students with learning disabilities in the Anderson Community Schools (ACS), Anderson, Indiana. She has been involved in the Thinking Mathematics (TM) project since it was first piloted in 1990. She conducts mathematics in-service programs to help ACS teachers implement the TM principles and the NCTM Standards.

**Carol Matern** is a teacher of students with learning disabilities in the Anderson Community Schools, Anderson, Indiana. She team-teaches forty-seven students with two colleagues at the intermediate level in a multiaged, fully inclusionary classroom. This classroom structure means that special education and regular-education students are enrolled in the same class for all subjects and are engaged in the same instructional activities.

is'll Kill
YOU!
FLAMMABLE
fire
extinguishe
...do you know where they
...how to use them?
TRIPLE T

18

# Spatial Sense and Competitive-Employment Options

## For Students with Mental Retardation

*Ann Massey*
*Mary Beth Noll*
*Joanne Stephenson*

THE National Council of Teachers of Mathematics (NCTM) emphasizes five general goals for all students in its *Curriculum and Evaluation Standards for School Mathematics* (1989): (1) learning to value mathematics, (2) becoming confident in one's own ability, (3) becoming a mathematical problem solver, (4) learning to communicate mathematically, and (5) learning to reason mathematically. How can these goals be realized for students with moderate and severe retardation?

A major theme of the NCTM's *Curriculum and Evaluation Standards for School Mathematics* (1989) is to empower *all* students to use mathematics in everyday life. Although the authors of this document clearly convey the message that mathematics is essential for all students, they also recognize that "students exhibit different talents, abilities, achievements, needs, and interests in relationship to mathematics" (p. 9). In other words, all students—even those with severe disabilities—can experience some success in meeting the general goals if curricula are sufficiently broad and flexible to allow for appropriately designed educational outcomes.

Parents and professionals agree that the most appropriate and desired educational outcome for students with moderate and severe retardation is the ability to participate meaningfully in the activities of daily living. Ultimately, this means living in the community and working in integrated competitive- employment settings. A learning environment that stresses the acquisition of functional skills is necessary if these students are to meet this desired educational outcome. The curriculum must be focused on the learning, performance, and participation that will enable them to function in targeted current and future environments (Brown et al. 1979; Lyon et al. 1990).

Follow-up studies have indicated that students with moderate and severe retardation have the highest rate of unemployment in the nation and often exit school only to have no work or to be placed in sheltered settings (Frank et al. 1990; Hasazi, Gordon, and Roe 1985; McDonnell, Hardman, and Hightower 1989). However, recent changes in traditional secondary-level vocational-preparation programs have resulted in significant increases in the numbers of individuals placed in competitive-employment settings in the community (Brown et al. 1987; McDonnell, Hardman, and Hightower 1989). These changes include provision of—

- age-appropriate, functional learning activities;
- community-based transition programs;
- supported employment programs (programs that employ persons with disabilities in which job coaches train the employees and work with them to ensure their success);
- integrated work options (jobs paying real wages in a "normal" workplace).

The success of curricular changes at the secondary school level has implications for the types of instructional programs implemented for students with moderate and severe mental disabilities at the elementary school level. Rather than address traditional developmental and academic skills, elementary school programs must devote resources to direct instruction in meaningful skills that will allow students to live and work in the community in which, ultimately, they will be expected to function (Brown et al. 1991; Westling and Floyd 1990).

The web of skills necessary to succeed in the community encompasses all aspects of an individual's life, but perhaps most important to functioning in the community is the job. To address the needs of future workers in the community, the mission of elementary school programs for youngsters with moderate and severe developmental disabilities must shift to the nurturing of employment-readiness skills, the assessment of employment strengths, and general preparation for employment settings.

## THE NEED FOR SPATIAL SENSE IN EMPLOYMENT SETTINGS

Successful experiences in many different areas of mathematics detailed in the NCTM *Curriculum and Evaluation Standards for School Mathematics* (1989) are foundational to enhancing employment options for individuals with moderate and severe learning disabilities. Number sense; spatial sense; time, measurement, and money concepts and skills; basic concepts enabling students to know which operation to use to solve a problem (perhaps with a calculator); and the recognition of a reasonable computational result are all essential to the majority of jobs and activities that can be performed by individuals with moderate and severe retardation.

Rather than address all aspects of a rich, broad mathematics curriculum for these students, however, this chapter will focus on the unique role of spatial sense in opening windows to competitive-employment options. The strategies for nurturing spatial sense presented in this chapter are designed for primary- and intermediate-level students with moderate to severe developmental disabilities, who are typically included in educational programs for the trainable mentally retarded.

## Spatial Sense

Spatial sense has been defined as "an intuitive feel for one's surroundings and the objects in them" (NCTM 1989, p. 49). "Without spatial sense and the vocabulary to describe relationships, we could not communicate about positions or the relationships of two or more objects. We could not give and receive directions for finding locations or completing simple tasks" (Shaw 1990, p. 4).

Spatial abilities involving body movement and eye-hand coordination are necessary for the majority of tasks required in school and employment settings, yet instruction in this area has often been ignored. For the learner with moderate and severe developmental disabilities, additional training may be required in order for some spatial skills to become automatic. Repetitive, visual, hands-on activities that assimilate task-related vocabulary must be provided as a part of daily instruction.

## Competitive-Employment Outcomes

A diverse range of competitive-employment opportunities are available in the community for individuals with moderate and severe retardation, including jobs in assembly, care provision, and industry, as well as numerous clerical and service positions. In order to illustrate how spatial sense is essential in job performance, several typical assembly jobs are presented and analyzed.

### *Assembly Work*

Assembly work is a typical competitive-employment option available to individuals with moderate or severe developmental disabilities. In assembly tasks, parts are positioned in relationship to each other. For example, a fiberboard-box stitcher uses an industrial stapling machine to staple fiberboard parts of boxes together.

Figure 18.1 illustrates the type of pattern a worker might follow to stitch the end of a fiberboard beer case. The movements from left to right, then down, across, and up can be very difficult to teach to the worker who is not accustomed to these directional movements. But once these few simple directions are mastered, the assembly can automatically be repeated over and over again.

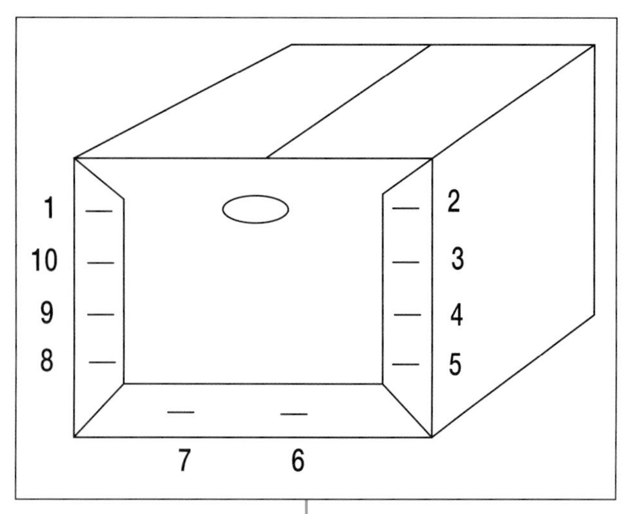


Fig. 18.1. A sample pattern for stitching the end of a fiberboard beer case

Pallet assembly, another type of assembly task, involves spatial judgment. Pallets are used to stack many kinds of stock, such as cases of food or soda, machine parts, or other materials, that must be transported by pallet jack to other locations in a warehouse or loaded onto a truck. Generally made of wood, they are assembled with stringers, deck boards, and nails according to specifications.

The specific dimensions and configurations are typically determined by the customer, but the assembly process is essentially the same. Deck boards are inserted into guides on a table, and stringers are evenly spaced across the top. Deck boards are nailed to the stringers using a pneumatic air fastener (nail gun). Each newly completed pallet is placed on top of the stack of finished pallets.

Pallet assembly involves numerous opportunities for spatial judgment. For example, the nail gun must be held in a straight-up-and-down position so that the nails go in straight. Stringers must be placed level, inside the guides, and flush to the end of the table. Deck boards must be evenly spaced, on top of the stringers, with the end boards flush with the edge of the table. Nailing begins at the left end and proceeds in a line to the other end. The center nails are done last.

Electronic assembly involves using hand tools and soldering irons to construct electronic components, subassemblies, and systems. This occupation requires the worker to follow not only the visual directives in production drawings or sample assemblies but also verbal instructions. The worker positions and aligns parts in a specified relationship to each other. Hand tools are used to crimp, bolt, solder, press-fit, or perform similar operations to secure parts in place. The worker mounts assembled components on panels and printed circuit boards or chases component wires and then connects them to printed circuit boards by soldering or other bonding techniques.

In the assembly of printed circuit boards and other electronic components, the worker must be able to follow a diagram and be able to run wires and other components in different directions. Examples include such directions as *to the right, up, down, to the bottom,* or *through the board.* Following a diagram requires the matching of the directionality on paper to the component being assembled.

If the learner has a thorough understanding of the positioning words before the process begins, the steps involving spatial judgment are mastered more easily. In order for students to acquire a basic understanding of spatial concepts, training must occur in the natural environment in a range of situations or conditions throughout the school career. The vocational-training experiences provided to a student should gradually build to full time as the students approach the completion of their schooling.

## DEVELOPING SPATIAL SENSE FOR EMPLOYMENT READINESS

How does one nurture the spatial abilities needed to enhance competitive-employment skills for students with moderate and severe retardation? As stated in NCTM's *Curriculum and Evaluation Standards for School Mathematics* (1989, p. 49), children should be provided "experiences that focus on geometric relationships; the direction, orientation, and perspectives of objects in space" to develop their spatial sense. Additionally, real-life activities should be an integral part of the classroom environment. This is particularly important for students with moderate or severe developmental disabilities, since generalizing classroom activities to the world in which they are used is difficult for them.

In order to enhance the generalization of vocational skills in the elementary school years, students should be involved to the maximum extent possible in meaningful, image-enhancing jobs and activities in classroom and school settings (Brown et al. 1987). That is, jobs and activities assigned to students with disabilities in the school environment should be those that nondisabled students also do, or would feel comfortable doing. The following sample jobs and activities can nurture the development of students' spatial skills in classroom and school settings in a long-range program leading to employment readiness.

### Readiness Activities for Assembly Jobs

Like the assembly jobs, school readiness activities run the gamut from simple to complex. The most relevant activities are those that involve the use of spatial sense to discriminate size and shape and to follow patterns. Children

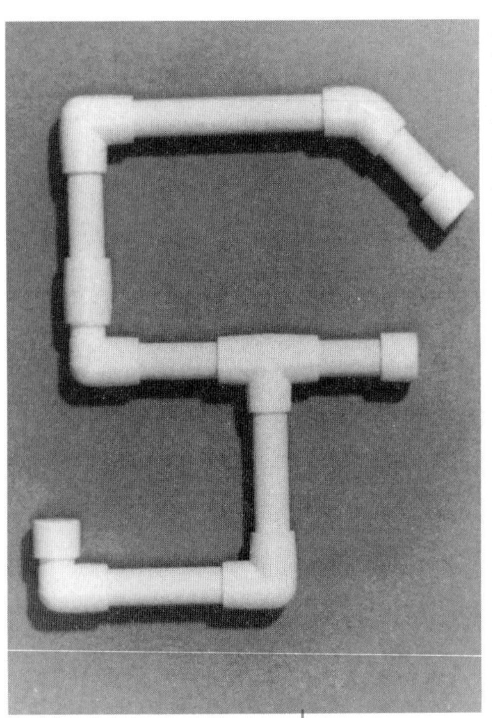

can begin in the early school grades by building with large blocks and, as their fine-motor skills develop, move to building with smaller blocks, such as Legos. In the elementary grades, they also can begin to follow a picture to make cars or airplanes with blocks. Students in junior and senior high school can learn to follow diagrams to connect plumbing pipes and elbows, nail boards, and wire boards.

A simple assembly task for primary school students is setting a table. This routine daily activity incorporates matching to a sample and using directionality with verbal prompts to teach independent table setting. Materials needed for teaching table setting are plates, glasses, knives, forks, spoons, napkins, and place mats. For instructional purposes, a variety of place mats should be prepared, with pictures of the items to be placed on the mats. At first, only a few pictures might be presented on a mat. For example, one place mat may contain only the picture of a plate; another, a picture of a plate and a fork; still another might picture a plate, knife, fork, spoon, and glass. The place mats can also reflect diverse cultures. For example, the Amish do not use knives, and both the fork and the spoon are placed to the right of the plate, whereas the steins (glasses) are to the left of and above the plate.

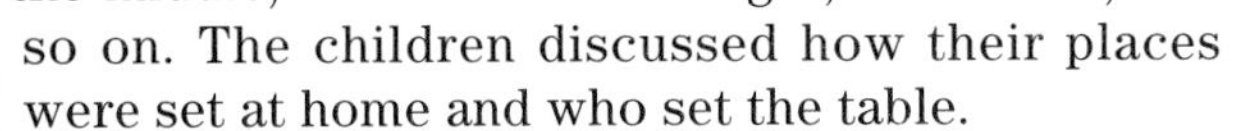

### Instructional Snapshot: Table Setting

One day at snack time, Ms. Setton asked the children to gather around the table. On the table was a place mat on which a napkin, fork, spoon, plate, and circle for a glass were drawn. She demonstrated putting a folded napkin, plate, knife, fork, and spoon in their appropriate places. She asked the children to name what was in the middle, what was on the right, on the left, and so on. The children discussed how their places were set at home and who set the table.

The next day at snack time, Ms. Setton gave each child a place mat with a picture of a plate on it and instructed the children to take a plate and match it to the picture of the plate on the place mat. Each day, one more picture was added to the place mat until the children were matching the plate, glass, knife, fork, and spoon to the appropriate pictures.

When the children had mastered matching, Ms. Setton used blank mats and gave verbal prompts. "Put the mat on the table in front of your chair. Take the plate and put it in the center of your mat. Put a fork to the left of the plate. Take a knife and put it to the right of the plate. Put a spoon to the right of the knife. Put the glass at the top of the knife."

Ms. Setton had recorded directions for table setting on cassette tapes. As a follow-up activity, the children were asked to listen to the tape and follow directions. They were challenged to memorize the table-setting directions. For practice, the children gave each other directions in setting a table. A concrete or pictorial model was always available for comparison.

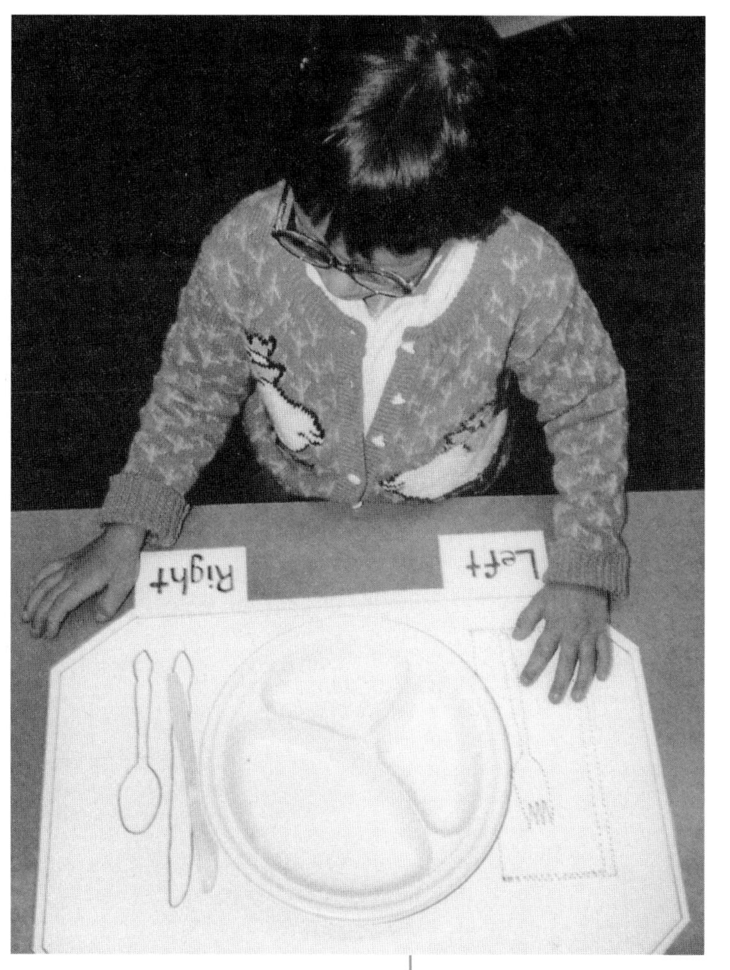


***Related Activities.*** A related seatwork activity that demands the recognition of shapes is putting together puzzles. Plywood board puzzles with removable geometric shapes are excellent for developing visualization abilities. Color matching, number matching, and color-to-number matching can be used to help students put the geometric pieces in the correct places. Students can be aided in assembling more difficult puzzles by matching an object to a picture. The age appropriateness of a puzzle should always be carefully considered before selecting it for instructional purposes. A wooden puzzle made from two or three large pieces depicting Sesame Street characters is not appropriate for adolescent students. Jigsaw puzzles would be a better alternative.

A more complex assembly task is that of collating student newspapers and student handouts. These jobs involve putting one paper on top of another in a specified order and stapling the papers together. Participation in these activities with nondisabled peers is recommended to further enhance the students' integration into the school community.

## Readiness Activities for Material-Handling Jobs

One everyday real-life activity for all children and adults involves going to the grocery store. For young students, playing store is a pretend activity that can be a rich mathematical experience. Many primary and elementary school classrooms have pretend stores that are used to teach money concepts and such skills as determining equivalent values for coins, counting to find total costs of purchases, and making change. Persons with moderate and severe retardation have a special need to become proficient in these basic activities involving money. However, employment opportunities for these individuals typically lie in jobs like stocking shelves, not in buying, selling, or making change. Pretend stores that are used to teach sorting and classifying and that develop spatial sense for directionality and orientation provide readiness for this type of job.

## Instructional Snapshot: Shelf Stocking

Mrs. Stockton is a teacher of a second-grade class that includes two students, John and Susan, who have moderate retardation and who are mainstreamed for specific periods of the day. She decided to use a grocery-store setting to teach spatial concepts. She sent letters to the parents of all the second-grade students requesting empty cereal boxes so that the children could set up a model section of the store. Three shelves had been cleared in the classroom for the project, and as the boxes arrived, the students sorted them by size and type of food.

During John and Susan's mathematics class, Mrs. Stockton put the boxes on the shelves with the brand name facing the front. Using two boxes, she asked such questions as, Which is on the right? Which is taller? Which is wider? Taking three boxes, she asked the children to point to the box in the middle. She put several boxes of two of the children's favorite cereals on the floor and asked them to choose a box and put it on the shelf in front of a box that matched it.

In extending this activity to incorporate other products, it was found that some labels were too difficult for nonreaders in the group to differentiate. For example, the labels on canned baby food looked too similar to one another (see fig. 18.2). Mrs. Stockton realized that she would have to help these students learn to match the jars by the bar code rather than by the name of the contents.

So for seatwork, Mrs. Stockton photocopied the bottoms of the boxes that showed the brand name, cereal name, and bar (UPC) code. She cut out the pictures of these bottoms for John and Susan to sort. She then made and cut out copies of only the bar codes (see fig. 18.3), which she put on cards so the children could play such matching games as bingo, fish, and concentration with the bar-code cards.

Finally, Mrs. Stockton taped bar codes to the shelves. The children's next task was to sort the boxes by bar code. To do so, they matched the bar codes on the boxes to the bar codes on the shelves and put the boxes in their appropriate places.

Readiness for a job as shelf stocker can be developed early. Similar activities for students in upper elementary school, junior high school, and senior high school should be increasingly complex and, whenever possible, should be conducted in community work sites.

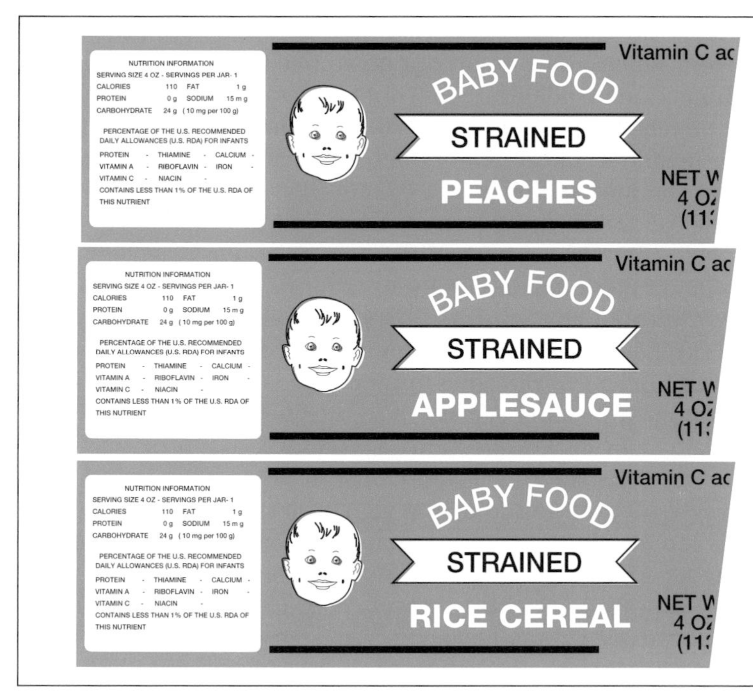


Fig. 18.2

***Related Activities.*** Activities that further nurture spatial sense include using directional cues to put toys on shelves (e.g., "Put the cars on the left and the dolls on the right"). Another activity could focus on stacking the dishes. As part of this activity, the teacher could ask the student if the stack will fit on the shelf (e.g., "Is the stack too high?").

## Readiness Activities for Messenger Jobs

Realization of one's position relative to space is an important component of spatial sense. For those with developmental disabilities, the school community and the community at large must be considered in addressing the student's self-perceptions in his or her environment. A community project such as the mapping investigation described below can be a vehicle for gaining an intuitive feel for one's surroundings, growing in spatial sense, and giving a sense of orientation and direction relative to oneself.

Fig. 18.3

*Kellogg's Just Right*, *Frosted Mini-Wheats*, and *Smacks* are registered trademarks, used with permission.

### Instructional Snapshot: Being a Messenger

Mr. Treadwright, a teacher in a primary school classroom for students with moderate and severe retardation in a small-town school, had become increasingly frustrated with the results of his efforts to nurture spatial sense in his students. Two pupils, Kate and Tom, have been only partially successful in completing assigned classroom mapping activities. Kate and Tom could not find their way around the school building without assistance. Their parents reported that although the students expressed their desire to walk the short distance to home from school, they could not do so because of their inability to find their way home independently. Ms. Globtrott, into whose classroom the students were mainstreamed for social studies, similarly reported that Kate and Tom were unable to participate effectively because of their lack of knowledge regarding community living. Mr. Treadwright thought that the students' lack of knowledge and inability to participate in the community—even at an awareness level—must be addressed in order for them to achieve the desired educational and functional outcomes.

Mr. Treadwright obtained the appropriate permissions from Kate and Tom's parents and purchased film for his camera. On a fine October day, he walked with Kate and Tom to the local shopping area on Main Street. Even though many preparatory trips had been taken to the front door of the school building, he continually asked such questions as, "Do we go left or right? Do we go up or down?" Once outside, at each decision point, they stopped to decide in which direction to turn or, at a stop sign, in which direction to look before crossing the street. Mr. Treadwright paid particular attention to the stop sign, discussing its color, configuration, and location.

When the group reached the business street, they stopped to talk about what they saw. Mr. Treadwright was pleased when Tom recognized the store where "you can buy tools" because he had been there with his dad. When Mr. Treadwright asked the children to identify the store next to the hardware store, Kate spoke eagerly about the beauty shop almost a block away.

When Mr. Treadwright asked if the signs for the stores were above or below the stores, he was met with silence. He took a picture of the line of stores along the street from where they were standing. They crossed the street; beginning at one end, they discussed each building: What goes on inside? Where do you go in? Point to the sign that says "In" or "Pull." He continued, asking frequent questions about the relationships of the stores to one another: "Which store is first? Which store is next to the department store? Which store is on the left? Which store are you standing in front of? Which store is the biggest?"

That evening, Mr. Treadwright returned to the village street and took pictures of each store they had discussed that day. The next day, Kate and Tom were shown the picture of the street. Mr. Treadwright then presented the individual pictures in order and invited Kate and Tom to talk about them. Their enthusiasm was evident as they recognized the places they had been.

Next, Mr. Treadwright gave each of the children the picture of the street and the stack of pictures of the individual stores. He asked them to put the stores in order, as they were in the picture. Then he mixed the pictures and had them repeat the lesson. Next, he removed some of the pictures and had Kate and Tom replace them. Each time the pictures were in place, he asked the same questions that he had asked during the previous day's walk. Return trips were made to the community to verify the results of the classroom lessons.

The ensuing lessons included tours of the bank, grocery store, beauty shop, and department store. Mr. Treadwright sketched the floor plans of the buildings as they discussed different aspects like the stores' location and functions, items within the stores, floor plans, shelving, and classification systems.

Mr. Treadwright walked frequently with Kate and Tom in the school building until they were able to follow his verbal directions. He and the students constructed a map of the layout of the school, and they put room numbers on the map. Eventually Kate and Tom were able to be messengers for Mr. Treadwright and to deliver materials to the school office and other classrooms.

With the cooperation of their parents, Kate and Tom were taught how to walk safely to and from school. Because of Mr. Treadwright's intervention,

Kate and Tom have gained an intuitive feel for their surroundings, the objects in them, and their personal relationship in space. With similar, ongoing experiences through senior high school, they will be ready for transitional jobs in the community. Job coaches may have to use mapping techniques similar to that employed by Mr. Treadwright, but Kate and Tom should learn quickly.

## CREATING WINDOWS OF OPPORTUNITY FOR COMPETITIVE EMPLOYMENT

Vocational experiences for students with moderate and severe retardation cover classroom and school jobs and community-based work experiences and jobs (Brown et al. 1987). The older these students are, or the closer they are to graduation, the greater the amount of time they should spend in vocational instruction in community-based training sites. However, if students are to be prepared to take advantage of vocational opportunities as part of the secondary school program and actually work as adults in integrated employment settings in the community, then training opportunities are an essential component of the elementary school curriculum.

Toward this end, the teachers in the classroom episodes in this chapter developed strategies to enhance the spatial skills of their students having moderate and severe retardation. The hands-on activities they designed focused on important spatial skills and incorporated essential spatial vocabulary into experiences related to their students' daily lives. The activities often afforded opportunities to interact meaningfully with peers in a natural, integrated environment.

This chapter highlighted how spatial-sense activities are an especially valuable and important component of the total mathematics program for students with moderate and severe retardation. The experiences and language developed through activities that promote spatial sense were directly related to the skills needed for employment. The students' successful mastery of the concepts, language, and skills developed through these activities nurtured their development of mathematical power in ways consistent with the vision of NCTM's *Curriculum and Evaluation Standards for School Mathematics* (1989).

### REFERENCES


Brown, Lou, Mary Beth Branston-McClean, Dian Baumgart, Lisbeth Vincent, Mary Falvey, and John Schroeder. "Using the Characteristics of Current and Subsequent Least Restrictive Environments in the Development of Curricular Content for Severely Handicapped Students." *AAESPH Review* 4 (Winter 1979): 407–24.

Brown, Lou, Patricia Rogan, Betsy Shiraga, Kathy Zanella-Albright, Kim Kessler, Fred Bryson, Patricia VanDeventer, and Ruth Loomis. *A Vocational Follow-up Evaluation of the 1984 to 1986 Madison Metropolitan School District Graduates with Severe Intellectual Disabilities.* Madison, Wis.: Madison Metropolitan School District, 1987. (Also available from the Association for Persons with Severe Handicaps.)

Brown, Lou, Patrick Schwarz, Alice Udvari-Solner, Elise F. Kampschroer, Fran Johnson, Jack Jorgensen, and Lee Gruenewald. "How Much Time Should Students with Severe Intellectual Disabilities Spend in Regular Education Classrooms and Elsewhere?" *Journal of the Association for Persons with Severe Handicaps* 16 (Spring 1991): 39–47.

Frank, Alan R., Patricia L. Sitlington, Linda Cooper, and Valerie Cool. "Adult Adjustment of Recent Graduates of Iowa Mental Disabilities Programs." *Education and Training in Mental Retardation* 25 (March 1990): 62–75.

Hasazi, Susan B., Lawrence R. Gordon, and Cheryl A. Roe. "Factors Associated with the Employment Status of Handicapped Youth Exiting High School from 1979 to 1983." *Exceptional Children* 57 (April 1985): 455–69.

Lyon, Steve R., Joseph W. Domaracki, Grace A. Lyon, and Susan Warsinske. *Preparation for Integrated Community Living and Employment: Curriculum and Program Development.* Harrisburg, Pa.: Mid-State Instructional Support Center, 1990.

McDonnell, John, Michael L. Hardman, and Julia Hightower. "Employment Preparation for High School Students with Severe Handicaps." *Mental Retardation* 27 (December 1989): 396–405.

National Council of Teachers of Mathematics. *Curriculum and Evaluation Standards for School Mathematics.* Reston, Va.: The Council, 1989.

Shaw, Jean M. "By Way of Introduction." *Arithmetic Teacher* 37 (February 1990): 4–5.

Westling, David L., and Jane Floyd. "Generalization of Community Skills: How Much Training Is Necessary?" *Journal of Special Education* 23 (Winter 1990): 386–406.

**Ann Massey** is an associate professor of mathematics education, Department of Mathematics, Indiana University of Pennsylvania, Indiana, Pennsylvania. She teaches mathematics content and methods courses to preservice students in special education, elementary education, and secondary mathematics education. She was a member of the Pennsylvania team for the NCTM project "Leading Mathematics Education into the 21st Century" to disseminate the Standards.

**Mary Beth Noll** is an assistant professor of special education at Saint Cloud State University, Saint Cloud, Minnesota. She is a former teacher of students with autism, mental retardation, and emotional and behavioral disorders and has taught in various educational programs in Kansas, Pennsylvania, and Minnesota.

**Joanne Stephenson** is the director of vocational supports for Parc-way Industries, the adult vocational services division of ARC-Allegheny in Pittsburgh, Pennsylvania. She is responsible for a range of community-integrated employment programs, vocational training, vocational evaluation, and case management. Formerly an education specialist, she has developed educational programs and implemented training programs for adults with disabilities.

19

# Sense Making in Middle School Mathematics

*Sandra K. Wilcox*
*Patricia Wagner*

RECENT calls to reform mathematics education for *all* children (National Council of Teachers of Mathematics [NCTM] 1989, 1991; National Research Council 1989) pose considerable challenges to classroom teachers—the primary agents of change in the nation's K–12 classrooms. The vision is of mathematics classrooms where students and teachers form a community of learners engaged in inquiry. In these classrooms, teachers pose interesting problems that lead to inventions by the learners. Students make conjectures, develop and use mathematical tools to investigate problems, and argue the reasonableness of the results. Assessment is an ongoing activity that teachers use to gain insights about students' understanding, to judge the usefulness of teaching practices and learning tasks, and to make decisions about where to go next.

Creating visions of classrooms that enable mathematical inquiry and reflection on teaching practices is one thing. Constructing classrooms that embody these visions in real schools with real students is quite another matter. Realizing the vision is especially challenging in special education classrooms.

This chapter describes our efforts to provide opportunities for classrooms of "learning disabled" students to investigate powerful mathematical ideas. (The majority of our students are African American, and many come from families receiving some form of public assistance. Most were identified as

---

For more than three years, we have been collaborating to make mathematical problem solving, reasoning, communication, and sense making important features in a classroom for "learning disabled" students. We collaborate in the context of the Professional Development School Partnerships established between the College of Education at Michigan State University and several school districts in the state to construct classroom and professional communities that support the empowerment of teachers and learners. We wish to acknowledge Anne Heidel, Elnora Crutchfield, Bill Rosenthal, and Sonia Marsalis, who have worked closely with us and our students and who contribute much to our collective efforts. Our work is supported, in part, by National Science Foundation grant no. MDR 9252881. The opinions expressed are those of the authors and not necessarily of the Foundation.

special-needs youngsters as early as first or second grade. Our choice to put the label in quotations reflects our own uncertainty about what it means to say that youngsters are disabled in mathematics. In fact, the more we work with students who are burdened with this label, the more we are coming to appreciate the many strengths they bring to the mathematics classroom that traditional curricula, modal teaching practices, and standardized testing programs simply fail to uncover.) In our work we have been guided by the following principles:

- *All* students should have opportunities to learn powerful mathematical ideas. This goal involves extending the breadth and depth of mathematical learning from the remediation of arithmetic skills to developing number and operation sense; developing spatial reasoning; identifying and using patterns; and using statistical methods to collect, describe, and analyze data.
- The mathematics classroom should be a place of active learning where students and teachers together are engaged in sense making.
- Mathematical investigations should lead to learners' inventions.
- Different approaches to problems should be encouraged and valued.
- Mathematical tasks should invite students to explore patterns embedded in problems and to generalize from the patterns, since "regularity is the essence of mathematics" (NCTM 1989, p. 60).
- The mathematics class should require students to communicate their understanding in multiple ways: using natural and symbolic language; using concrete, pictorial, numeric, algebraic, geometric, graphical, and spatial representations.
- Assessment of what students are coming to understand should be an ongoing activity that teachers use to make instructional decisions.

The following sections present three teaching and learning episodes that exemplify two of these principles: (1) looking for, reasoning about, and generalizing from patterns; and (2) making sense of, and communicating about, mathematics. These classroom events and examples of students' work support our claim that if given the chance, "learning disabled" students can develop a set of mathematical tools and a disposition to engage with others in problem solving *and* problem posing.

## Lessons from Perimeters: Making Sense of Patterns

The students are working on a problem that involves adding tiles to the arrangement in figure 19.1. Their task is to add tiles to make a figure with a perimeter of 18 (Shroyer and Fitzgerald 1986, p. 18). Adding tiles is constrained in the following ways: Each tile that is added must completely touch the entire edge of adjacent tiles and no holes are allowed. The additions shown in figure 19.2 are not allowed.

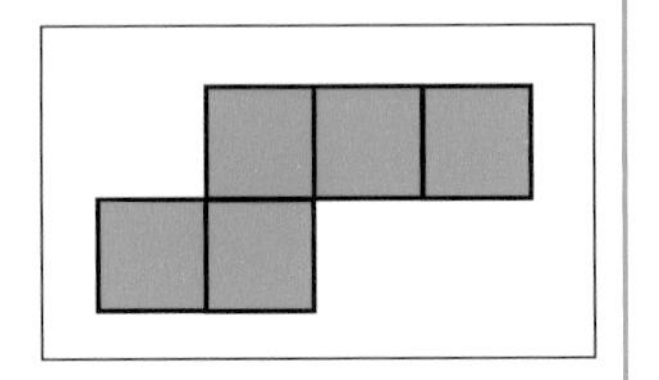

Fig. 19.1

Each student has a bag of one-inch-square tiles. Some of the students find one figure with a perimeter of 18 and then quit working. We encourage them

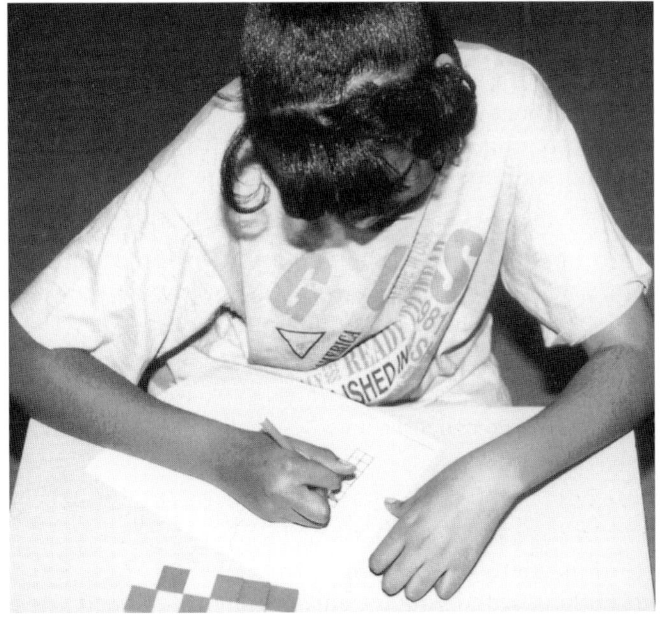

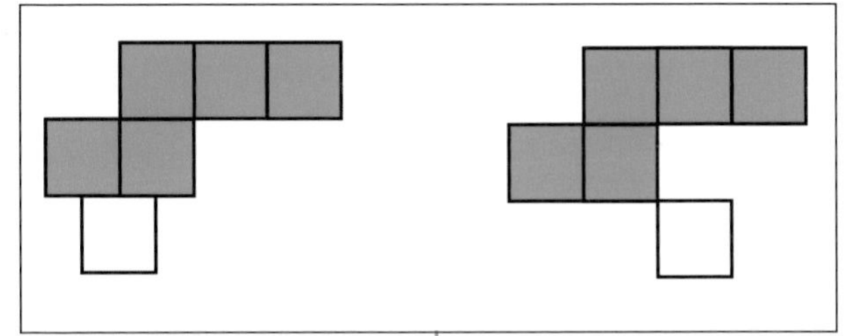

Fig. 19.2

to find another figure. They are to make a drawing on centimeter grid paper for each figure they construct.

We observe that three students build their figure on top of the grid paper and then trace around it to make the drawing. We show them how to use one square on the grid paper to stand for one square tile and tell them their drawing will look like a smaller version of the figure they make with the tiles. Two students quickly draw an outline of their figure and then draw in the edges of the individual tiles.

Chiwanna tries the same strategy but hesitates, not sure how long to draw the top edge of her figure. She continues with the general outline of her figure but ends up with a drawing that shows more squares on the grid than tiles in her figure. We suggest that moving from left to right, she try outlining each square separately. This tack seems to work well for her, and she continues, successfully drawing other figures as she constructs them with her tiles.

We give the students a transparency of the grid and ask them to draw one or more of their figures to show on the overhead projector. After the students have shared their figures with the class, we remark that not everyone used the same number of tiles. The students pick up on this observation, then go back and count the number of tiles in each figure, noting who used the most and who used the least.

*Teacher:* This is pretty interesting. I have a challenge for you. What is the *greatest* number of tiles you can add and still have a perimeter of 18?

There is a flurry of activity as the students try various ways of adding tiles and count the perimeter.

By the end of the class, one student has made a $4 \times 5$ rectangle with twenty tiles. Since no one has used more than twenty, the students quickly conclude that twenty is the greatest possible number of tiles. We warn them that finding twenty tiles does not necessarily mean that there cannot be more. They need to figure out a way to convince us and each other that twenty is the greatest.

As Bobby packs up his books, he adds with great delight, "I see another challenge: What's the *smallest* number of tiles you can add?"

Over the next two days, the youngsters investigate the two challenges. To get them started, we suggest that they investigate if some patterns emerge in how the perimeter changes when a single tile is added. They find that the perimeter either increases by 2 or it does not change at all. At this point no one has discovered that the perimeter can also decrease by 2.

We notice as we observe the students that they recount the perimeter each time they add a tile.

*Teacher:* I want you to go back to the original figure. Without counting around your figure, what do you think will happen to the perimeter if you add one tile to the far-left end of the figure?

*Orlando:* You get three more.

*Kevin:* Yah, three more.

*Teacher:* Why is that?

*Orlando:* Because you added three sides (*pointing to the three exposed edges of the added tile*).

*Bobby:* That isn't right. I counted and you get only two more, fourteen.

*Teacher:* Kevin, you and Orlando think it gets bigger by three, but Bobby says it gets bigger only by two. See if you can make sense of this.

We leave them for a while to puzzle over this inconsistency. Fairly soon, Bobby and Orlando figure out that they have actually covered the side of one tile in the original figure. Their way of describing the situation is that they are adding three edges but losing one, so the result is an increase of two. We ask them to share their finding with the rest of the class.

Finally, we return to their earlier conjecture that twenty is the greatest number of tiles that can be used. We ask if anyone can make a convincing argument that twenty is the greatest. Bobby is quick to volunteer. He goes to the overhead projector, builds a $3 \times 5$ rectangle, and then adds one tile to begin a fourth row.

*Bobby:* (*Pointing with his finger as he counts*) See, the perimeter is eighteen now. I can put a tile here (*in the corner on the fourth row*), but it's still eighteen 'cause I added two sides. But I took these two sides away, so it didn't change anything. And I can fill this row up and still keep the eighteen. But if I add one more up here (*starting a fifth row*), this makes twenty 'cause I added three but took away one. So the most tiles is twenty.

This slice of the classroom dialogue illustrates our intent to push beyond simply finding and describing patterns. We want students to be able to *reason* why the perimeter changes as it does. We want students to *make sense of patterns* and be able to *generalize from them*. Finding a pattern can be a powerful problem-solving tool. Students who make sense of a pattern begin to understand a problem situation and can use it to justify their results to others.

Making sense in this classroom has two aspects. The first entails the students' making sense of the mathematics in problem situations. The second aspect involves our making sense of what the students are saying and doing. We are continually surprised by the diverse and unexpected ways in which students work on problems. Oftentimes solutions appear at first glance not to make much sense at all. But we have learned that much of the time the students have some reasonable, albeit unconventional, ways of presenting their solutions to a problem, as the following classroom event illustrates.

## Does Everyone Take the Same Number of Paces to Walk the Same Distance?

The students have been working on a unit in which they learn to use statistical methods to collect, describe, and analyze data. Earlier investigations have introduced students to line plots, bar graphs, and coordinate graphs as ways of representing data they have collected. The topic of this investigation is the distance students travel in school in a typical day (Connected Mathematics Project 1992, p. 25). The following problem guides the investigation.

> Does it take everyone the same number of paces to travel the same distance?

The task is to decide on a common route that each of the students travels, gather data from classmates about the number of paces required to travel this distance, and make a display of the data. The students decide on a route from their classroom to the cafeteria and back again. The eight students leave the room and after several minutes return with the count of their paces.

The data are collected and displayed on the overhead projector (see table 19.1). The students get graph paper and set about making displays of the class's data. It appears that everyone has decided to make a bar graph.

Table 19.1

| | |
|---|---|
| Robert | 151 |
| Winston | 147 |
| Natasha | 178 |
| Monica | 157 |
| Kenyatta | 151 |
| Raymond | 159 |
| Montel | 146 |
| Kenneth | 154 |

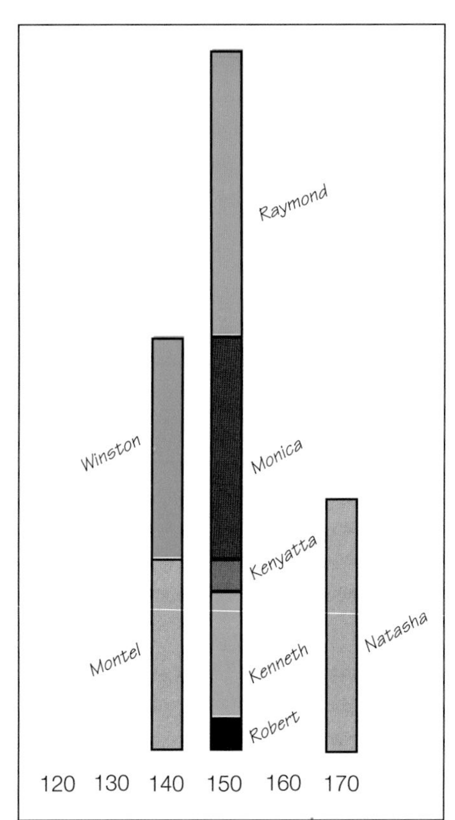


Fig. 19.3

Winston turns in the display shown in figure 19.3. Each "bar" is actually one or more bars stacked on top of each other at 140, 150, or 170. Each section is shaded with a color, and the names of the students are written beside the sections. Initially we are puzzled about what Winston has done. But when we ask him to explain his work, he has a perfectly reasonable explanation for his treatment of the data.

Pointing to the two sections above 140, he explains that the first section is six squares high for the 146 paces it took Montel. The section stacked on top of Montel's is seven squares high for Winston's own 147 paces. Above 150 he has a bar one square high for Robert's 151 paces, a section four squares high for Kenneth's 154, another section one square high for Kenyatta's 151, a section seven squares high for Monica's 157, and a last section nine squares high for Raymond's 159.

We ask about the names that are written on the paper. He points to the bar for Montel and Montel's name, both of which are in brown pencil. We ask about two other sections of the graph that are shaded with brown. He points to two other names written with brown pencil and says, "I didn't have enough colors to do 'em all different."

After class, we talk more about Winston's work. We realize that although he has created an unconventional display, he has invented a representation that makes sense to him and to us. We also realize that his display is a kind of graphical representation of a stem-and-leaf plot. We make a note to come back to Winston's unconventional bar graph later in the unit to make a connection to stem-and-leaf plots. We have come to realize that part of helping students to make sense of mathematics is allowing for and expecting students' uncommon ways of working with problems and presenting results. We work hard to understand these unique situations from the student's perspective.

## How Long Does It Take To Get To School?

The class has been working on an investigation to determine about how long it takes students to get to school (Connected Mathematics Project 1992, pp. 33–34). Their textbook includes a map of a school neighborhood with the homes of several children marked. The scale on the map is 1 cm = 1 km. The textbook also includes a data table that indicates the time it takes each of the children to get to school. Our students measured the distance for each child from home to school. Then they compiled the data shown in table 19.2.

Using the information in table 19.2, the class has been discussing two questions: (1) Which students live closest to school and do they take the shortest amount of time to get to school? (2) Which students live farthest from school and do they take the longest to get to school? Then a question is posed that students cannot answer simply by finding data in the table.

*Teacher:* If a student lives twelve kilometers from the school, how long do you think it will take the student to travel to school?

Table 19.2

| Student's Name | Distance in Kilometers | Time in Minutes |
|---|---|---|
| Anna | 8 | 14 |
| Cary | 9 | 12 |
| Eve | 6 | 5 |
| Fern | 6 | 9 |
| George | 6 | 10 |
| Hal | 9 | 24 |
| Jon | 9 | 16 |
| Kate | 7 | 18 |

*Raymond:* Six and a half minutes. (*Raymond has a habit of taking a stab at an answer without much thought.*)

Two other students answer quickly, "Seven minutes" and "Eight minutes."

*Raymond:* What time does the person go to school?

*Teacher:* (*Appearing a bit surprised*) Why does *that* matter?

*Winston:* It does! I think between twelve and fifteen minutes.

*Teacher:* What information did you use?

Winston refers to Anna, who traveled 8 km in 14 min, and Hal, who traveled 9 km in 24 min. But then he revises his answer.

*Winston:* He's gonna take longer 'cause he has to walk more kilometers.

*Raymond:* It depends on how fast he walks. I think it will take about twelve to sixteen minutes.

*Kenneth:* I say twenty-one minutes.

*Teacher:* Why do you think that, Kenneth?

Kenneth goes immediately to the table, pointing out that Jon takes 16 min to go 9 km and Cary takes 12 min to go 9 km.

It is interesting here that all the times the students are giving are greater than 12, which is the least time corresponding to a distance of 9 km (the maximum distance in the table). At the same time, the responses are all less than 24, which is the greatest time corresponding to a distance of 9 km, although Winston's last comment may indicate that he is thinking the time should be more than 24 min.

*Teacher:* Look at your data table. What about Hal? He also has to go nine kilometers but it takes him twenty-four minutes.

On the overhead projector, the teacher writes the information shown in figure 19.4. The students look at it for a moment, and then Monica speaks up.

| | |
|---|---|
| 9 km (24 min) | 12–16 min—Raymond |
| | 12–15 min—Winston |
| 12 km | |
| | 21 min—Kenneth |

Fig. 19.4

*Monica:* Maybe it's twenty-six or twenty-seven minutes.

*Teacher:* (*Pointing to the chart*) What does this tell you?

*Monica:* That he gets up earlier.

*Winston:* (*Revising his earlier estimate*) Maybe it takes him thirty-two minutes.

*Raymond:* I think the person who took twenty-four minutes was riding a bike—he cheated.

*Teacher:* But why does this person take *longer*?

*Raymond:* That person is bigger.

It is time for the class to end. An unresolved question remains, to be taken up the next day.

One aspect of building a community of learners is establishing classroom patterns that enable children to share through language "things seen and significances perceived" (Schwab 1975, p. 32). The challenge to the teacher is to give students ownership of their ideas while offering prompts, questions, or counterexamples for incorrect approaches. This requires, in part, making sense, in the moment-to-moment conversation, of what students do and do not consider relevant in a situation.

In the event just described, Raymond, Winston, and Monica make comments related to when students get up in the morning and when they go to school. On the surface, these comments may seem irrelevant to the question at hand. However, in the students' world, everyone must get to school at the same time. Therefore, if some students take longer to get to school, it makes sense that they need to get up earlier in the morning.

We were uncertain about whether and to what extent other factors appeared relevant to the students. They seem to hold a simplistic, intuitive notion that a longer distance corresponds to more time. Their conception seems to be a nearly linear relation between distance and time. At the same time, there is a hint of a competing explanation, as evidenced by Raymond's comment "It depends on how fast he walks." We are also uncertain about their sense of the distances in the table. Initially, Winston and Raymond refer to the students as walking to school. Yet a close look at the distances and corresponding times in the table reveals that some students could not be walking to school. For example, it is not possible for Cary to walk a distance of nine kilometers in twelve minutes. We suspect that the students are not considering what would be reasonable walking times for particular distances.

Our analysis of this event and the partial understandings we think students have suggest some further instructional tasks. We need to orchestrate the conversation in ways that reintroduce and give legitimacy to Raymond's insight. We need to challenge the assumptions our students may have about how the people in this problem get to school. Identifying some well-known landmarks in the community and their metric distances from our school could be useful in our investigation. Students could then estimate the time it would take to get to school from various starting points if they walked, rode a bike, or took the school bus. This approach might challenge the students' views and lead to a more robust understanding of the complexities of this problem.

## NURTURING POTENTIAL

Our experiences have convinced us that given opportunities, support, and time to develop competence and confidence in new areas, "learning disabled" students *can* pursue mathematical investigations in geometry, probability, number theory, and statistics. We have tried to shape our teaching practices in ways that we believe will nurture this potential in youngsters with special needs. We aim to create a learning environment in which our students feel safe to take risks in mathematics. We endeavor to create a learning community where all of us—students and teachers—share the responsibility for learning.

When our students encounter challenging problems, we try to be guides by contributing prompts, questions, counterexamples, or strategies for thinking about a problem rather than by telling them how to solve the problem. We want our students to make sense and have ownership of mathematical ideas. Attempting to reconstruct the mathematics classroom in ways that empower both teachers and learners brings challenges and uncertainty. But the growth we see in our students convinces us that we are doing something right—for them and for us.

## REFERENCES


Connected Mathematics Project. *About Us*. East Lansing, Mich.: Michigan State University, 1992.

National Council of Teachers of Mathematics. *Curriculum and Evaluation Standards for School Mathematics*. Reston, Va.: The Council, 1989.

———. *Professional Standards for Teaching Mathematics*. Reston, Va.: The Council, 1991.

National Research Council. Mathematical Sciences Education Board. *Everybody Counts: A Report to the Nation on the Future of Mathematics Education.* Washington, D.C.: National Academy Press, 1989.

Schwab, Joseph. "Learning Community." *The Center Magazine* 8 (1975): 30–44.

Shroyer, Janet, and William Fitzgerald. *Mouse and Elephant: Measuring Growth.* Menlo Park, Calif.: Addison-Wesley Publishing Co., 1986.

**Sandra Wilcox** is an assisant professor in the Department of Teacher Education, Michigan State University. She is interested in teaching mathematics for understanding; the social, political, economic, and organizational contexts in which the reform of mathematics education is situated; and race, gender, class, and educational equity. She directs two NSF-funded projects on alternative assessments in mathematics.

**Patricia Wagner** is a mathematics teacher of students with learning disabilities at Holmes Middle School, Flint, Michigan. She is interested in developing mathematical-reasoning skills through writing; exploring teaching strategies that promote active student participation; using multiple representations to investigate mathematical situations; and making mathematics relevant to adolescents' lives.

20

# Inquiry and Problem Solving in a General Mathematics Classroom

*Steven A. Kirsner*
*Sandra Callis Bethell*

THIS chapter describes one high school teacher's attempt to change her mathematics teaching in ways that are consistent with the *Standards* documents of the National Council of Teachers of Mathematics (NCTM 1989, 1991). The setting is Holt High School, one of several Professional Development Schools established in collaboration with Michigan State University to begin to create conditions that support the quality of learning envisioned in the NCTM *Standards* (1989, 1991) and to prepare prospective teachers better to assume responsibility for ensuring quality teaching. The school encompasses grades 10–12.

As a result of their experiences over a two-year period in a classroom containing many special education students, the teacher (Bethell) and the university collaborator (Kirsner) argue that it is possible to create a classroom environment that produces meaningful mathematical learning. Our purpose is not to support the idea that special education students should be in general mathematics classes but rather to show that given the opportunity to learn more powerful mathematics than strictly computational skills, special education students show the ability to think creatively and to express themselves clearly.

A vision of active mathematics learning requires classroom teaching that differs from the norm of broadcasting mathematics as a fixed body of rules and procedures (National Research Council 1989). The *Curriculum and Evaluation Standards for School Mathematics* (NCTM 1989) calls

Bethell wishes to dedicate this chapter to the fond memory of Steve Kirsner, who was killed in an automobile accident during the period in which this chapter was being written. His kind words, sense of humor, and respectful attitude contributed to a wonderful collegial relationship that had a great impact on Bethell's perspective and practice. She also wishes to acknowledge gratefully the support and assistance of Michael Sedlak and Daniel Chazan from Michigan State University in reacting to final drafts of this chapter.

for students to examine, transform, apply, prove, communicate, and solve mathematical problems and concepts. These behaviors exist in an environment envisioned by the *Professional Standards for Teaching Mathematics* (NCTM 1991, p. 57):

> The teacher of mathematics should create a learning environment that fosters the development of each student's mathematical power by—
>
> - providing and structuring the time necessary to explore sound mathematics and grapple with significant ideas and problems; ...
> - providing a context that encourages the development of mathematical skill and proficiency;
> - respecting and valuing students' ideas, ways of thinking, and mathematical dispositions;
>
> and by consistently expecting and encouraging students to—
>
> - work independently or collaboratively to make sense of mathematics;
> - take intellectual risks by raising questions and formulating conjectures;
> - display a sense of mathematical competence by validating and supporting ideas with mathematical argument.

In documenting Bethell's attempt to create such an environment in her tenth-grade classroom, we will pay particular attention to students' learning as well as to the conditions that promote or limit meaningful mathematical learning among students. We include an interview with one student to contrast individual assessment with group assessment and to portray the complex learning that occurred in this special education student.

## THE GENERAL MATHEMATICS CLASS

The class highlighted in this chapter was team taught by Bethell and a special education teacher, Margaret Lamb. Bethell and Lamb planned the units of instruction together, with Bethell taking primary responsibility for curriculum and Lamb providing insight into students' learning. Lamb was a source of inspiration and ideas about how diverse students could best access the mathematical ideas. She consistently modeled risk-taking behavior for the students as she asked questions and posed conjectures in the class.

Sixteen students were enrolled in the class. They included sophomores, juniors, and seniors. Eleven of the students had been classified special education students, either while they were in the class or shortly before. Students were extremely diverse in their perspectives on and ability to do mathematics and in their views of themselves as mathematics learners. Virtually all the students in the class had encountered some degree of failure and frustration during their previous school mathematics experiences.

Bethell was especially intent on instilling in her students a sense of self-efficacy in mathematics. Her conception of doing mathematics corresponds to the first four standards of the *Curriculum and Evaluation Standards for School Mathematics* (NCTM 1989)—mathematics as problem solving, reasoning, communicating, and making connections. Above all, she wanted all her students to "make sense of mathematical ideas."

In order to create an environment in which her students could be successful, Bethell employed a blend of rather eclectic teaching activities. Mathematics instructors typically draw on many teaching strategies. Whereas some may rely heavily on manipulatives, others emphasize students' discussion and communication of mathematical ideas. Journal writing, group work, and cooperative learning are common pedagogical strategies that support more meaningful communication. Traditional teaching, of course, relies on demonstrations by the teacher and on individual students' seatwork.

Bethell emphasized these types of teaching strategies about equally. She did not use a textbook in her class. She relied on materials she had collected from coursework and workshops; wrote her own activities, worksheets, and tests; and occasionally consulted with other mathematics educators.

During the period focused on in this chapter, the students were engaged in a unit on probability. Although the *Curriculum and Evaluation Standards for School Mathematics* recommends this topic for all grade levels, most of the students in the class—typical of K–12 students (Shaughnessy 1992)—had never been exposed to this branch of mathematics.

The prescribed district curriculum for general mathematics did not include probability. Disagreements about policy were avoided by simply framing the ten-week probability curriculum in the context of helping students learn the computational skills that the prescribed curriculum dictated. Specifically, Bethell justified this unit on the grounds that it helped students learn to understand and compute with fractions—objectives that occupied an important place in the adopted curriculum.

As part of the unit, the students explored the meaning of probability, the concepts of experimental and theoretical probability, the meaning and usefulness of simulating experiments, and the concepts of independent and dependent probability. To help students master these understandings, Bethell relied on her entire repertoire of pedagogical strategies. The following problem is illustrative of the way in which central questions were used to encourage students to think about ways to model problems, communicate their ideas, and justify their solutions.

## MONTY'S PROBLEM

Kirsner introduced Bethell to this problem, which is based on "Let's Make a Deal," hosted by Monty Hall, because of its intriguing counterintuitive solution. Although Kirsner was not suggesting it for use in her class, Bethell thought it was so interesting that she immediately wanted to pose it to her students. Shaughnessy and Dick (1991) frame the problem this way (p. 252):

> During a certain game show, contestants are shown three doors. One of the doors has a big prize behind it, and the other two have junk behind them. The contestants are asked to pick a door, which remains closed to them. Then the game show host, Monty, opens one of the other two doors and reveals the contents to the contestant. Monty always chooses a door with a gag gift behind it. The contestants are then given the option to stick with their original choice or switch to the unopened door.

The problem is, If you were a contestant, would you stick with your original door or switch to the other door, or does it even matter what you decide (see fig. 20.1)?

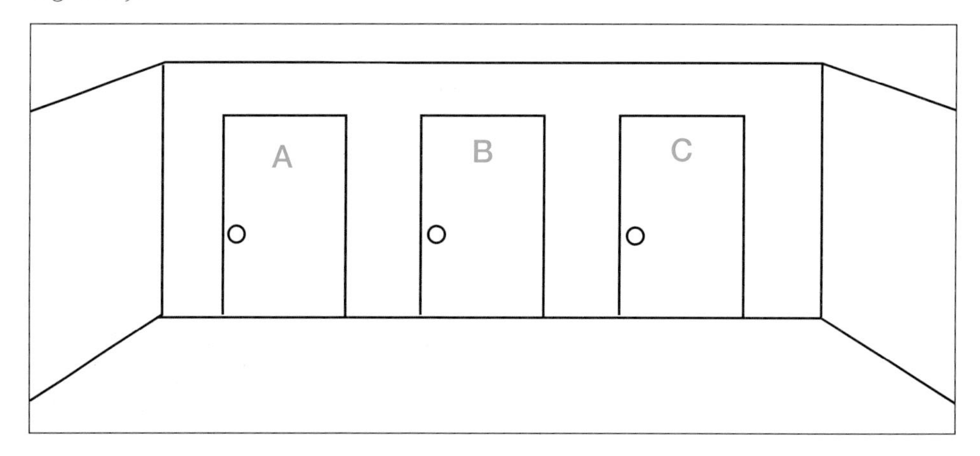


Fig. 20.1

When Kirsner and Bethell first discussed this problem, Bethell refused to accept Kirsner's conclusion that her chances are doubled if she chooses the other door. She suggested that it doesn't matter whether she stays or switches; after one door has been eliminated as a winning one, two equally likely chances remain: the original choice and the remaining door. Bethell's response was typical of "people across the whole range of mathematical expertise, from novices to teachers to research mathematicians" (Shaughnessy and Dick 1991, p. 252). She decided to see what the students would think.

When Bethell posed the problem to her students, she told them that she and Kirsner had different opinions about the most advantageous strategy for the contestants. The students predictably agreed with Bethell, who asked them how they could prove to Kirsner that he was wrong. A number of students responded that they could simulate the game, and they proceeded to work in pairs to do so.

Before long, students observed that the contestants were winning more often when they switched than when they stayed with their original selection. Not totally convinced, the students decided to alter the manner in which the simulations were conducted. They decided to conduct a larger number of simulations and to do them more systematically.

Bethell kept track of the simulation responses that were generated by the class. As each pair of students reported the results of their simulations (i.e., either stay or switch), Bethell tallied the results on the chalkboard. By the end of the class, the students were convinced that it was advantageous to switch doors because this strategy wins the more desirable prize twice as often as the other strategy. They were also perplexed, and expressed curiosity about why this outcome occurred.

The next time Kirsner observed Bethell's class, the students were eager to present their results. They also wanted an explanation for their counterintuitive findings. Kirsner did his best to offer different explanations for the clear advantages of the switch strategy. He was not quite sure whether the students understood or accepted his explanations. However, it seemed clear that the students' simulations had convinced them that the switch strategy was best.

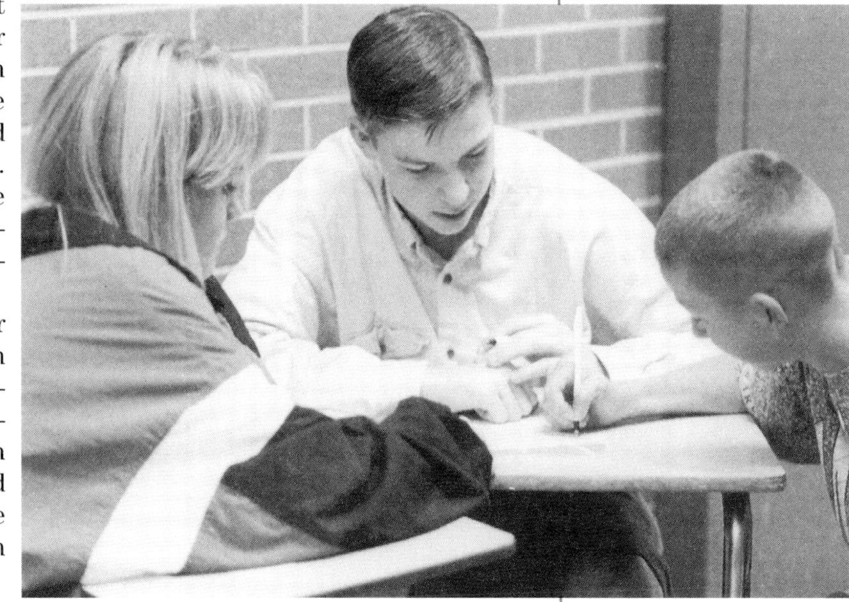

Over the next few days, other adults, including Lamb and a team of three regular-education preservice teachers, discussed the problem with the students. In each instance, the students explained that the switch strategy was the preferred one, and a discussion ensued about why this was so.

## ASSESSMENT AT THE CLASSROOM AND THE INDIVIDUAL LEVEL

Bethell evaluated herself and her students in several ways. She found it important to reflect on classroom discourse to evaluate the class as a whole. She discussed her evidence with Lamb, Kirsner, and other colleagues in the mathematics department.

Although most students had trouble making sense of *why* the switch strategy was advantageous in Monty's problem, they clearly accepted that it was. They demonstrated an understanding of the concept, strategy, and usefulness of simulating experiments. They accepted the evidence yielded by the simulations and rejected their original, strongly held guess about a desired strategy. Moreover, they conducted simulations and discussed the evidence in an environment that closely approximates the visions of the *Professional Standards for Teaching Mathematics* (NCTM 1991).

Bethell also considered it essential to elicit information from individual students. Her interviews reveal the complex learning that had occurred and reinforced the value of student interviews for assessment. The following interview exemplifies one student's thinking about the general mathematics class, the probability unit and Monty's problem in particular, and herself as a mathematics learner.

### Penelope

Penelope is a sophomore and a special education student. Her interview data reveal a surprising level of mathematics learning. Although she frequently seems disengaged from classroom activities and often expresses verbally that she is not smart enough to understand the subject matter, Penelope's interview responses demonstrate a relatively sophisticated understanding of probability. Her case illustrates how difficult it is to discern students' learning accurately. Excerpts of the interview follow. Kirsner is designated by *S. K.* and Penelope by *P.* Penelope begins by talking about never before having been given an opportunity to study important mathematics and about how different and positive this mathematics class has been.

*S.K.:* OK. Can you tell me a little bit about this year's math class. Is it a lot different from or different at all from previous math classes, and if so how? What has it been like?

*P.:* Yeah. It's different. I like this math class. And I like Miss Bethell. She's a good teacher. It's really better than my other math classes because the school I used to go to, when they had special ed, they always pulled us out of the class.... We don't feel self-confidence in ourself. We feel a little bit [more] stupid than other people. They just pull us out. And here they don't have to pull us out. It's not like we're in a special class ... and plus, I get a better grade in this math class than I did last year. Even though it's basically the same, it's just that I learn more here than I did over there because over there all I learned is all that addition. Here I learned geometry and how to do fractions, which I had no idea how to do, and I didn't know how to do geometry. And I learned probability and all this stuff that I never learned when I was in special ed. So all the stuff I learned was basic over there in special ed, in that other school. And they didn't show me probability, geometry, or fractions or how to use fractions.... They didn't show me all that. So therefore I was behind.

I always thought I was stupid in mathematics. And Miss Bethell thinks I'm very smart in it. I always felt bad because I didn't know all the stuff that everybody else knew, and now I know that stuff that everybody else knows now. And I'm really enjoying being in this class. And it just did me good. And now I don't feel stupid. I feel kind of smart. I mean because I know stuff that my aunt doesn't know and she wants me to come over and teach her how to do this stuff. And when I take prealgebra, I'm going to take algebra 1. I mean, next year with Miss Bethell because we heard she might be teaching algebra 1 again.... She wants me to come over and teach, help her learn stuff like that. So I kind of feel great about myself.

The seriousness and sophistication with which Penelope was responding surprised us. In class she regularly acted as if she was uninterested in virtually all academic tasks, spending considerable time walking around the room and often asking to put her head down because she had a headache. Yet she made the following responses in the interview:

*S.K.:* Can you tell me either what probability is about or give me an example or tell me why it might be important to study probability?

*P.:* Well. I have it written down what probability is about. See if I can find it.

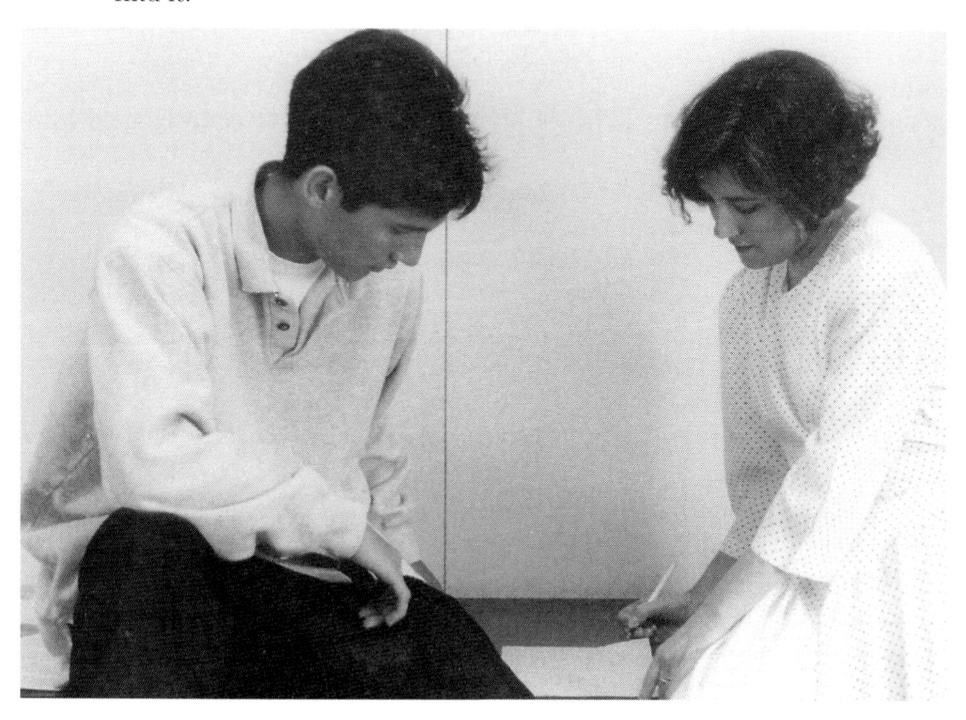

*S.K.:* Sure. Or if you can think about an example.

*P.:* Probability. Ah. Here it is. It's a chance that something will happen. It is expressed as a ratio.... Like if you flip a quarter you can, you have either a chance of getting a heads or a tails. The probability is one-half. Because there's two sides, one with a heads and one with a tails. So actually there's a one-half. You can get one heads if you flip and one tails if you flip so there's actually one-half. That's the probability of getting it. Or you can get—this is where I get stuck sometimes. With that one thing that you showed us. Like that ABC thing. You know.

*S.K.:* Oh, yeah. Where there are three doors: A, B, and C. And there's a prize behind one of them.

*P.:* Yep. There could be—I don't really know quite what you said, but I remember there was a one-third chance of being answered right. And there's a two-thirds chance if you switch.

*S.K.:* This was the problem. The problem was ABC. And say behind one of these is a thousand-dollar bill. And you pick one.

*P.:* A

*S.K.:* (*Pointing to door C*) And then I tell you it's not behind this one. Should you stay with A or should you switch to B?

*P.:* I'll switch because you get more of a chance of getting the answer right if you switch. Which my teacher, at first she thought that it'd be the same. You'd get a half-and-half chance. But you don't because you started out with three. And then when they ask you if it wasn't this one, you could change. You have the chance of changing or not changing. And so you have a better chance of changing than you do by staying.

Shaughnessy (1992, p. 485) uses four categories to characterize people's probabilistic conceptions:

1. *Non-statistical*.... Responses based on beliefs, ... no attention to or awareness of chance or random events
2. *Naive-statistical*.... Mostly experientially based ... responses; some understanding of chance and random events
3. *Emergent-statistical*.... Ability to apply normative models to simple problems; recognition that there is a difference between intuitive beliefs and a mathematized model ...
4. *Pragmatic-statistical*.... An in-depth understanding of mathematical models of chance

We believe that Penelope, who had never before studied probability and who refers to herself as "stupid," is a student in the emergent-statistical stage. According to Shaughnessy (1992, p. 486), such students—

> are under the influence of didactical [instructional] interventions, and their conceptions are changing. They are beginning to see the difference between degree of belief, and a mathematical model of a sample space. They are becoming able to

> apply normative models to an ever widening range of ... settings, but initially they may still be subject to falling back upon the familiar, causal or heuristic explanations when confronted with an unfamiliar type of task. Their ... conceptions are developing but are still unstable.

Penelope's interview demonstrates that she falls into this category. At the very least, she has clearly moved out of the nonstatistical category. This effect is not insignificant for a ten-week intervention, and it is unlikely to have occurred in the context of traditional instruction in probability.

## CONCLUSION

We have attempted to demonstrate that it is possible for a high school mathematics teacher to construct a learning environment that promotes the type of student learning described in the NCTM *Standards* documents (1989, 1991). Bethell's collaboration with Lamb and Kirsner; the goals of the Professional Development project for mathematics; and the thoughtful discussion and inquiry with other professionals in the school building supported the development of the nurturing environment.

The learning community that developed in Bethell's classroom grew out of her commitment to the philosophy and ideas in the *Curriculum and Evaluation Standards for School Mathematics* (NCTM 1989) and the *Professional Standards for Teaching Mathematics* (NCTM 1991) as well as her attitude of acceptance and respect for diverse students and their ideas. Her attitude and commitment found expression as she took risks by creatively reframing her district's curriculum, incorporating basic skills into more powerful topics, posing authentic questions to which she herself did not know the "correct answer," allowing the students to develop exploratory strategies and activities, and assessing students and her teaching by reflecting on whole-class discussions and individual interviews.

These strategies, rather than a particular unit or set of activities, gave the students' voices a forum in the classroom. This forum allowed them to understand mathematics conceptually, to find meaning in mathematical activities, and to see themselves as capable mathematics learners.

## REFERENCES


National Council of Teachers of Mathematics. *Curriculum and Evaluation Standards for School Mathematics*. Reston, Va.: The Council, 1989.

———. *Professional Standards for Teaching Mathematics*. Reston, Va.: The Council, 1991.

National Research Council. Mathematical Sciences Education Board. *Everybody Counts: A Report to the Nation on the Future of Mathematics Education*. Washington, D.C.: National Academy Press, 1989.

Shaughnessy, J. Michael. "Research in Probability and Statistics: Reflections and Directions." In *Handbook of Research on Mathematics Teaching and Learning: A Project of the National Council of Teachers of Mathematics*, edited by Douglas A. Grouws, pp. 465–94. New York: Macmillan Publishing Co., 1992.

Shaughnessy, J. Michael, and Thomas Dick. "Monty's Dilemma: Should You Stick or Switch?" *Mathematics Teacher* 84, (April 1991): 252–56.



At the time of his death in December 1992, **Steven Kirsner** was completing his doctoral dissertation, "Teaching Mathematics for Understanding and School Restructuring," at Michigan State University's (MSU) College of Education. Kirsner's work with the Michigan Partnership for New Education, the National Center for Research on Teacher Learning, and the U.S. Office of Educational Research and Improvement led to the publication of several articles on the teaching and learning of mathematics and the production of a widely used documentary film on mathematics teaching. Prior to attending MSU, Kirsner was and award-winning mathematics teacher in Dade County, Florida.

**Sandra Callis Bethell** is a mathematics teacher at Holt High School and an assistant instructor at Michigan State University, where she coteaches mathematics methods courses. She has participated in research projects with the Institute for Research on Teaching and the National Center for Research on Teacher Learning at Michigan State University. Her presentations and publications reflect her interest in providing access for all students to meaningful mathematics learning.

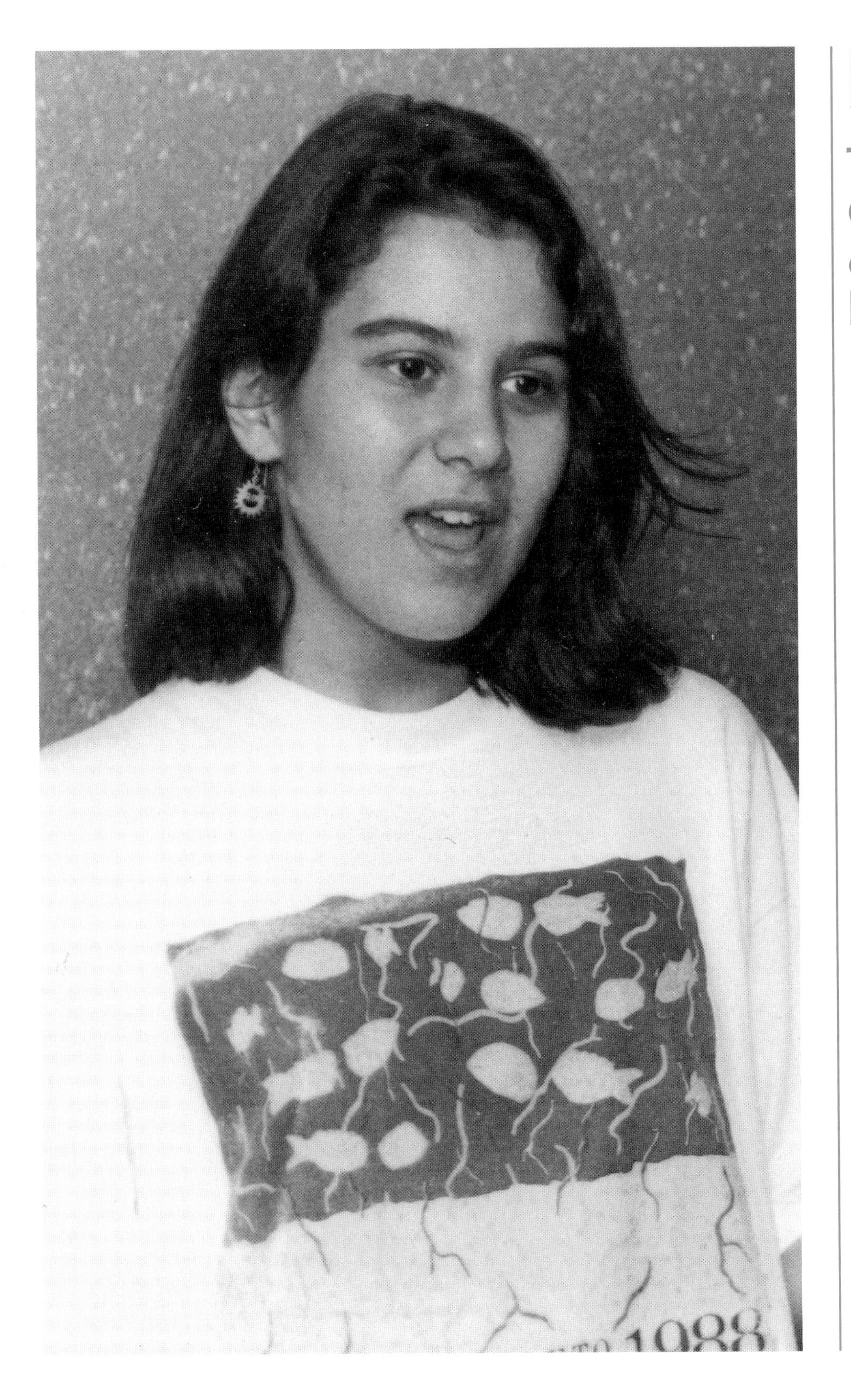

# Part 4

## The Challenge and the Promise

POLO
Kansas

21

# Opening the Windows of Opportunity: The Challenge of Implementing Change

*Linda P. Ware*

THE authors in each of the previous chapters have provided examples and evidence to substantiate the bold declaration by the National Council of Teachers of Mathematics (NCTM) that "*all* students can learn to think mathematically" (NCTM 1991, p. 21). Despite the simplicity of this conviction, it signals a fundamental reform of *what* mathematics is taught, *how* it is taught, and for *whom* instruction should be provided. NCTM (NCTM 1989, 1991; Meiring et al. 1992) has set the direction for rethinking the teaching and learning of school mathematics that will open windows of opportunity for all students, but as the authors in this text have described, the gains will be particularly significant for students with special needs.

Recognizing the potential of all learners in the call for reform in school mathematics is a significant and welcome change in the current dual system of education in which the separation rather than the inclusion of students has been standard practice. Well-established structures currently divide general education and special education into tracks, each sustained by separate personnel, programs, resources, and students. Thus the inclusion of all learners suggests a major reconceptualization of schooling beyond reform of the mathematics curriculum per se. The vision put forth by NCTM is a powerful one that cannot be realized without a greater understanding of the barriers to the successful implementation of the *Curriculum and Evaluation Standards for School Mathematics* (1989) and the *Professional Standards for Teaching Mathematics* (1991)—hereafter referred to as the *Standards*—in both general and special education.

Recognizing the potential of all learners in the call for reform in school mathematics is a significant and welcome change in the current dual system of education in which the separation rather than the inclusion of students has been standard practice.

This chapter describes some of the barriers that pose the greatest challenge to realizing the vision of the *Standards*. Through the collective efforts

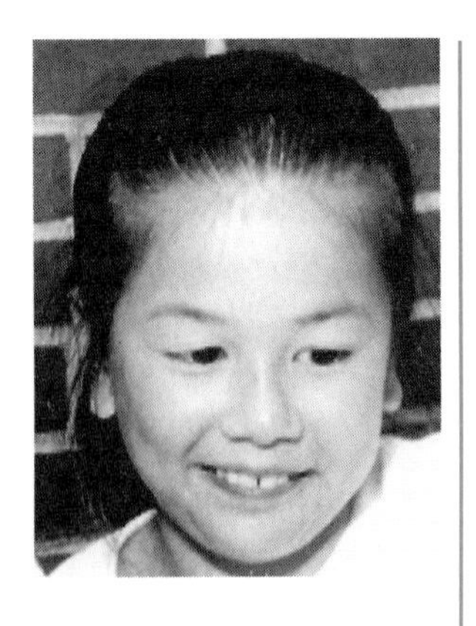

of general and special educators, a unique opportunity exists to transform these barriers into bridges that will in fact insure enriched learning opportunities for all students.

The chapter is divided into four sections. Barriers to *what* is taught and *how* it is taught in both general and special education will be detailed in sections 1 and 2. In section 3, a discussion of *who* is taught emphasizes the inherent barriers for special-needs learners. Finally, in section 4, specific recommendations are outlined for bridging the chasm between general and special education through the implementation of the *Standards* in one unified educational system.

## IMPLEMENTING THE *STANDARDS:* REFORMING *WHAT* IS TAUGHT

Among the many components of curriculum reform proposed by NCTM, perhaps the most provocative has been the recommendation for a reduced emphasis on arithmetic computation and memorization of complex algorithms in favor of an increased emphasis on the meaning and understanding of appropriate operations and procedures. Romberg (1992, p. 433) has proposed that

> the aims of teaching mathematics need to include the empowerment of learners to create their own mathematical knowledge; the reshaping of mathematics, at least in school, to give all groups more access to its concepts and to the wealth and power its knowledge brings; and bringing the social contexts in which mathematics is used and practiced into the classroom—the implicit values of mathematics need to be squarely faced by students. What it means when mathematics is seen in this way is that students need to study it in living contexts that are meaningful and relevant to them—contexts that include their languages, cultures, and everyday lives as well as their school-based experiences.

Forged of learners' reflection on, and discourse about, real-world problems, the curriculum reform proposed by NCTM would alter what is taught while it reshapes the meaning of mathematics. As a consequence, the traditional objective assumptions about the nature of reality and the development of knowledge would give way to a more subjective interpretation and an emphasis on individuals and their own interpretation or creation of knowledge. This shift from objectivism to subjectivism represents a radical departure from the conventional behavioristic learning theory that has driven instruction in schools.

### The Influence of Behaviorism

One of the principal barriers to the successful implementation of the *Standards* is the reluctance of educators to abandon the long-established strictures of behavioristic traditions that have influenced teaching, learning, and educational measurement throughout this century (see, e.g., Resnick and Klopfer [1989]; Schoenfeld [1987]; Schubert [1986]). Reflecting behaviorism, learning has traditionally been cast as a hierarchy of components, each of which is mastered in a logical fashion.

Learning and teaching structures in general education have been typified by specified curriculum objectives, rote responses, and worksheets of problem sets. Similarly, in special education, learning and teaching are typified by individualized education programs comprising specified behavioral objectives and students' mastery of discrete skills tied to worksheet exercises, often taught in isolation from any meaningful context.

Although learning in both general and special education is viewed as the mastery of a hierarchy of knowledge, additional constraints are imposed on special education students. Behavioristic assessment practices influence the determination of students' eligibility for special education services, and a linear approach to instruction is characteristic of the long-standing focus in special education on the remediation of diagnosed weaknesses and deficits. This diagnostic-prescriptive, deficit-driven approach to teaching, learning, and assessment reinforces the belief that the mastery of skills takes precedence over understanding (Heshusius 1982; Poplin 1984).

Because students with special needs require more time to master "the very large number of very small steps" needed to demonstrate competence, they typically receive instruction outside the regular classroom. Despite recent criticism from the general education community (see, e.g., Gardner [1991]; Resnick and Klopfer [1989]; Resnick [1989]; Schoenfeld [1987]; Shepard [1991]; and Sternberg [1990]) and equal criticism from the special education community (see, e.g., Heshusius [1989]; Iano [1987]; Poplin [1984]; Skrtic [1991]), behavioristic learning theory remains firmly entrenched in education.

In contrast, the curriculum recommended by the *Standards* adopts the view of reality as a personal construction of complex interactions between the environment and the learner and thus rejects behaviorism in favor of constructivism. Based on the work of Piaget, constructivist learning theory assumes activity and reflection by the learner. Thus, when the *Standards* stress the increased use of concrete materials and the construction of models, the goal is to provide students the opportunity to construct their own meaning and knowledge about mathematics through activity and reflection. Likewise, when the *Standards* stress communication, the goal is to promote verbal and written expression by students as a consequence of structured opportunities for activity and reflection in mathematics.

The curriculum recommended by the *Standards* adopts the view of reality as a personal construction of complex interactions between the environment and the learner and thus rejects behaviorism in favor of constructivism.

## A Constructivist Perspective

Curriculum reform grounded in constructivist learning theory is inspired and ambitious. For many educators it represents a wholly new learning theory and poses formidable challenges to well-established teaching and learning structures. Learning time, subject matter, instructional materials, and the classroom and school environment each require review and revision to accommodate a constructivist perspective to teaching and learning mathematics.

In the interests of time, logic once dictated that the school day be divided into six or seven equal periods, rather than three or four blocks of time, to accommodate the presentation and mastery of individual curriculum components. As a consequence, instruction, instructional materials, and the learning environment have been characterized by efficiency and management. Constrained by this structure, teachers have often been influenced by two considerations:

- How quickly can the material be covered?
- How quickly will the students demonstrate mastery?

Concerns like these affect instruction in ways that present a barrier to the reforms outlined by the *Standards*. The use of concrete materials, interactive learning activities, and ongoing student-directed learning projects are time-intensive processes that deviate from the time-efficient tasks of the context-free memorization of algorithms, the drill and practice of facts, and the daily routine presentation of a two-page spread of textbook problems.

The management of learning in special education is similar to that in the general education classroom, despite the broad range of curricula presented to accommodate the diverse range of students' ability levels. Students are assigned to a resource room for specified periods with the expectation that they will progress through modified curricular activities in a stepwise approach. Drill-and-practice skills development and grade-level workbook or textbook activities form the core of special-needs students' daily work.

Thus, the requirements of a constructivist learning environment cannot readily be accommodated in either general or special education as each is currently structured. In order to minimize frustration and failure by both students and teachers, an activity-based, hands-on curriculum assumes that through restructuring, teachers will set aside larger *blocks* of time, rather than traditional *periods* of time, for learning to accommodate exploration and investigation.

Teachers need to set aside larger *blocks* of time, rather than traditional *periods* of time, for learning to accommodate exploration and investigation.

The proposed shift to a structure in which blocks of time replace periods of time has many implications. Whenever blocks of time replace class periods, small groups of students working together become a structural necessity. In the transformation from a "class" to a "community" structure, students assume greater control and responsibility for their own learning and that of their peers. Cooperative learning becomes more than *what* children do in school; instead, it reflects *how* children learn in school as well as in life.

For this reason, it is essential that these changes be viewed in the context of a learning theory that encourages individual reflection and interaction

among learners. It then becomes more meaningful to emphasize “discussing, describing, and demonstrating [mathematical knowledge as a] social process” (NCTM 1989, p. 214). As a consequence, activity and reflection become less contrived and a more natural response to learning.

An understanding of the shift from behavioristic to constructivist learning theory is central to the successful implementation of mathematical reform according to the *Standards*. Without such an understanding, the necessary reappropriation of learning time will not occur, and the ineffective organization of time will pose a significant barrier to the implementation of the *Standards*.

In the following section, these arguments will be extended in a discussion of how teachers have been encouraged to teach in the spirit of the *Standards*. After ineffective classroom structures and the inefficient organization of instructional time, it will be argued, the greatest barrier to the implementation of NCTM's recommendations is the current organization of schools.

In the transformation from a "class" to a "community" structure, students assume greater control and responsibility for their own learning and that of their peers.

## REALIZING THE *STANDARDS*: REFORMING *HOW* TEACHERS TEACH

The NCTM *Standards* documents (1989, 1991) and the accompanying K–12 Addenda Series (1991–94) represent the efforts of many individuals who have collaborated to produce a vision for the improvement of mathematics learning and teaching in America's schools. Increased opportunities for professional collaboration and improved collegiality among mathematics

Increased opportunities for professional collaboration and improved collegiality among mathematics professionals are viewed as essential to the successful implementation of the *Standards*.

professionals are viewed as essential to the successful implementation of the *Standards*. The *Professional Teaching Standards* urges that school administrators allow "time for teachers to plan, to reflect, to help each other improve instruction; time for professional development; and time to interact with the community" (NCTM 1991, p. 181).

We cannot assume, however, that such interactions occur under the current system. With its hierarchical authority and well-established power structures, the traditional organization of schools undermines meaningful collaboration in both daily operations and daily interactions (Ware 1993).

Numerous individuals and educational commissions, including both the Holmes Group (1986) and the Carnegie Task Force (Carnegie Foundation 1986), have long recognized that increased professionalism among teachers is essential to educational reform. Nonetheless, the goal remains illusive. NCTM recognizes that although the process of developing a collaborative ethic may be slow, it is clearly essential to the implementation of the *Standards*. For a more complete discussion of mathematics teachers' progress toward this goal, see Noddings (1992).

Certainly, a danger exists in assuming that teachers will naturally shift into the new roles demanded by a collaborative ethic; the anticipation of reaching this goal is as likely to overwhelm teachers as to empower them. On this point, Apple (1992) has cautioned that given the dramatic changes in both teaching and learning set forth in the *Standards*, the potential exists for teachers to experience a "chronic sense of work overload" (p. 426)—a phenomenon he has described as "intensification" (Apple 1988). He cites intensification as one of the unanticipated negative outcomes of site-based reform, wherein pressures on teachers become so great as to undermine their meaningful inclusion in the change process.

NCTM has attempted to minimize this pressure on teachers. Such publications as those in the *Curriculum and Evaluation Standards for School Mathematics* Addenda Series (1991, 1992, 1993) have been developed to give teachers "direction for moving toward excellence in teaching mathematics" (NCTM 1991, p. 7) as well as a basis for professional discourse. Through the use of annotated vignettes of model teaching in accordance with the *Standards*, activities and instruction are presented that discourage the "tell, show, and do" model common to most mathematics classrooms in favor of increased student activity and nonroutine, open-ended problem solving.

Activities and instruction discourage the "tell, show, and do" model in favor of increased student activity and nonroutine, open-ended problem solving.

By creating a problem-rich learning environment, teachers can observe and listen to students in the process of developing mathematical understanding. Thus, teachers are encouraged to develop their monitoring skills in order to gain information and improve student learning and simultaneously to reflect on their own teaching and its effect on student learning.

## Professional Development as a Dynamic Process

NCTM contends that through professional development and professional dialogue teachers can develop skills to improve mathematics content and pedagogy. The *Professional Teaching Standards* (NCTM 1991) suggests that professional development assumes a dynamic and continuous process in which teachers' growth is "deeply embedded in their philosophies of learning, their attitudes and beliefs about learners and mathematics, and their willingness to make changes in how and what they teach" (p. 125).

Given the significance of the shift from behavioristic to constructivist practices, it is essential that educators articulate a learning theory to guide their own beliefs and practices. Through professional development and dialogue, educators can begin to address that which was previously identified as a barrier to the implementation of the *Standards:* unreflective conformance to the strictures of behavioristic learning theory.

Further, because constructivism does not offer "pedagogical recipes or convenience" (Davis, Maher, and Noddings 1990, p. 188), the value of professional dialogue is that it will emerge in concert with actual instructional exploration and experimentation by both students and teachers. Just as reflection and communication among students are less contrived and more meaningful in conjunction with concrete learning activities, dialogue among teachers will be less contrived if it occurs in conjunction with meaningful, theory-based curriculum reform. Thus, reflection is to activity as theory is to practice.

## Professional Dialogue and Collaboration

Collaboration is premised on the ability of professionals to work together to identify common problems and through collective reflection, to devise, test, and revise solutions. Collaboration requires a school climate wherein accountability and responsibility are shared by all. However, because schools are based on *specialization* rather than *collaboration*, individuals generally work alone; maintain responsibility for some, but not all, students; and rarely integrate instructional content. When a problem emerges that defies a simple solution, the specialist who is presumed to have the necessary skills is assigned the problem, which reinforces the isolation of both students and teachers (Skrtic and Ware 1992).

Collaboration is premised on the ability of professionals to work together to identify common problems and through collective reflection, to devise, test, and revise solutions.

The most visible structures that result from specialization include special, remedial, and compensatory education. Each of these separate systems would be a structural impossibility if schools were organized for collaboration rather than specialization. The current system is itself a barrier that discourages the development of a collaborative ethic wherein participants can share common problems, derive common solutions, or accept responsibility for all students. In short, the very structure of schools impedes innovation, professional development, and collaboration (see, e.g., Elmore and

McLaughlin [1988]; Rosenholtz [1989]; Sarason [1990]; Skrtic [1991]; Skrtic and Ware [1992]).

The barriers thus far identified represent well-established structures that many educators may find impossible to reconsider or revise. Mathematics education, however, "exists as part of a larger curriculum and needs to be thought about and integrated into the larger picture of policies and practices of transformations in the entire school itself" (Apple 1992, p. 429). Behavioristic learning theory, specified learning periods, and specialization are no mere stage props in the drama of American education. Each component could easily be labeled the *sine qua non* of the system, but when they are lumped together and labeled as "barriers" to the successful implementation of reform, the challenge to change these structures becomes formidable.

Add to the agenda for reform the belief expressed by NCTM that the *Standards* aim to provide enriched learning experiences for all students, and change becomes even more overwhelming. A discussion follows of the barriers to including all learners in this reform, with particular emphasis on those in special education.

## INCLUDING ALL LEARNERS IN A SINGLE SYSTEM: REFORMING *WHO* IS TAUGHT

Despite NCTM's ambitious inclusion of all learners in the proposed reform, the fact is that all learners are not included in most mathematics classrooms. The response to diversity in students' abilities has traditionally been exclusion and placement in special, remedial, or compensatory education programs. More recently, programs for at-risk students and other such targeted populations as inner-city, black, male youth have also been created with specialization as a major component.

Until the educational system welcomes diversity as a strength rather than a weakness, the *Standards* will not be successfully implemented.

The widely held belief that providing separate programs and services is in the best interest of the child is somewhat suspect, given that this practice clearly serves to perpetuate the current system. The deficiences of the current system will be confirmed once teachers and schools attempt to alter the structure to create an environment where all students can experience success. Until the educational system welcomes diversity as a strength rather than a weakness, the *Standards* will not be successfully implemented.

We have elaborated on the meaning of the phrase *all learners* in describing practices that meet the needs of students in gifted, special, remedial, and compensatory education. The policy remains unclear, however, on where this revised curriculum will be furnished to all students. The *Standards* recommend that through "differentiation" of content, teachers can address learner diversity. However, until the setting for this instruction is clarified, differentiation amounts to no more than continued exclusion.

If the organization of schools remains unchanged, special education students will continue to receive instruction in the special education classroom from special education teachers in a segregated setting. The NCTM *Standards* documents included *all* students in reform, and subsequent messages have been more explicit. *A Core Curriculum* (Meiring et al. 1992), for example, presents a common core of topics for grades 9–12 that all students should have the opportunity to learn. This volume recommends several cur-

ricular models for organizing a nonelitist curriculum, including a "crossover" model that represents a "first step away from current multitrack programs that offer advanced mathematics for a few and minimal mathematics for the majority" (p. vi). The activities endorsed emphasize cooperative group work and are designed to accommodate a broad range of learners through differentiation in learning outcomes.

In this document, NCTM reiterates the vision of the *Standards* as promoting an "increased sense of community" (p. 114) among teachers. The criticism of the tradition in mathematics in which specialized instruction reinforces an elitist curriculum includes its impeding the emergence of cooperative communities (p. 114):

> Instituting a core program will bring department members into closer working relationships: more sharing of knowledge, more working together, and more vesting in expertise and dissemination. Individuals will need to give up some autonomy in making teaching decisions and engage in more cooperative planning, program assessment, and analyses of data. Those who have the knowledge and expertise needed to bring new topics to more diverse populations of students will need to share it .

NCTM's vision for change clearly exceeds the depth and breadth of any previously attempted curriculum reform. In view of the barriers that have

In view of the barriers, restructuring becomes integral to the successful implementation of the *Standards*.

been outlined, restructuring becomes integral to the successful implementation of the *Standards*. After a decade of national debate on reform and restructuring, the *Standards* hold much promise for creating change not only in the curriculum but in all facets of schooling.

The final section of this chapter offers a unique opportunity for those who have urged fundamental school reform and restructuring in both general and special education to ally themselves with the proponents of the *Standards*. Through collective action, the barriers can be transformed into bridges that will ensure enriched learning opportunities for all students in one unified system.

## REMOVING THE BARRIERS TO THE *STANDARDS*: PROVIDING FOR ALL LEARNERS

The three fundamental barriers to the implementation of the *Standards* for both general and special education are these:

- The inability of educators to abandon the strictures of behaviorism
- The inability of educators to structure blocks of time
- The inability of schools to initiate systemwide change from specialization to collaboration as the preferred organizational configuration

The proponents of the Regular Education Initiative (REI), who call for special education reform, and general educators who call for restructuring of the current system encounter similar obstacles. If removing these barriers cleared the way for a nonelitist mathematics curriculum, the *Standards* would have a greater potential for implementation.

### Coalescing Reform Strands

#### *The Regular Education Initiative*

In 1986 when Madeline Will, then assistant secretary for the Office of Special Education and Rehabilitative Services (OSERS), challenged general and special educators to renew their commitment to the education of children with learning problems, the Education of the Handicapped Act (EHA) (P.L. 94-142) had been in effect for little more than ten years. Will proposed the Regular Education Initiative (REI) in response to the empirical data that questioned the efficacy of the dual system of educational services (see, e.g., Heller, Holtzman, and Messick [1982]; Wang, Reynolds, and Walberg [1987]).

In the years since, REI proponents have recognized the need to unite with general educators to change the current educational system and ensure that it becomes "both one and special for all students" (Lipsky and Gartner 1989, p. 74). For a more comprehensive analysis of the current reevaluation of special education, see Hehir and Latus (1992).

#### *General Education Reform*

Similarly during the 1980s, an unprecedented groundswell of reform and restructuring proposals emerged in general education (Bacharach 1990;

Cuban 1989; Raywid 1990). The issues raised included equity versus excellence, top-down versus grass-roots change, increased centralization versus decentralization, the empowerment of teachers, authentic assessment, block scheduling versus Carnegie units, and parental choice. Despite their pet buzzwords and the usual arguments, most reformers can agree that "extensive improvement is imperative" (Raywid 1990, p. 143).

Although many suggestions for reform addressed the structures of schools, very few addressed teaching and learning. We anticipate that the decade of the '90s and the dawning of the twenty-first century will bring a new awareness of, and emphasis on, teaching and learning influenced by earlier reforms and responding to the need to organize schools better for instruction and ongoing curriculum reform (see, e.g., Cuban [1989, 1990]; Guskey [1990]; Raywid [1990]).

In a proposal to address the need for attention to the curriculum, Sykes (1990) has suggested the need for "norms of conduct and standards of practice" (p. 367). His proposal is strikingly similar to the documents produced by NCTM. Sykes explains (p. 367):

> By norms of conduct I mean a range of moral and aesthetic judgments governing the role responsibilities of teachers. By standards of practice I mean the technical and procedural rules and guidelines that direct instruction. Such norms and standards can rise only out of dialogue and reflection based on the close observation of teaching.

Sykes further encourages teachers to "experiment with activities unfamiliar to themselves and to their students" (p. 368) as a means to improve teaching and learning. Even though no specific reference is made to the *Standards*, it is clear that if this experimentation were to occur in the context of a reform initiative, dialogue and reflection among teachers and other invested professionals would become necessary.

## Reshaping the Barriers as Bridges

Working in tandem, teachers have the opportunity to learn from one another as they learn together.

Consider the potential opportunity for general and special education teachers to experiment and explore together to create a curriculum consistent with the framework proposed by the *Standards*. Working in tandem, teachers would have the opportunity to learn from one another as they learn together. As a matter of policy, special education teachers currently instruct a broad range of learners, and their expertise could contribute to the development of a curriculum with differentiated learning outcomes. Given the numerous challenges posed by the implementation of a constructivist curriculum in mathematics, a structure wherein two teachers rather than one facilitate learning has much promise. Working cooperatively, the general and special education teachers could realistically accommodate a wider range of student needs with concrete activities and model-based instruction.

This proposal differs from many of the current models of in-class support prompted by the "collaboration movement" that have emerged in the field of special education (see, e.g., Curtis and Meyers [1988]; Idol, Paolucci-Whitcomb, and Nevin [1986]; Pugach and Johnson [1988]). Central to this proposal is its focus on the *development* of a specific curriculum for *all*. Rather than adapt the existing curriculum to accommodate special-needs learners, this proposal calls for the development of a wholly new curriculum that centers on the learner and de-emphasizes the role of the teacher as the authority or single source of information.

Beginning with a structure wherein both teachers assume the responsibility for all learners will eliminate the problem of "terrain" that has impeded the success of collaboration between general and special educators (see, e.g., Pugach and Johnson [1989]; Wesson [1991]; Phillips and McCullough [1990]; Ysseldyke, Thurlow, Wortuba, and Nania [1990]). The fundamental reform proposed by NCTM would enable teachers to begin this collaborative effort as equal partners so that neither professional could assume dominance over the other.

A new professional culture might prosper in the absence of a structure that reinforces elitism and separatist practices among professionals. Classrooms would become communities of learners—creative problem solvers involved in the process of creating knowledge—and schools could become communities of teachers who themselves are creative problem solvers, actively involved in the process of creating curriculum and collaboratively solving problems as they arise.

Implementing the *Standards* provides a unique opportunity to merge the policy and practice of special and general education. Success will be manifest in the implementation of a novel curriculum grounded in the belief that all students can learn. This proposal represents two dramatic shifts in the current operation of schools. The first shift is the structural change from the present dual system of education to the proposed unified system. The second is somewhat more subtle, resulting from the cultural change prompted by the shift from behaviorism to constructivism. Both pose a unique set of issues with which schools must contend, and both are ambitious and long-overdue reforms that can, through collective efforts, be accomplished in tandem. An obvious and necessary starting point is to explicitly include special education teachers in the professional development and dialogue endorsed by the NCTM for implementation of the *Standards*.

## FROM VISION TO REALIZATION

This chapter began with the restatement of the simple belief that has inspired NCTM and its visionary development of the *Standards* and related curriculum efforts. Since the publication of the documents, efforts have been made by numerous educators, alone and collectively, to translate this belief into practice. Their inchoate and uncertain attempts have been refined over time, and as demonstrated in this chapter, the challenge of the project continues to entice and include a broad range of educators.

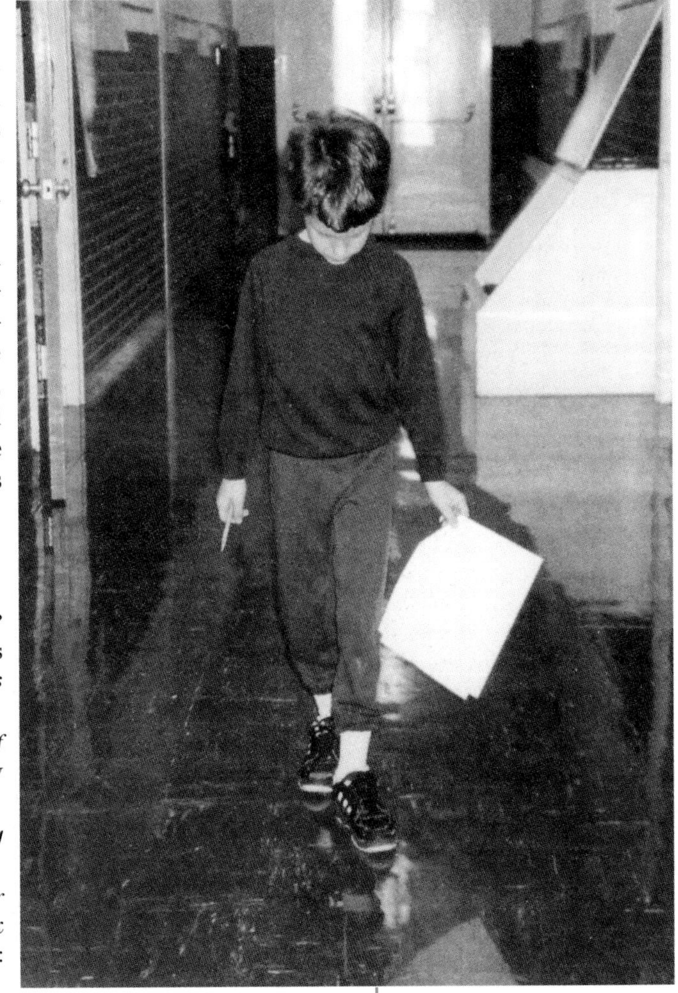

As contact is renewed among professionals and teaching and learning are reconsidered, new environments for mathematics instruction will be created. Our next steps to reform may prove more difficult than the first ones. It is clear, however, that the inspired community has grown, and progress will continue. In the words of the African American poet Audre Lorde, "Our visions begin with our desires" (Tate 1983, p. 107).

## REFERENCES


Apple, Michael W. "Do the *Standards* Go Far Enough? Power, Policy, and Practice in Mathematics Education." *Journal for Research in Mathematics Education* 23 (November 1992): 412–31.

———. *Teachers and Texts: A Political Economy of Class and Gender Relations in Education.* New York: Routledge, 1988.

Bacharach, Samuel B. *Education Reform: Making Sense of It All.* Boston: Allyn & Bacon, 1990.

Carnegie Foundation. *A Nation Prepared: Teachers for the Twenty-first Century: The Report of the Task Force on Teaching as a Profession.* New York: Carnegie Corporation, 1986.

Cuban, Larry. "The 'At-Risk' Label and the Problem of Urban School Reform." *Phi Delta Kappan* 70 (1989): 780–84, 799–81.

———. "Four Stories about National Goals for American Education." *Phi Delta Kappan* 72 (December 1990): 264–71.

Curtis, Michael J., and Joel Meyers. "Consultation: A Foundation for Alternative Services in the Schools." In *Alternative Educational Delivery Systems: Enhancing Instructional Options for All Students*, edited by Janet L. Graden, Joseph E. Zins, and Michael J. Curtis, pp. 35–48. Washington, D.C.: National Association of School Psychologists, 1988.

Davis, Robert B., Carolyn A. Maher, and Nel Noddings. "Suggestions for the Improvement of Mathematics Education." In *Constructivist Views on the Teaching and Learning of Mathematics*, *Journal for Research in Mathematics Education* Monograph No. 4, edited by Robert B. Davis, Carolyn A. Maher, and Nel Noddings, pp. 187–91. Reston, Va.: National Council of Teachers of Mathematics, 1990.

Elmore, Richard F., and Milbery W. McLaughlin. *Steady Work: Policy, Practice, and the Reform of American Education.* Santa Monica, Calif.: Rand Corporation, 1988.

Gardner, Howard. *The Unschooled Mind: How Children Think and How Schools Should Teach.* New York: Basic Books, 1991.

Guskey, Thomas R. "Integrating Innovation." *Educational Leadership* 47 (February 1990): 11–15.

Hehir, Thomas, and Thomas Latus. *Special Education at the Century's End: Evolution of Theory and Practice since 1970.* Cambridge: Harvard University Press, 1992.

Heller, Kirby, Wayne Holtzman, and Samuel Messick. *Placing Children in Special Education: A Strategy for Equity.* Washington, D.C.: National Academy of Science Press, 1982.

Heshusius, Lous. "At the Heart of the Advocacy Dilemma: A Mechanistic World View." *Exceptional Children* 49 (August/September 1982): 6–13.

———. "The Newtonian-Mechanistic Paradigm, Special Education, and Contours of Alternatives: An Overview." *Journal of Learning Disabilities* 22 (September 1989): 403–15.

Holmes Group. *Tomorrow's Teachers.* East Lansing, Mich.: Holmes Group, 1986.

Iano, Richard P. "Rebuttal: Neither the Absolute Certainty of Prescriptive Law nor a Surrender to Mysticism." *Remedial and Special Education* 7 (January/February 1987): 50–61.

Idol, Lorna, Phyllis Paolucci-Whitcomb, and Ann Nevin. *Collaborative Consultation.* Austin, Tex.: Pro-Ed, 1986.

Lipsky, Dorothy K., and Alan Gartner, eds. *Beyond Separate Education: Quality Education for All.* Baltimore: Paul H. Brookes Publishing Co., 1989.

Meiring, Steven P., Rheta N. Rubenstein, James E. Schultz, Jan de Lange, and Donald L. Chambers. *A Core Curriculum: Making Mathematics Count for Everyone.* Reston, Va.: National Council of Teachers of Mathematics, 1992.

National Council of Teachers of Mathematics. *Curriculum and Evaluation Standards for School Mathematics.* Reston, Va.: The Council, 1989.

———. *Curriculum and Evaluation Standards for School Mathematics* Addenda Series. K–6 series, edited by Miriam A. Leiva; 5–8 series, edited by Frances R. Curcio; 9–12 series, edited by Christian R. Hirsch. Reston, Va.: The Council, 1991–94.

———. *Professional Standards for Teaching Mathematics.* Reston, Va.: The Council, 1991.

Noddings, Nel. "Professionalization and Mathematics Teaching." In *Handbook of Research on Mathematics Teaching and Learning*, edited by Douglas A. Grouws, pp. 197–208. New York: Macmillan Publishing Co., 1992. (Available also from the NCTM.)

Phillips, Vicki, and Laura McCullough. "Consultation-Based Programming: Instituting the Collaborative Ethic in Schools." *Exceptional Children* 56 (January 1990): 291–304.

Poplin, Mary S. "Toward an Holistic View of Persons with Learning Disabilities." *Learning Disability Quarterly* 7 (Fall 1984): 290–94.

Pugach, Marlene, and Lawrence J. Johnson. "The Challenge of Implementing Collaboration between General and Special Education." *Exceptional Children* 56 (November 1989): 232–35.

———. "Peer Collaboration." *Teaching Exceptional Children* 21 (1988): 75–77.

Raywid, Mary Anne. "The Evolving Effort to Improve Schools: Pseudo-Reform, Incremental Reform, and Restructuring." *Phi Delta Kappan* 72 (October 1990): 139–43.

Resnick, Lauren B. "Developing Mathematical Knowledge." *American Psychologist* 44 (February 1989): 162–69.

Resnick, Lauren B., and L. E. Klopfer. *Toward the Thinking Curriculum: Current Cognitive Research*. Alexandria, Va.: Association for Supervision and Curriculum Development, 1989.

Romberg, Thomas A. "Further Thoughts on the *Standards:* A Reaction to Apple." *Journal for Research in Mathematics Education* 23 (November 1992): 432–37.

Rosenholtz, Susan J. *Teachers' Workplace: The Social Organization of Schools*. White Plains, N.Y.: Longman, 1989.

Sarason, Seymour. *The Predictable Failure of Educational Reform*. San Francisco: Jossey-Bass Publishers, 1990.

Schoenfeld, Alan H., ed. *Cognitive Science and Mathematics Education*. Hillsdale, N.J.: Lawrence Erlbaum Associates, 1987.

Schubert, William H. *Curriculum: Perspective, Paradigm, and Possibility*. New York: Macmillan, 1986.

Shepard, Lorrie A. "Psychometricians' Beliefs about Learning." *Educational Researcher* 20 (October 1991): 2–16.

Skrtic, Thomas. *Behind Special Education: A Critical Analysis of Professional Culture and Schools Organization*. Denver: Love Publishing, 1991.

Skrtic, Thomas, and Linda Ware. "Reflective Teaching and the Problem of School Organization." In *Teacher Personal Theorizing: Connecting Curriculum Practice, Theory, and Research*, edited by E. Wayne Ross, Jeffrey W. Cornett, and Gail McCutcheon, pp. 207–18. New York: State University of New York Press, 1992.

Sternberg, Robert J. "Thinking Styles: Keys to Understanding Student Performance." *Phi Delta Kappan* 71 (January 1990): 366–71.

Sykes, Gary. "Creative Teaching: Evaluating the Learning Process instead of Just the Outcome." In *Education Reform: Making Sense of It All*, edited by S. B. Bacharach, pp. 362–69. Boston: Allyn & Bacon, 1990.

Tate, Claudia, ed. *Black Women Writers at Work*. New York: Continuum Publishing Group, 1983.

Wang, Margaret C., Maynard C. Reynolds, and Herbert J. Walberg. *Handbook of Special Education: Research and Practice*. Vol. 1, *Learner Characteristics and Adaptive Education*. Oxford: Pergamon Press, 1987.

Ware, Linda. "Contextual Barriers to Collaboration." *Journal of Educational and Psychological Consultation*. Forthcoming.

Wesson, Caren L. "Curriculum-Based Measurement and Two Models of Follow-up Consultation." *Exceptional Children* 57 (December/January 1991): 246–56.

Ysseldyke, James E., Martha L. Thurlow, Joseph W. Wortuba, and Paula A. Nania. "Instructional Arrangements: Perceptions from General Education." *Teaching Exceptional Children* 22 (Summer 1990): 4–8.

**Linda Ware,** an experienced teacher, is a doctoral candidate in the Department of Special Education at the University of Kansas, Lawrence, Kansas. She is employed by the Northeast Kansas Education Service Center and serves as a consultant for a variety of teacher-training projects in mathematics, curriculum development, and special education inclusion.

# Appendix A

# Recommendations for Curriculum Planning from the NCTM: Teaching Mathematics to Empower ALL Students, K–12

## RECOMMENDATIONS FOR CURRICULUM PLANNING

In its *Curriculum and Evaluation Standards for School Mathematics*, the National Council of Teachers of Mathematics (1989) has set the tone for framing a quality mathematics program for all students, K–12. Summary information from this document is presented below.

## CURRICULUM STANDARDS FOR GRADES K–4

STANDARD 1: Mathematics as Problem Solving

In grades K–4, the study of mathematics should emphasize problem solving so that students can—

- use problem-solving approaches to investigate and understand mathematical content;
- formulate problems from everyday and mathematical situations;
- develop and apply strategies to solve a wide variety of problems;
- verify and interpret results with respect to the original problem;
- acquire confidence in using mathematics meaningfully.

STANDARD 2: Mathematics as Communication

In grades K–4, the study of mathematics should include numerous opportunities for communication so that students can—

- relate physical materials, pictures, and diagrams to mathematical ideas;
- reflect on and clarify their thinking about mathematical ideas and situations;
- relate their everyday language to mathematical language and symbols;
- realize that representing, discussing, reading, writing, and listening to mathematics are a vital part of learning and using mathematics.

STANDARD 3: Mathematics as Reasoning

In grades K–4, the study of mathematics should emphasize reasoning so that students can—

- draw logical conclusions about mathematics;
- use models, known facts, properties, and relationships to explain their thinking;
- justify their answers and solution processes;
- use patterns and relationships to analyze mathematical situations;
- believe that mathematics makes sense.

STANDARD 4: Mathematical Connections

In grades K–4, the study of mathematics should include opportunities to make connections so that students can—

- link conceptual and procedural knowledge;
- relate various representations of concepts or procedures to one another;
- recognize relationships among different topics in mathematics;
- use mathematics in other curriculum areas;
- use mathematics in their daily lives.

STANDARD 5: Estimation

In grades K–4, the curriculum should include estimation so students can—

- explore estimation strategies;
- recognize when an estimate is appropriate;
- determine the reasonableness of results;
- apply estimation in working with quantities, measurement, computation, and problem solving.

STANDARD 6: Number Sense and Numeration

In grades K–4, the mathematics curriculum should include whole number concepts and skills so that students can—

- construct number meanings through real-world experiences and the use of physical materials;
- understand our numeration system by relating counting, grouping, and place-value concepts;
- develop number sense;
- interpret the multiple uses of numbers encountered in the real world.

STANDARD 7: Concepts of Whole Number Operations

In grades K–4, the mathematics curriculum should include concepts of addition, subtraction, multiplication, and division of whole numbers so that students can—

- develop meaning for the operations by modeling and discussing a rich variety of problem situations;
- relate the mathematical language and symbolism of operations to problem situations and informal language;
- recognize that a wide variety of problem structures can be represented by a single operation;
- develop operation sense.

STANDARD 8: Whole Number Computation

In grades K–4, the mathematics curriculum should develop whole number computation so that students can—

- model, explain, and develop reasonable proficiency with basic facts and algorithms;
- use a variety of mental computation and estimation techniques;
- use calculators in appropriate computational situations;
- select and use computation techniques appropriate to specific problems and determine whether the results are reasonable.

STANDARD 9: Geometry and Spatial Sense

In grades K–4, the mathematics curriculum should include two- and three-dimensional geometry so that students can—

- describe, model, draw, and classify shapes;
- investigate and predict the results of combining, subdividing, and changing shapes;
- develop spatial sense;
- relate geometric ideas to number and measurement ideas;
- recognize and appreciate geometry in their world.

STANDARD 10: Measurement

In grades K–4, the mathematics curriculum should include measurement so that students can—

- understand the attributes of length, capacity, weight, mass, area, volume, time, temperature, and angle;
- develop the process of measuring and concepts related to units of measurement;
- make and use estimates of measurement;
- make and use measurements in problem and everyday situations.

STANDARD 11: Statistics and Probability

In grades K–4, the mathematics curriculum should include experiences with data analysis and probability so that students can—

- collect, organize, and describe data;
- construct, read, and interpret displays of data;
- formulate and solve problems that involve collecting and analyzing data;
- explore concepts of chance.

STANDARD 12: Fractions and Decimals

In grades K–4, the mathematics curriculum should include fractions and decimals so that students can—

- develop concepts of fractions, mixed numbers, and decimals;
- develop number sense for fractions and decimals;
- use models to relate fractions to decimals and to find equivalent fractions;
- use models to explore operations on fractions and decimals;
- apply fractions and decimals to problem situations.

STANDARD 13: Patterns and Relationships

In grades K–4, the mathematics curriculum should include the study of patterns and relationships so that students can—

- recognize, describe, extend, and create a wide variety of patterns;
- represent and describe mathematical relationships;
- explore the use of variables and open sentences to express relationships.

# SUMMARY OF CHANGES IN CONTENT AND EMPHASIS FOR MATHEMATICS IN GRADES K–4

## Increased Attention

### *Number*

- Number sense
- Place-value concepts
- Meaning of fractions and decimals
- Estimation of quantities

### *Operations and Computation*

- Meaning of operations
- Operation sense
- Mental computation
- Estimation and the reasonableness of answers
- Selection of an appropriate computational method
- Use of calculators for complex computation
- Thinking strategies for basic facts

### *Geometry and Measurement*

- Properties of geometric figures
- Geometric relationships
- Spatial sense
- Process of measuring
- Concepts related to units of measurement
- Actual measuring
- Estimation of measurements
- Use of measurement and geometry ideas throughout the curriculum

### *Probability and Statistics*

- Collection and organization of data
- Exploration of chance

### *Patterns and Relationships*

- Pattern recognition and description
- Use of variables to express relationships

## Decreased Attention

### *Number*

- Early attention to reading, writing, and ordering numbers symbolically

### *Operations and Computation*

- Complex paper-and-pencil computations
- Isolated treatment of paper-and-pencil computations
- Addition and subtraction without renaming
- Isolated treatment of division facts
- Long division
- Long division without remainders
- Paper-and-pencil fraction computation
- Use of rounding to estimate

### *Geometry and Measurement*

- Primary focus on naming geometric figures
- Memorization of equivalencies between units of measurement

## Increased Attention

### *Problem Solving*

- Word problems with a variety of structures
- Use of everyday problems
- Applications
- Study of patterns and relationships
- Problem-solving strategies

### *Instructional Practices*

- Use of manipulative materials
- Cooperative work
- Discussion of mathematics
- Questioning
- Justification of thinking
- Writing about mathematics
- Problem-solving approach to instruction
- Content integration
- Use of calculators and computers

## Decreased Attention

### *Problem Solving*

- Use of clue words to determine which operation to use

### *Instructional Practices*

- Rote practice
- Rote memorization of rules
- One answer and one method
- Use of work sheets
- Written practice
- Teaching by telling

## CURRICULUM STANDARDS FOR GRADES 5–8

STANDARD 1: Mathematics as Problem Solving

In grades 5–8, the mathematics curriculum should include numerous and varied experiences with problem solving as a method of inquiry and application so that students can—

- use problem-solving approaches to investigate and understand mathematical content;
- formulate problems from situations within and outside mathematics;
- develop and apply a variety of strategies to solve problems, with emphasis on multistep and nonroutine problems;
- verify and interpret results with respect to the original problem situation;
- generalize solutions and strategies to new problem situations;
- acquire confidence in using mathematics meaningfully.

STANDARD 2: Mathematics as Communication

In grades 5–8, the study of mathematics should include opportunities to communicate so that students can—

- model situations using oral, written, concrete, pictorial, graphical, and algebraic methods;
- reflect on and clarify their own thinking about mathematical ideas and situations;
- develop common understandings of mathematical ideas, including the role of definitions;
- use the skills of reading, listening, and viewing to interpret and evaluate mathematical ideas;
- discuss mathematical ideas and make conjectures and convincing arguments;
- appreciate the value of mathematical notation and its role in the development of mathematical ideas.

STANDARD 3: Mathematics as Reasoning

In grades 5–8, reasoning shall permeate the mathematics curriculum so that students can—

- recognize and apply deductive and inductive reasoning;
- understand and apply reasoning processes, with special attention to spatial reasoning and reasoning with proportions and graphs;
- make and evaluate mathematical conjectures and arguments;
- validate their own thinking;
- appreciate the pervasive use and power of reasoning as a part of mathematics.

STANDARD 4: Mathematical Connections

In grades 5–8, the mathematics curriculum should include the investigation of mathematical connections so that students can—

- see mathematics as an integrated whole;
- explore problems and describe results using graphical, numerical, physical, algebraic, and verbal mathematical models or representations;
- use a mathematical idea to further their understanding of other mathematical ideas;
- apply mathematical thinking and modeling to solve problems that arise in other disciplines, such as art, music, psychology, science, and business;
- value the role of mathematics in our culture and society.

STANDARD 5: Number and Number Relationships

In grades 5–8, the mathematics curriculum should include the continued development of number and number relationships so that students can—

- understand, represent, and use numbers in a variety of equivalent forms (integer, fraction, decimal, percent, exponential, and scientific notation) in real-world and mathematical problem situations;
- develop number sense for whole numbers, fractions, decimals, integers, and rational numbers;
- understand and apply ratios, proportions, and percents in a wide variety of situations;
- investigate relationships among fractions, decimals, and percents;
- represent numerical relationships in one- and two-dimensional graphs.

STANDARD 6: Number Systems and Number Theory

In grades 5–8, the mathematics curriculum should include the study of number systems and number theory so that students can—

- understand and appreciate the need for numbers beyond the whole numbers;
- develop and use order relations for whole numbers, fractions, decimals, integers, and rational numbers;
- extend their understanding of whole number operations to fractions, decimals, integers, and rational numbers;
- understand how the basic arithmetic operations are related to one another;
- develop and apply number theory concepts (e.g., primes, factors, and multiples) in real-world and mathematical problem situations.

STANDARD 7: Computation and Estimation

In grades 5–8, the mathematics curriculum should develop the concepts underlying computation and estimation in various contexts so that students can—

- compute with whole numbers, fractions, decimals, integers, and rational numbers;

- develop, analyze, and explain procedures for computation and techniques for estimation;
- develop, analyze, and explain methods for solving proportions;
- select and use an appropriate method for computing from among mental arithmetic, paper-and-pencil, calculator, and computer methods;
- use computation, estimation, and proportions to solve problems;
- use estimation to check the reasonableness of results.

STANDARD 8: Patterns and Functions

In grades 5–8, the mathematics curriculum should include explorations of patterns and functions so that students can—

- describe, extend, analyze, and create a wide variety of patterns;
- describe and represent relationships with tables, graphs, and rules;
- analyze functional relationships to explain how a change in one quantity results in a change in another;
- use patterns and functions to represent and solve problems.

STANDARD 9: Algebra

In grades 5–8, the mathematics curriculum should include explorations of algebraic concepts and processes so that students can—

- understand the concepts of variable, expression, and equation;
- represent situations and number patterns with tables, graphs, verbal rules, and equations and explore the interrelationships of these representations;
- analyze tables and graphs to identify properties and relationships;
- develop confidence in solving linear equations using concrete, informal, and formal methods;
- investigate inequalities and nonlinear equations informally;
- apply algebraic methods to solve a variety of real-world and mathematical problems.

STANDARD 10: Statistics

In grades 5–8, the mathematics curriculum should include exploration of statistics in real-world situations so that students can—

- systematically collect, organize, and describe data;
- construct, read, and interpret tables, charts, and graphs;
- make inferences and convincing arguments that are based on data analysis;
- evaluate arguments that are based on data analysis;
- develop an appreciation for statistical methods as powerful means for decision making.

STANDARD 11: Probability

In grades 5–8, the mathematics curriculum should include explorations of probability in real-world situations so that students can—

- model situations by devising and carrying out experiments or simulations to determine probabilities;
- model situations by constructing a sample space to determine probabilities;
- appreciate the power of using a probability model by comparing experimental results with mathematical expectations;
- make predictions that are based on experimental or theoretical probabilities;
- develop an appreciation for the pervasive use of probability in the real world.

STANDARD 12: Geometry

In grades 5–8, the mathematics curriculum should include the study of the geometry of one, two, and three dimensions in a variety of situations so that students can—

- identify, describe, compare, and classify geometric figures;
- visualize and represent geometric figures with special attention to developing spatial sense;
- explore transformations of geometric figures;
- represent and solve problems using geometric models;
- understand and apply geometric properties and relationships;
- develop an appreciation of geometry as a means of describing the physical world.

STANDARD 13: Measurement

In grades 5–8, the mathematics curriculum should include extensive concrete experiences using measurement so that students can—

- extend their understanding of the process of measurement;
- estimate, make, and use measurements to describe and compare phenomena;
- select appropriate units and tools to measure to the degree of accuracy required in a particular situation;
- understand the structure and use of systems of measurement;
- extend their understanding of the concepts of perimeter, area, volume, angle measure, capacity, and weight and mass;
- develop the concepts of rates and other derived and indirect measurements;
- develop formulas and procedures for determining measures to solve problems.

## SUMMARY OF CHANGES IN CONTENT AND EMPHASIS FOR MATHEMATICS IN GRADES 5–8

### Increased Attention

*Problem Solving*

- Pursuing open-ended problems and extended problem-solving projects
- Investigating and formulating questions from problem situations
- Representing situations verbally, numerically, graphically, geometrically, or symbolically

*Communication*

- Discussing, writing, reading, and listening to mathematical ideas

*Reasoning*

- Reasoning in spatial contexts
- Reasoning with proportions
- Reasoning from graphs
- Reasoning inductively and deductively

*Connections*

- Connecting mathematics to other subjects and to the world outside the classroom
- Connecting topics within mathematics
- Applying mathematics

*Number/Operations/Computation*

- Developing number sense
- Developing operation sense
- Creating algorithms and procedures
- Using estimation both in solving problems and in checking the reasonableness of results
- Exploring relationships among representations of, and operations on, whole numbers, fractions, decimals, integers, and rational numbers
- Developing an understanding of ratio, proportion, and percent

### Decreased Attention

*Problem Solving*

- Practicing routine, one-step problems
- Practicing problems categorized by types (e.g., coin problems, age problems)

*Communication*

- Doing fill-in-the-blank worksheets
- Answering questions that require only yes, no, or a number as responses

*Reasoning*

- Relying on outside authority (teacher or an answer key)

*Connections*

- Learning isolated topics
- Developing skills out of context

*Number/Operations/Computation*

- Memorizing rules and algorithms
- Practicing tedious paper-and-pencil computations
- Finding exact forms of answers
- Memorizing procedures, such as cross-multiplication, without understanding
- Practicing rounding numbers out of context

## Increased Attention

### *Patterns and Functions*

- Identifying and using functional relationships
- Developing and using tables, graphs, and rules to describe situations
- Interpreting among different mathematical representations

### *Algebra*

- Developing an understanding of variables, expressions, and equations
- Using a variety of methods to solve linear equations and informally investigate inequalities and nonlinear equations

### *Statistics*

- Using statistical methods to describe, analyze, evaluate, and make decisions

### *Probability*

- Creating experimental and theoretical models of situations involving probabilities

### *Geometry*

- Developing an understanding of geometric objects and relationships
- Using geometry in solving problems

### *Measurement*

- Estimating and using measurement to solve problems

### *Instructional Practices*

- Actively involving students individually and in groups in exploring, conjecturing, analyzing, and applying mathematics in both a mathematical and real-world context
- Using appropriate technology for computation and exploration
- Using concrete materials
- Being a facilitator of learning
- Assessing learning as an integral part of instruction

## Decreased Attention

### *Patterns and Functions*

- Topics seldom in the current curriculum

### *Algebra*

- Manipulating symbols
- Memorizing procedures and drilling on equation solving

### *Statistics*

- Memorizing formulas

### *Probability*

- Memorizing formulas

### *Geometry*

- Memorizing geometric vocabulary
- Memorizing facts and relationships

### *Measurement*

- Memorizing and manipulating formulas
- Converting within and between measurement systems

### *Instructional Practices*

- Teaching computations out of context
- Drilling on paper-and-pencil algorithms
- Teaching topics in isolation
- Stressing memorization
- Being the dispenser of knowledge
- Testing for the sole purpose of assigning grades

## CURRICULUM STANDARDS FOR GRADES 9–12

STANDARD 1: Mathematics as Problem Solving

In grades 9–12, the mathematics curriculum should include the refinement and extension of methods of mathematical problem solving so that all students can—

- use, with increasing confidence, problem-solving approaches to investigate and understand mathematical content;
- apply integrated mathematical problem-solving strategies to solve problems from within and outside mathematics;
- recognize and formulate problems from situations within and outside mathematics;
- apply the process of mathematical modeling to real-world problem situations.

STANDARD 2: Mathematics as Communication

In grades 9–12, the mathematics curriculum should include the continued development of language and symbolism to communicate mathematical ideas so that all students can—

- reflect upon and clarify their thinking about mathematical ideas and relationships;
- formulate mathematical definitions and express generalizations discovered through investigations;
- express mathematical ideas orally and in writing;
- read written presentations of mathematics with understanding;
- ask clarifying and extending questions related to mathematics they have read or heard about;
- appreciate the economy, power, and elegance of mathematical notation and its role in the development of mathematical ideas.

STANDARD 3: Mathematics as Reasoning

In grades 9–12, the mathematics curriculum should include numerous and varied experiences that reinforce and extend logical reasoning skills so that all students can—

- make and test conjectures;
- formulate counterexamples;
- follow logical arguments;
- judge the validity of arguments;
- construct simple valid arguments;

and so that, in addition, college-intending students can—

- construct proofs for mathematical assertions, including indirect proofs and proofs by mathematical induction.

STANDARD 4: Mathematical Connections

In grades 9–12, the mathematics curriculum should include investigation of the connections and interplay among various mathematical topics and their applications so that all students can—

- recognize equivalent representations of the same concept;
- relate procedures in one representation to procedures in an equivalent representation;
- use and value the connections among mathematical topics;
- use and value the connections between mathematics and other disciplines.

STANDARD 5: Algebra

In grades 9–12, the mathematics curriculum should include the continued study of algebraic concepts and methods so that all students can—

- represent situations that involve variable quantities with expressions, equations, inequalities, and matrices;
- use tables and graphs as tools to interpret expressions, equations, and inequalities;
- operate on expressions and matrices, and solve equations and inequalities;
- appreciate the power of mathematical abstraction and symbolism;

and so that, in addition, college-intending students can—

- use matrices to solve linear systems;
- demonstrate technical facility with algebraic transformations, including techniques based on the theory of equations.

STANDARD 6: Functions

In grades 9–12, the mathematics curriculum should include the continued study of functions so that all students can—

- model real-world phenomena with a variety of functions;
- represent and analyze relationships using tables, verbal rules, equations, and graphs;
- translate among tabular, symbolic, and graphical representations of functions;
- recognize that a variety of problem situations can be modeled by the same type of function;
- analyze the effects of parameter changes on the graphs of functions;

and so that, in addition, college-intending students can—

- understand operations on, and the general properties and behavior of, classes of functions.

STANDARD 7: Geometry from a Synthetic Perspective

In grades 9–12, the mathematics curriculum should include the continued study of the geometry of two and three dimensions so that all students can—

- interpret and draw three-dimensional objects;
- represent problem situations with geometric models and apply properties of figures;
- classify figures in terms of congruence and similarity and apply these relationships;
- deduce properties of, and relationships between, figures from given assumptions;

and so that, in addition, college-intending students can—

- develop an understanding of an axiomatic system through investigating and comparing various geometries.

STANDARD 8: Geometry from an Algebraic Perspective

In grades 9–12, the mathematics curriculum should include the study of the geometry of two and three dimensions from an algebraic point of view so that all students can—

- translate between synthetic and coordinate representations;
- deduce properties of figures using transformations and using coordinates;
- identify congruent and similar figures using transformations;
- analyze properties of Euclidean transformations and relate translations to vectors;

and so that, in addition, college-intending students can—

- deduce properties of figures using vectors;
- apply transformations, coordinates, and vectors in problem solving.

STANDARD 9: Trigonometry

In grades 9–12, the mathematics curriculum should include the study of trigonometry so that all students can—

- apply trigonometry to problem situations involving triangles;
- explore periodic real-world phenomena using the sine and cosine functions;

and so that, in addition, college-intending students can—

- understand the connection between trigonometric and circular functions;
- use circular functions to model periodic real-world phenomena;
- apply general graphing techniques to trigonometric functions;
- solve trigonometric equations and verify trigonometric identities;
- understand the connections between trigonometric functions and polar coordinates, complex numbers, and series.

STANDARD 10: Statistics

In grades 9–12, the mathematics curriculum should include the continued study of data analysis and statistics so that all students can—

- construct and draw inferences from charts, tables, and graphs that summarize data from real-world situations;
- use curve fitting to predict from data;
- understand and apply measures of central tendency, variability, and correlation;
- understand sampling and recognize its role in statistical claims;
- design a statistical experiment to study a problem, conduct the experiment, and interpret and communicate the outcomes;
- analyze the effects of data transformations on measures of central tendency and variability;

and so that, in addition, college-intending students can—

- transform data to aid in data interpretation and prediction;
- test hypotheses using appropriate statistics.

STANDARD 11: Probability

In grades 9–12, the mathematics curriculum should include the continued study of probability so that all students can—

- use experimental or theoretical probability, as appropriate, to represent and solve problems involving uncertainty;
- use simulations to estimate probabilities;
- understand the concept of a random variable;
- create and interpret discrete probability distributions;
- describe, in general terms, the normal curve and use its properties to answer questions about sets of data that are assumed to be normally distributed;

and so that, in addition, college-intending students can—

- apply the concept of a random variable to generate and interpret probability distributions including binomial, uniform, normal, and chi square.

STANDARD 12: Discrete Mathematics

In grades 9–12, the mathematics curriculum should include topics from discrete mathematics so that all students can—

- represent problem situations using discrete structures such as finite graphs, matrices, sequences, and recurrence relations;
- represent and analyze finite graphs using matrices;
- develop and analyze algorithms;
- solve enumeration and finite probability problems;

and so that, in addition, college-intending students can—

- represent and solve problems using linear programming and difference equations;
- investigate problem situations that arise in connection with computer validation and the application of algorithms.

STANDARD 13: Conceptual Underpinnings of Calculus

In grades 9–12, the mathematics curriculum should include the informal exploration of calculus concepts from both a graphical and a numerical perspective so that all students can—

- determine maximum and minimum points of a graph and interpret the results in problem situations;
- investigate limiting processes by examining infinite sequences and series and areas under curves;

and so that, in addition, college-intending students can—

- understand the conceptual foundations of limit, the area under a curve, the rate of change, and the slope of a tangent line, and their applications in other disciplines;
- analyze the graphs of polynomial, rational, radical, and transcendental functions.

STANDARD 14: Mathematical Structure

In grades 9–12, the mathematics curriculum should include the study of mathematical structure so that all students can—

- compare and contrast the real number system and its various subsystems with regard to their structural characteristics;
- understand the logic of algebraic procedures;
- appreciate that seemingly different mathematical systems may be essentially the same;

and so that, in addition, college-intending students can—

- develop the complex number system and demonstrate facility with its operations;
- prove elementary theorems within various mathematical structures, such as groups and fields;
- develop an understanding of the nature and purpose of axiomatic systems.

## SUMMARY OF CHANGES IN CONTENT AND EMPHASIS FOR MATHEMATICS IN GRADES 9–12

### Increased Attention

#### *Algebra*

- The use of real-world problems to motivate and apply theory
- The use of computer utilities to develop conceptual understanding
- Computer-based methods such as successive approximations and graphing utilities for solving equations and inequalities
- The structure of number systems
- Matrices and their applications

#### *Geometry*

- Integration across topics at all grade levels
- Coordinate and transformation approaches
- The development of short sequences of theorems
- Deductive arguments expressed orally and in sentence or paragraph form
- Computer-based explorations of 2-D and 3-D figures
- Three-dimensional geometry
- Real-world applications and modeling

#### *Trigonometry*

- The use of appropriate scientific calculators
- Realistic applications and modeling
- Connections among the right triangle ratios, trigonometric functions, and circular functions
- The use of graphing utilities for solving equations and inequalities

### Decreased Attention

#### *Algebra*

- Word problems by type, such as coin, digit, and work
- The simplification of radical expressions
- The use of factoring to solve equations and to simplify rational expressions
- Operations with rational expressions
- Paper-and-pencil graphing of equations by point plotting
- Logarithm calculations using tables and interpolation
- The solution of systems of equations using determinants
- Conic sections

#### *Geometry*

- Euclidean geometry as a complete axiomatic system
- Proofs of incidence and betweenness theorems
- Geometry from a synthetic viewpoint
- Two-column proofs
- Inscribed and circumscribed polygons
- Theorems for circles involving segment ratios
- Analytic geometry as a separate course

#### *Trigonometry*

- The verification of complex identities
- Numerical applications of sum, difference, double-angle, and half-angle identities
- Calculations using tables and interpolation
- Paper-and-pencil solutions of trigonometric equations

Increased Attention

*Functions*

- Integration across topics at all grade levels
- The connections among a problem situation, its model as a function in symbolic form, and the graph of that function
- Function equations expressed in standardized form as checks on the reasonableness of graphs produced by graphing utilities
- Functions that are constructed as models of real-world problems

*Statistics*

*Probability*

*Discrete Mathematics*

*Instructional Practices*

- The active involvement of students in constructing and applying mathematical ideas
- Problem solving as a means as well as a goal of instruction
- Effective questioning techniques that promote student interaction
- The use of a variety of instructional formats (small groups, individual explorations, peer instruction, whole-class discussions, project work)
- The use of calculators and computers as tools for learning and doing mathematics
- Student communication of mathematical ideas orally and in writing
- The establishment and application of the interrelatedness of mathematical topics
- The systematic maintenance of student learnings and embedding review in the context of new topics and problem situations
- The assessment of learning as an integral part of instruction

Decreased Attention

*Functions*

- Paper-and-pencil evaluation
- The graphing of functions by hand using tables of values
- Formulas given as models of real-world problems
- The expression of function equations in standardized form in order to graph them
- Treatment as a separate course

*Instructional Practices*

- Teacher and text as exclusive sources of knowledge
- Rote memorization of facts and procedures
- Extended periods of individual seatwork practicing routine tasks
- Instruction by teacher exposition
- Paper-and-pencil manipulative skill work
- The relegation of testing to an adjunct role with the sole purpose of assigning grades

## RECOMMENDATIONS FOR TEACHING AND LEARNING MATHEMATICS IN GRADES K–12

Assuming that teachers are the key figures in changing the ways in which mathematics is taught and learned in schools so that all students are mathematically empowered, the National Council of Teachers of Mathematics (NCTM) through its *Professional Standards for Teaching Mathematics* (1991, p. 1) challenges teachers to be more proficient in—

- selecting mathematical tasks to engage students' interests and intellect;
- providing opportunities to deepen their understanding of the mathematics being studied and its applications;
- orchestrating classroom discourse in ways that promote the investigation and growth of mathematical ideas;
- using, and helping students use, technology and other tools to pursue mathematical investigations;
- seeking, and helping students seek, connections to previous and developing knowledge;
- guiding individual, small-group, and whole-class work.

In particular, six Standards for Teaching Mathematics, summarized below, are presented by NCTM in its 1991 document.

STANDARD 1: Worthwhile Mathematical Tasks

The teacher of mathematics should pose tasks that are based on—

- sound and significant mathematics;
- knowledge of students' understandings, interests, and experiences;
- knowledge of the range of ways that diverse students learn mathematics;

and that

- engage students' intellect;
- develop students' mathematical understandings and skills;
- stimulate students to make connections and develop a coherent framework for mathematical ideas;
- call for problem formulation, problem solving, and mathematical reasoning;
- promote communication about mathematics;
- represent mathematics as an ongoing human activity;
- display sensitivity to, and draw on, students' diverse background experiences and dispositions;
- promote the development of all students' dispositions to do mathematics.

STANDARD 2: The Teacher's Role in Discourse

The teacher of mathematics should orchestrate discourse by—

- posing questions and tasks that elicit, engage, and challenge each student's thinking;

- listening carefully to students' ideas;
- asking students to clarify and justify their ideas orally and in writing;
- deciding what to pursue in depth from among ideas that students bring up during a discussion;
- deciding when and how to attach mathematical notation and language to students' ideas;
- deciding when to provide information, when to clarify an issue, when to model, when to lead, and when to let a student struggle with a difficulty;
- monitoring students' participation in discussions and deciding when and how to encourage each student to participate.

STANDARD 3: Students' Role in Discourse

The teacher of mathematics should promote classroom discourse in which students—

- listen to, respond to, and question the teacher and one another;
- use a variety of tools to reason, make connections, solve problems, and communicate;
- initiate problems and questions;
- make conjectures and present solutions;
- explore examples and counterexamples to investigate a conjecture;
- try to convince themselves and one another of the validity of particular representations, solutions, conjectures, and answers;
- rely on mathematical evidence and argument to determine validity.

STANDARD 4: Tools for Enhancing Discourse

The teacher of mathematics, in order to enhance discourse, should encourage and accept the use of—

- computers, calculators, and other technology;
- concrete materials used as models;
- pictures, diagrams, tables, and graphs;
- invented and conventional terms and symbols;
- metaphors, analogies, and stories;
- written hypotheses, explanations, and arguments;
- oral presentations and dramatizations.

STANDARD 5: Learning Environment

The teacher of mathematics should create a learning environment that fosters the development of each student's mathematical power by—

- providing and structuring the time necessary to explore sound mathematics and grapple with significant ideas and problems;
- using the physical space and materials in ways that facilitate students' learning of mathematics;

- providing a context that encourages the development of mathematical skill and proficiency;
- respecting and valuing students' ideas, ways of thinking, and mathematical dispositions;

and by consistently expecting and encouraging students to—

- work independently or collaboratively to make sense of mathematics;
- take intellectual risks by raising questions and formulating conjectures;
- display a sense of mathematical competence by validating and supporting ideas with mathematical argument.

STANDARD 6: Analysis of Teaching and Learning

The teacher of mathematics should engage in ongoing analysis of teaching and learning by—

- observing, listening to, and gathering other information about students to assess what they are learning;
- examining effects of the tasks, discourse, and learning environment on students' mathematical knowledge, skills, and dispositions;

in order to—

- ensure that every student is learning sound and significant mathematics and is developing a positive disposition toward mathematics;
- challenge and extend students' ideas;
- adapt or change activities while teaching;
- make plans, both short- and long-range;
- describe and comment on each student's learning to parents and administrators, as well as to the students themselves.

## REFERENCES


National Council of Teachers of Mathematics. *Curriculum and Evaluation Standards for School Mathematics.* Reston, Va.: The Council, 1989.

———. *Professional Standards for Teaching Mathematics.* Reston, Va.: The Council, 1991.

# Appendix B

## Students with Special Needs: Categories of Identification Established by Public Laws

*Ruth E. Downs*
*Janice L. Matthew*
*Marilyn L. McKinney*

The categories and definitions associated with special education that are presented in this appendix cannot begin to describe the unique strengths and weaknesses of each individual child. However, they will provide regular classroom teachers with a resource of special education terminology and serve as a review of the terminology for special education teachers.

These definitions are summarized from (1) the Individuals with Disabilities Education Act of 1991 (IDEA), which was enacted on 7 October 1991; (2) "Education of Handicapped Children, Implementation of Part B of the Education of the Handicapped Act" (EHA) (Department of Health, Education, and Welfare, Office of Education 1977), which was signed on 23 August 1977; and (3) other public documents. Information on Attention Deficit Disorder is summarized from the legal memorandum issued 16 September 1991 by the United States Department of Education Office of Special Education and Rehabilitative Services and from written and verbal information obtained from the Education Committee of Children with Attention Deficit Disorders (1988).

### DEFINITIONS RELATED TO SPECIAL EDUCATION

**Attention Deficit Disorder (ADD)**. ADD is a syndrome legally classified as an "other health impairment" (Davila, Williams, and MacDonald 1991) that is characterized by serious and persistent difficulties in these specific areas:

- Attention span
- Impulse control
- Hyperactivity (sometimes)

ADD is a chronic disorder that can begin in infancy and extend through adulthood while having negative effects on a child's life at home and school and in the community. It is estimated that 3 to 5 percent of our school-aged population is affected by ADD (Davila, Williams, and MacDonald 1991), a condition that previously fell under the headings "learning disabled," "brain damaged," "hyperkinetic," or "hyperactive." The newer term, Attention Deficit Disorder, was introduced to describe more clearly the characteristics of these children. The two types of Attention Deficit Disorder are described below.

*Attention Deficit Hyperactivity Disorder (ADHD).* To be diagnosed as having ADHD, a child must display for six months or more at least eight of the following characteristics before the age of seven:

1. Fidgets, squirms, or seems restless
2. Has difficulty remaining seated
3. Is easily distracted
4. Has difficulty awaiting turn
5. Blurts out answers
6. Has difficulty following instructions
7. Has difficulty sustaining attention
8. Shifts from one uncompleted task to another
9. Has difficulty playing quietly
10. Talks excessively
11. Interrupts or intrudes on others
12. Does not seem to listen
13. Often loses materials necessary for tasks
14. Frequently engages in dangerous actions

*Undifferentiated Attention Deficit Disorder.* In this form of ADD, the primary and most significant characteristic is inattentiveness; hyperactivity is *not* present. Nevertheless, these children manifest problems with organization and distractibility, and they may be seen as quiet or passive in nature. It is speculated that Undifferentiated ADD is currently underdiagnosed, since these children tend to be easily overlooked in the classroom. Thus these children may be at a higher risk for academic failure than those with Attention Deficit Hyperactivity Disorder.

**Autism**. Autism is a severe neurological disorder characterized by qualitative distortions in the development of cognitive, language, social, or motor skills. Symptoms are typically manifested before three years of age, are not usual for any stage of child development, and must include two or more of the following:

1. The impairment of reciprocal social interaction
2. The impairment of communication and imaginative activity, including verbal and nonverbal skills
3. A markedly restricted repertoire of activities and interests, often including stereotypical motor or verbal behavior and resistance to change
4. Abnormal or inconsistent responses to sensory stimuli in one or more of the following areas: sight, hearing, pain, balance, smell, taste, posture, and motor behavior

**Gifted and talented.** The 1972 U.S. Office of Education report (Marland 1972) defines *gifted and talented* in this way:

> Gifted and talented children are those identified by professionally qualified persons who by virtue of outstanding abilities are capable of high performance. These are children who require differentiated educational programs and services beyond those normally provided by the regular school program in order to realize their contribution to self and society. Children capable of high performance include those with demonstrated achievement and/or potential in any of the following areas:
>
> 1. General intellectual ability
> 2. Specific academic aptitude
> 3. Creative or productive thinking
> 4. Leadership ability
> 5. Visual and performing arts
> 6. Psychomotor ability

In 1978, the U.S. Congress revised Marland's definition to exclude psychomotor ability (Davis and Rimm 1989, pp. 11, 12):

> The gifted and talented are children and, whenever applicable, youth who are identified at the pre-school, elementary or secondary level as possessing demonstrated or potential abilities that give evidence of high performance capability in areas such as intellectual, creative, specific academic or leadership ability or in the performing and visual arts, and who by reason thereof require services or activities not ordinarily provided by the school.

**Hearing impairments, including deafness.** Hearing impairments are hearing losses that interfere with the development of the communication process and result in students' failure to achieve their educational potential.

**Mental disability.** Mental disability (sometimes termed *mental retardation*) involves impaired mental development that adversely affects a person's educational performance. Individuals with mental disabilities exhibit significantly impaired adaptive behavior in learning, maturation, or social adjustment as a result of subaverage intellectual functioning.

**Orthopedic impairments**. Orthopedic impairments are disabilities that adversely affect performance. They can result from a congenital anomaly (e.g., clubfoot, absence of some member, or absence of a portion of the skull or skin), a disease (e.g., poliomyelitis, bone tuberculosis, or Hurler Syndrome), or other causes (e.g., cerebral palsy, amputations, or fractures or burns that cause contractures).

**Other health impairments.** Other health impairments are chronic or acute health problems that cause a person to exhibit limited strength, vitality, or alertness. These conditions, which adversely affect a child's educational performance, include heart conditions, spina bifida, tuberculosis, rheumatic fever, nephritis, asthma, sickle-cell anemia, hemophilia, epilepsy, environmental illnesses (e.g., lead poisoning), leukemia, and diabetes. As noted above, ADD and its variations are now included in this category (Davila, Williams, and MacDonald 1991).

**Serious emotional disturbance.** Persons with a serious emotional disturbance exhibit over a long time and to a marked degree one or more of the following characteristics that adversely affect educational performance: an inability to learn that cannot be explained by intellectual, sensory, or health factors; an inability to build or maintain satisfactory interpersonal relationships with peers and teachers; inappropriate types of behavior or feelings under normal circumstances; a general pervasive mood of unhappiness or depression; and a tendency to develop physical symptoms or fears associated with personal or school problems. This category does not include students who are socially maladjusted, unless they are also seriously emotionally disturbed.

**Special education.** Special education is instruction specially designed to meet the needs of an exceptional student.

**Specific learning disability.** A specific learning disability is a disorder in one or more of the basic psychological processes involved in understanding or in using spoken or written language that can manifest itself in an imperfect ability to listen, think, speak, read, write, spell, or do mathematical calculations. Such disorders include such conditions as perceptual disabilities, brain injury, minimal brain disfunction, dyslexia, and developmental aphasia. The term does not apply to children who have learning problems that are primarily the result of visual, hearing, or motor disabilities; of mental disabilities; of emotional disturbances; or of environmental, cultural, or economic disadvantages.

**Speech and language disabilities.** Speech and language disabilities are impairments of language, voice, fluency, or articulation that are not due to sensory impairment or developmental delay but that are present to such a degree that academic achievement is affected and the condition is significantly disabling.

**Traumatic brain injury.** Traumatic brain injury is a moderate to severe injury to the brain resulting in severe behavior and learning disorders. Persons whose behavior and learning disorders are primarily the result of visual, hearing, or motor impairments; mental disabilities; emotional factors; or environmental disadvantages are not neurologically impaired and therefore do not fall in this category.

**Visual impairments, including blindness.** Visual impairments are losses of vision that adversely affect educational performance.

## REFERENCES


Davila, Robert R., Michael L. Williams, and John T. MacDonald. *Memorandum on Clarification of Policy to Address the Needs of Children with Attention Deficit Disorders within General and/or Special Education.* Washington, D.C.: U.S. Department of Education, Office of Special Education and Rehabilitative Services, 1991.

Davis, Gary, A., and Sylvia Rimm. *Education of the Gifted and Talented.* 2nd ed. Englewood Cliffs, N.J.: Prentice Hall, 1989.

Department of Health, Education, and Welfare. Office of Education. "Education of Handicapped Children, Implementation of Part B of the Education of the Handicapped Act." *Federal Register* 42, no. 163, 23 August 1977.

Education Committee of Children with Attention Deficit Disorders (CHADD). *Attention Deficit Disorders: A Guide for Teachers.* Plantation, Fla.: CHADD, 1988.

*Individuals with Disabilities Education Act, 10, U.S. Code.* Chapter 33, as amended by PL 101-476, sec. 1401(a)(20) (1991).

Marland, S. P., Jr. *Education of the Gifted and Talented.* Vol. 1, *Report to the Congress of the United States by the U.S. Commissioner of Education.* Washington, D.C.: U.S. Government Printing Office, 1972.

# Appendix C

## Individualized Education Plan (IEP) for Edward and Mathematics Individualized Learning Plan (MILP) for Lisa

*Carole Greenes*
*Frank Garfunkel*
*Melissa DeBussey*

SPED 766
PAGE 1 of 10

DEPARTMENT OF EDUCATION
INDIVIDUALIZED EDUCATIONAL PLAN

PLAN COVERS THE PERIOD FROM: 9/91 TO 6/92
MEETING DATE: 5-17-91
TYPE OF MEETING (CHECK ONE):
Initial Evaluation: ____
Review I: ____
Review II: ____
Re-evaluation: ____
DATE 3-YEAR RE-EVALUATION DUE: 5/94

ETHNIC CODE: ____
IDENTIFICATION NUMBER: ____
SCHOOL: Devotion
GRADE: 7
STUDENT'S DOMINANT LANGUAGE: English
LANGUAGE OF HOME: English
PROTOTYPE: 502.3
LIASION NAME: ____
LIASION POSITION: ____

STUDENT NAME: Edward BIRTHDATE: ____
PARENT(S) NAME(S): ____
ADDRESS: ____

HOME TELEPHONE NUMBER: ____
WORK TELEPHONE NUMBER: ____

TEAM PARTICIPANTS

| SIGNATURE(if in attendance) | NAME | ROLE/ASSESSMENT RESPONSIBILITY |
|---|---|---|
| | | Chairperson / Guidance Counselor |
| | | Chapter I Reading |
| | | Father |
| | | mother |
| | | Examiner - Intern School Psychologist |
| | | Teacher - Grade 5 |
| | | Educational Diagnostician |

REV. 9/86

SPED 766
PAGE 2 of 10

**STUDENT PROFILE, including but not limited to the child's performance level, measurable physical constraints on such performance, and learning style:**

Eddie is a 13-year-old 5th grader who has been retained twice since he started school. Although he has the ability to succeed, his emotional concerns, including low self-esteem, and his weak skills in language arts and math have played a major role in his poor performance.

**SPECIAL EDUCATION SERVICE DELIVERY:**

| TYPE OF SERVICE | FOCUS ON GOAL NUMBERS | TYPE OF SETTING C | SG | I | R | SE | LOCATION | PERSONNEL | DATE SERVICE BEGINS | FREQUENCY AND DURATION OF SERVICE PER DAY/WEEK | TOTAL HOURS PER WEEK |
|---|---|---|---|---|---|---|---|---|---|---|---|
| Learning Center | 1-4 | | X | X | | X | Devotion | Sped Staff | 9/91 | 10 periods @ 40min | 400 min |
| | | | | | | | | | | | |
| | | | | | | | | | | | |
| | | | | | | | | | | | |
| | | | | | | | | | | | |
| | | | | | | | | | | | |
| TOTAL HOURS OF SPECIAL EDUCATION SERVICE DELIVERY PER WEEK | | | | | | | | | | | $6\frac{2}{3}$ |

C=CLASS; SG=SMALL GROUP; I=INDIVIDUAL
R=REGULAR; SE=SPECIAL EDUCATION

**CRITERIA FOR MOVEMENT TO NEXT LESS RESTRICTIVE PROTOTYPE, including entry skills to be met by the student and accommodations to be made in the regular or special education program:**

When the goals of this plan have been met, Eddie will be moved to the next less restrictive prototype.

**ANNUAL/DAILY DURATION OF PROGRAM:** ____Days per Year ________Hours per Day

**TRANSPORTATION PLAN:**

____Regular Transportation
____Parent-Provided Transportation with Reimbursement
____Special Transportation as follows:____________________

REV. 9/86

On the whole, Eddie has worked hard in 5th grade. He does best in situations where he can be given a lot of one-on-one attention in order to prevent him from slipping on his responsibilities. Often he works quickly and won't check his work for errors. He has difficulty organizing, but once he's shown a strategy, he can internalize it and use it when he chooses to do so. Unfortunately, at times, he can get depressed and stay that way for a long time, and his production is seriously affected.

Eddie picks up a lot by listening. He has low vision in one eye and needs to wear his glasses in school, but he often won't do so.

Eddie's difficulties in school have been further compounded this year by his poor attendance. This will definitely need to be monitored next year.

The TEAM feels that it would be in Eddie's best interest to promote him to grade 7 beginning in September. He will need continuous monitoring to ensure that he knows what his assignments are and that he is organized for action. He will also need continued support in both language arts and math in order to strengthen his skills.

REV. 9/86

ADDITIONAL INFORMATION: including a description of the child's participation in regular education and physical education, recommendations regarding state mandated testing programs, a description of the program for transitioning from private to public school, and other applicable information:

Eddie will participate in a small-group math class taught by a member of the special education staff. He will receive a second period of English instead of a foreign language. He will miss social studies in order to attend one period a day in the Learning Center.

__Need for Continuing Services:__ For students __two years prior to graduation or age 22__, the TEAM has determined that there IS, ___ IS NOT a need for continuing services to be provided by a human service agency.

__Discipline:__ The student's handicapping condition requires modification of the rules and regulations outlined in the student handbook. __✓__ No ______ Yes If YES, describe modifications below.

__Graduation/Diploma:__ For students __fourteen years or older__, the TEAM has determined that the student _____ IS expected to graduate in ____________________ (month/year), _______ IS NOT expected to graduate.

If the student is expected to graduate, the criteria for graduation and the plan for meeting those criteria are noted below:

REV. 9/86

| PRIORITY NUMBER | CURRENT PERFORMANCE LEVELS | GENERAL STUDENT CENTERED GOALS | TEACHING APPROACH AND METHODOLOGY; MONITORING AND EVALUATION TECHNIQUES; SPECIALIZED EQUIPMENT AND MATERIALS |
|---|---|---|---|
| 1 | Although Eddie can read a 5th-grade passage silently with enough comprehension to answer some multiple-choice questions correctly, he misreads some words and has difficulty summarizing in his own words.<br>CAT scores:<br>Vocabulary 5.2, Comprehension 4.0, Total 4.6, Year 5/90<br>Vocabulary 4.7, Comprehension 4.4, Total 4.6, Year 5/91<br>Eddie did not wear eyeglasses during testing, finished each section early, and refused to check his work. | Eddie will continue to improve his reading skills | use of classroom materials and supplementary high-interest material.<br>One-on-one and small-group instruction.<br>Evaluation by teacher-made and standardized tests. |
| 2 | While Eddie is capable of writing a well-organized paragraph or essay, his written language is filled with punctuation and spelling errors. He tends to spell phonetically. Usually, he will not proofread his work unless someone sits with him. | Eddie will improve his written language skills. | Emphasis on proofreading of regular education written assignments; use of typewriter, computer, and word processor. |

REV. 9/86

| PRIORITY NUMBER | CURRENT PERFORMANCE LEVELS | GENERAL STUDENT CENTERED GOALS | TEACHING APPROACH AND METHODOLOGY; MONITORING AND EVALUATION TECHNIQUES; SPECIALIZED EQUIPMENT AND MATERIALS |
| --- | --- | --- | --- |
| 3 | As measured by the California Achievement Tests, Eddie's math skills are approximately one year below grade level. While his work is average on concepts of fractions, decimals and percents he has problems with subtraction, multiplication, division, measurement and problem solving. | Eddie will strengthen his ability to perform the four basic mathematical operations and to solve word problems. | Small group and individual approach. Reinforcement of math facts through auditory channel, possibly using a tape recorder. Emphasis on careful re-checking of work. Evaluation by teacher-made and standardized tests. |
| 4 | Eddie's absenses have played a part in his poor school performance. He works well in structured situations and can internalize strategies to handle his classwork and homework although he doesn't always employ them. | Eddie will improve his study skills. | Capitalize on Eddie's like of structure. Communication between regular education teachers, specialists, and home is essential for Eddie to stay on top of his homework. Evaluation by observation of behavior and grades. |
| | | | |
| | | | |
| | | | |
| | | | |

SPED 766
PAGE 7 of 10

PLAN COVERS THE PERIOD FROM: 9/91 TO: 6/92

| GOAL # | OBJ # | SPECIFIC STUDENT CENTERED GOALS | QUARTERS DURING WHICH OBJECTIVES WILL BE ADDRESSED 1 Date: FR: 9/91 TO: 11/91 | 2 Date: FR: 11/91 TO: 2/92 | 3 Date: FR: 3/92 TO: 6/92 | 4 Date: FR: TO: |
|---|---|---|---|---|---|---|
| 1 | 1 | Eddie will continue to improve his decoding skills. | ✓ | ✓ | ✓ | |
| 1 | 2 | Eddie will improve his structural analysis skills in order to | ✓ | ✓ | ✓ | |
| | | be able to read more multisyllabic words. | | | | |
| 1 | 3 | Eddie will broaden his knowledge of prefixes, suffixes, and | ✓ | ✓ | ✓ | |
| | | root words. | | | | |
| 1 | 4 | Eddie will broaden his vocabulary. | ✓ | ✓ | ✓ | |
| 1 | 5 | Eddie will demonstrate an increasing ability to draw | ✓ | ✓ | ✓ | |
| | | conclusions from what he has read. | | | | |
| 2 | 1 | Eddie will expand his writing by adding details. | ✓ | ✓ | ✓ | |
| 2 | 2 | Eddie will compose, edit, and revise compositions. | ✓ | ✓ | ✓ | |
| 2 | 3 | Eddie will demonstrate increased ability to use capitals, | ✓ | ✓ | ✓ | |
| | | periods, and commas correctly. | | | | |
| 2 | 4 | Eddie will continue to improve his ability to spell common | ✓ | ✓ | ✓ | |
| | | sight words. | | | | |
| 3 | 1 | Eddie will practice borrowing and regrouping for subtraction | ✓ | ✓ | ✓ | |
| 3 | 2 | Eddie will practice subtraction involving zeros. | ✓ | ✓ | ✓ | |
| 3 | 3 | Eddie will practice two-digit by two-digit | ✓ | ✓ | ✓ | |
| | | multiplication. | | | | |
| 3 | 4 | Eddie will improve his ability to solve a variety of division | ✓ | ✓ | ✓ | |

REV. 9/86

PLAN COVERS THE PERIOD FROM: 9/91 TO: 6/92

| GOAL # | OBJ # | SPECIFIC STUDENT CENTERED GOALS | QUARTERS DURING WHICH OBJECTIVES WILL BE ADDRESSED | | | |
|---|---|---|---|---|---|---|
| | | | 1 | 2 | 3 | 4 |
| | | | Date: FR: 9/91 TO: 11/91 | Date: FR: 12/91 TO: 2/92 | Date: FR: 3/92 TO: 6/92 | Date: FR: TO: |
| | | problems. | | | | |
| 3 | 5 | Eddie will display a growing ease with word problems. | ✓ | ✓ | ✓ | |
| 4 | 1 | Eddie will attend school each day, unless he is sick. | ✓ | ✓ | ✓ | |
| 4 | 2 | Eddie will keep a daily assignment notebook. | ✓ | ✓ | ✓ | |
| 4 | 3 | Eddie will devise a daily schedule for the completion | ✓ | ✓ | ✓ | |
| | | of his homework assignments. | | | | |
| 4 | 4 | Eddie will complete and hand in all homework. | ✓ | ✓ | ✓ | |
| 4 | 5 | When given a long-range assignment, Eddie will be able | ✓ | ✓ | ✓ | |
| | | to divide the various steps necessary to complete the task | | | | |
| | | into easily manageable chunks. | | | | |
| | | | | | | |
| | | | | | | |
| | | | | | | |
| | | | | | | |
| | | | | | | |
| | | | | | | |
| | | | | | | |
| | | | | | | |
| | | | | | | |

REV. 9/86

SPED 766
PAGE 9 of 10

## STATE MANDATED AND LOCAL TESTING PROGRAMS

STUDENT: Edward ____________ GRADE: 7 ______ SCHOOL: Devotion ____________________

State mandated testing in ______________________________ will be given in grade ____________ in ____________.
(Basic Skills/Curriculum Assessment) (Month/Year)

Your child, ______________________________, is eligible to participate in the testing program, which consists of tests in the following areas: ______________________________________________________________

______________________________________________________________

You should know that although testing will be appropriate for most special needs students, modifications to the testing procedures may be made when necessary. These modifications would be similar to those which are usually made in the child's instructional program.

Under Massachusetts law, parents may excuse their child from taking all or part of the test. If you do not want your child to participate, please check the appropriate space(s), sign and date this form, and return it to:

______________________________ by ______________________________.
(Name) (Date)

IF YOU DO NOT RESPOND IN WRITING ON THIS FORM, TESTING WILL BE ADMINISTERED ON THE SCHEDULED DATE.

______ I do not want my child to participate in the state testing program described above.

______ I do not want my child to participate in the following areas of the state testing program described above:
______________________________________________________________

______________________________________________________________

______ I do not want my child to participate in the local testing program with modification (i.e. untimed).

______ I do not want my child to participate in the local testing program described above.

______________________________ ______________________
(Parent's signature) (Date)

REV. 9/86

SPED 766
PAGE 10 of 10

**RESPONSE TO EDUCATIONAL PLAN:** Parent(s)/Guardian/Surrogate Parent/Student over age 18.

In the space below, check the option(s) of your choice, sign and date this form, and make comments you wish. You may request an independent evaluation under the following circumstances: if you postpone a decision, if you reject the plan in full, if you reject the finding of no special needs, if you reject the plan in part.

OPTION CHOICES

_____ I accept the educational plan in full.

_____ I accept the finding of no special needs.

_____ I postpone a decision until the completion of an independent evaluation.

_____ I request an independent evaluation.

_____ I reject the educational plan in full.

_____ I reject the finding of no special needs.

_____ I reject the following portions of the educational plan with the understanding that the portions which I accept will be implemented immediately.

______________________________
______________________________
______________________________

Signature:______________________________ Date:______________________________
(Parent/Guardian/Surrogate Parent/Student Over Age 18)

Comments:______________________________
______________________________
______________________________
______________________________

I certify that the goals in this plan are those recommended by the TEAM and the indicated services will be provided.

Principal:______________________________ Special Education Administrator:______________________________
(Signature/Date) (Signature/Date)

If placement outside the local education agency is recommended, I certify that ______________________________ is able to provide the services stated in this plan.
(Facility Name/Address)

Director of Accepting Facility:______________________________
(Signature/Date)

REV. 9/86

STUDENT NAME: Lisa
DATE: September 8, 1992
PLAN COVERS PERIOD FROM Sept. 1992 TO May 1993
STUDENT'S DOMINANT LANGUAGE: English
PARENTS NAME:
ADDRESS:

BIRTHDATE: February 12, 1986
SCHOOL: School District
GRADE: 2
LANGUAGE OF HOME: Spanish/English
PHONE:
WORK PHONE NUMBER:

**MATHEMATICS INDIVIDUALIZED LEARNING PLAN**

TEAM MEETING:

| SIGNATURE | NAME | ROLE/ASSESSMENT RESPONSIBILITY |
|---|---|---|
| ______ | ______ | Math Teacher ______ |
| ______ | ______ | Classroom Teacher—Grade 2 ______ |
| ______ | ______ | Classroom Teacher—Grade 1 ______ |
| ______ | ______ | Talented and Gifted Teacher ______ |
| ______ | ______ | Parent ______ |
| ______ | ______ | Parent ______ |
| ______ | ______ | ______ |

I certify that the goals stated in this plan are those recommended by the team and that the indicated services will be provided.

PRINCIPAL:

MATH COORDINATOR/TEACHER:

CLASSROOM TEACHER:
(if different from math teacher)

SPECIAL EDUCATION ADMINISTRATOR/
TALENTED AND GIFTED ADMINISTRATOR:

STUDENT:

PARENT (optional):

STUDENT PROFILE: Lisa is an engaging, well-adjusted 6-year, 7-month-old girl entering the second grade in the _______ School District. She is the youngest of three children. Lisa socializes easily with children of her own age as well as those who are older. Lisa's siblings' friends regularly include her in their activities.

Lisa is very articulate in both verbal and written expression. She converses easily with adults and with peers. Lisa writes in full sentences, using capital letters at the beginning of a sentence and the appropriate punctuation at the end of the sentence. Her stories are creative and written with a level of sophistication expected of a third grader. Lisa is a voracious reader and has a special interest in dinosaurs. Lisa's mother reports that Lisa reads her grade-3 sibling's books and attempts to read her grade-5 sibling's books. She also often attempts their homework. On the California Test of Basic Skills, Lisa tests in the mid-third-grade range for reading. She independently reads chapter books with a third-grade readability level, as well as information books, including the World Book Encyclopedia, and books of poetry.

Lisa enjoys creating and telling stories to her friends and family. She is also artistically talented. She illustrates much of her written work.

Lisa scored a 137 verbal, a 135 performance, and a 136 combined score on the Wechsler Intelligence Scale for Children—Revised. Lisa's scores for this intelligence test place her in the Superior range.

**CURRENT MATHEMATICS PERFORMANCE LEVELS:** Achievement tests, Portfolio Assessment, and Observations at the end of grade 1 demonstrate that Lisa has met all the Grade-1 Mathematics Objectives as well as many of the Grade-2 Mathematics Objectives.

Lisa can add, subtract, multiply, and divide basic facts, as evidenced by her performance on the California Test of Basic Skills.

She computes basic facts in addition, subtraction, multiplication, and division using skip counting and chips and has memorized the majority of facts. She can solve one-step story problems involving single-digit addition, subtraction, multiplication, and division. Lisa can use numbers and symbols to write equations for computation story problems involving basic facts.

Using base-ten blocks and place-value cards, Lisa is able to illustrate the expanded form of two- and three-digit numbers. Lisa is able to use base-ten blocks to model addition and subtraction of multidigit numbers. Lisa is able to add and subtract multidigit numbers with a calculator.

Lisa is able to find information in horizontal and vertical bar graphs and tables with two columns of data, as well as record results of measurements and experiments. She can read a Fahrenheit thermometer.

Lisa likes to solve number-puzzle problems and will persevere for lengthy periods of time until she is able to solve the problems.

**MATHEMATICS GOALS:** Lisa has mastered the basic computational skills, data organization skills, and reasoning processes that will be taught in the second grade. To capitalize on Lisa's strengths in mathematics, writing, and art, Lisa's math curriculum will focus on creating and illustrating story problems that involve computation with multidigit numbers and the development of various logical-reasoning strategies for solving noncomputational problems.

Lisa will need to participate in the regular class discussions involving comparison of numbers, estimation, addition and subtraction of money, rounding of numbers, measurement involving the English and metric systems, geometry, and fractions.

Lisa's enrichment will occur primarily in the regular grade-2 classroom. Since she is a self-starter, much of Lisa's mathematics enrichment will occur as individual investigations. Lisa will be encouraged to assist her classmates with their mathematics. She will be grouped with third and fourth graders for various after-school integrated projects (e.g., the grade-3 study of prehistoric animals and the grade-4 exploration of the solar system). In addition, Lisa will maintain a daily Math Log of in-class and homework investigations, comments, and questions. Lisa will be encouraged to use word-processing and drawing programs.

Lisa's performance and progress will be assessed through analysis of Math Log entries, observation of her presentations and work with peers, evaluation of her projects, and results of teacher-student interviews and teacher-designed achievement tests.

| OBJ # | CURRICULUM OBJECTIVES | SCHOOL QUARTER | | | |
|---|---|---|---|---|---|
| | | 1 | 2 | 3 | 4 |
| 1 | Given a mathematical story problem to solve, Lisa will draw a picture or construct a model to solve the problem. | X | X | X | X |
| 2 | Given a one-step story problem involving multidigit addition, subtraction, multiplication, or division, Lisa will identify the operation needed to solve the problem. | X | X | X | X |
| 3 | Given a one-step story problem involving multidigit addition, subtraction, multiplication, and division, Lisa will use numbers and symbols to write an equation to solve the problem and record the equation. | X | X | X | X |
| 4 | Given a two-step story problem involving addition, subtraction, multiplication, and/or division, Lisa will identify the two steps of the story problem and the sequence in which the steps should be performed and give a rationale for the sequence. | X | X | X | X |
| 5 | Given a two-step story problem involving addition, subtraction, multiplication, and/or division, Lisa will use numbers and symbols to write an equation or equations to solve the problem. | X | X | X | X |
| 6 | Given a set of multidigit numbers, Lisa will use the numbers to create, record, and illustrate story problems and solve the problems. | X | X | X | X |
| 7 | Given a map, Lisa will locate information in the map. | | | | X |
| 8 | Given a map, Lisa will create and illustrate story problems that use the data in the map and solve the problems. | | | | X |
| 9 | Given information in tables, Lisa will use the data to solve one-step problems. | X | X | | |
| 10 | Given information in tables, Lisa will use the data to solve two-step problems. | | X | X | X |
| 11 | Given information in tables, Lisa will use the data to create, record, and illustrate story problems and solve the problems. | | | X | X |
| 12 | Given information in pictographs, Lisa will use the data to solve one-step story problems. | X | X | X | X |

| OBJ # | CURRICULUM OBJECTIVES | SCHOOL QUARTER | | | |
|---|---|---|---|---|---|
| | | 1 | 2 | 3 | 4 |
| 13 | Given information in pictographs, Lisa will use the data to solve multistep story problems. | | | X | X |
| 14 | Given information in pictographs, Lisa will use the data to create and record story problems and solve the problems. | | | X | X |
| 15 | Lisa will make organized lists to solve problems and describe the organization techniques. | | | X | X |
| 16 | Lisa will use the guess-and-check strategy to solve problems and describe the strategy. | | | X | X |
| 17 | Lisa will identify and generalize shape and number patterns and describe the patterns. | | | X | X |
| 18 | Given pairs of related numbers, Lisa will describe the rule that relates the numbers. | | X | X | X |
| 19 | Lisa will create, record, and illustrate multiclue logic problems about numbers and shapes and describe how she constructed the problems. | X | X | X | X |
| 20 | Given a spinner, Lisa will perform and record results of probability experiments with the spinner and describe the experiments. | | X | X | X |
| 21 | Given a pair of dice, Lisa will perform and record results of probability experiments involving numbers on the dice and describe the experiments. | | | | X |
| 22 | Given a bag of red, green, and blue chips, Lisa will perform and record results of probability experiments involving the chips and describe the experiments. | | | X | X |
| 23 | Given a probability situation involving two outcomes, Lisa will identify and provide a rationale for the more likely of two outcomes. | | | X | X |
| 24 | Given the first quadrant of the coordinate plane, Lisa will identify the coordinates of a point. | | | X | X |
| 25 | Given the coordinates of a point in the first quadrant, Lisa will graph the point. | | | X | X |

| OBJ # | PEDAGOGICAL APPROACH | INSTRUCTIONAL MATERIALS | ASSESSMENT STRATEGIES |
|---|---|---|---|
| 1 | Individual investigation | *LEGO/LOGO* (LEGO)<br>Computer<br>*TOPS Problem Solving Deck A* (Dale Seymour Publications) | Assessment of Math Log entries |
| 2 | Individual investigation<br>Lisa paired with a student who needs extra help | *Fitz and Martha at Blue Falls High* (video) (Tom Snyder Productions)<br>*Getting Started with Story Problems* (DLM)<br>*How to Solve Story Problems* (DLM)<br>VCR and TV | Assessment of Math Log entries<br>Teacher interview |
| 3 | Individual investigation | *Fitz and Martha at Blue Falls High* (video) (Tom Snyder Productions)<br>*TOPS Communication Deck I* (Dale Seymour Publications)<br>VCR and TV | Assessment of Math Log entries |
| 4 | Individual investigation | *Fitz and Martha at Blue Falls High* (video) (Tom Snyder Productions)<br>*How to Solve 1 and 2 Step Story Problems* (DLM)<br>*TOPS Problem Solving Deck A* (Dale Seymour Publications)<br>*TOPS Calculator Problem Deck III* (Dale Seymour Publications)<br>VCR and TV | Assessment of Math Log entries<br>Teacher-designed achievement test<br>Teacher interview |
| 5 | Individual investigation | *Thinker Math (Grades 3–4)* (Creative Publications)<br>*TOPS Communication Deck I* (Dale Seymour Publications) | Assessment of Math Log entries |
| 6 | Individual investigation<br>Presentation to peers | Computer<br>*Math By-Lines (Grades 3–6)* (Dale Seymour Publications)<br>*Thinker Math (Grades 3–4)* (Creative Publications)<br>*TOPS Communication Deck I* (Dale Seymour Publications)<br>Word-processing and drawing programs (IBM) | Assessment of Math Log entries<br>Evaluation of book of problems written, illustrated, and solved by Lisa<br>Peer solution of problems |

| OBJ # | PEDAGOGICAL APPROACH | INSTRUCTIONAL MATERIALS | ASSESSMENT STRATEGIES |
|---|---|---|---|
| 7 | Individual investigation<br>Small-group (accelerated students) investigation<br>Parent participation—Lisa plans a family trip | Map of Chelsea<br>Map of Massachusetts<br>Map of the United States | Teacher-designed achievement test |
| 8 | Individual investigation<br>Presentation to peers | Map of Chelsea<br>Map of Massachusetts<br>Map of the United States | Assessment of Math Log entries<br>Evaluation of book of problems written, illustrated, and solved by Lisa |
| 9 | Individual investigation<br>Lisa paired with a student who needs extra help | *Developing Skills with Tables and Graphs: Book A* (Dale Seymour Publications) | Teacher-designed achievement test |
| 10 | Individual investigation | *Developing Skills with Tables and Graphs: Book A* (Dale Seymour Publications) | Teacher-designed achievement test |
| 11 | Individual investigation<br>Presentation to peers | Computer<br>*Developing Skills with Tables and Graphs: Book A* (Dale Seymour Publications)<br>*TOPS Communication Deck I* (Dale Seymour Publications)<br>Word-processing and drawing programs (IBM) | Assessment of Math Log entries<br>Evaluation of book of problems written, illustrated, and solved by Lisa<br>Peer solution of problems |
| 12 | Individual investigation | *TOPS Problem Solving Deck A* (Dale Seymour Publications) | Teacher-designed achievement test |
| 13 | Individual investigation | *TOPS Problem Solving Deck A* (Dale Seymour Publications) | Teacher-designed achievement test |
| 14 | Individual investigation | Computer<br>*TOPS Communication Deck III* (Dale Seymour Publications)<br>Word-processing and drawing programs (IBM) | Assessment of Math Log entries<br>Evaluation of book of problems written, illustrated, and solved by Lisa<br>Peer solution of problems |

| OBJ # | PEDAGOGICAL APPROACH | INSTRUCTIONAL MATERIALS | ASSESSMENT STRATEGIES |
|---|---|---|---|
| 15 | Individual investigation<br>Parent participation—Lisa makes family food-shopping list organized by aisles in food store | *Reach—Living Things: Petunias, Potatoes, Pets, and People* (Dale Seymour Publications)<br>*Reach—That's Entertainment: Shows, Sports, Stumpers, and Stories* (Dale Seymour Publications)<br>*Used Numbers—Sorting: Groups and Graphs* (Dale Seymour Publications) | Assessment of Math Log entries<br>Teacher interview |
| 16 | Individual investigation<br>Small-group (accelerated students) investigation | Computer<br>*Reach—Living Things: Petunias, Potatoes, Pets, and People* (Dale Seymour Publications)<br>*Reach—That's Entertainment: Shows, Sports, Stumpers, and Stories* (Dale Seymour Publications)<br>*TOPS Problem Solving Deck A* (Dale Seymour Publications)<br>*The Pond* (IBM) (Sunburst Communications) | Assessment of Math Log entries<br>Teacher interview |
| 17 | Individual investigation<br>Small-group (accelerated students) investigation | Attribute Blocks (Dale Seymour Publications)<br>Computer<br>*Logix* (Mondia)<br>Hundreds board<br>Pattern Blocks (Dale Seymour Publications)<br>*Pattern Block Activities* (Dale Seymour Publications)<br>*The Pond* (IBM) (Sunburst Communications)<br>*Winkies World of Patterns* (IBM) (Sunburst Communications) | Assessment of Math Log entries<br>Teacher-designed achievement test<br>Teacher interview |
| 18 | Individual investigation | *Developing Skills with Tables and Graphs: Book A* (Dale Seymour Publications) | Assessment of Math Log entries<br>Teacher-designed achievement test |
| 19 | Individual investigation<br>Small-group (accelerated students) investigation<br>Presentation to peers | Attribute Blocks (Dale Seymour Publications)<br>Computer<br>Hundreds board<br>*Logic Problems for Primary People* (Creative Publications)<br>*Logix* (Mondia)<br>*Master Mind* (Dale Seymour Publications)<br>*Playing with Logic* (Dale Seymour Publications)<br>*The Pond* (IBM) (Sunburst Communications) | Assessment of Math Log entries<br>Evaluation of book of problems written, illustrated, and solved by Lisa<br>Peer solution of problems |

| OBJ # | PEDAGOGICAL APPROACH | INSTRUCTIONAL MATERIALS | ASSESSMENT STRATEGIES |
|---|---|---|---|
| 20 | Small-group (accelerated students) investigation | Spinners<br>*Probability Jobcards Intermediate* (Creative Publications) | Observation<br>Evaluation of probability game involving a spinner, designed by Lisa, that includes directions, rules, and a written explanation of the outcomes of the game |
| 21 | Small-group (accelerated students) investigation | Dice<br>*Probability Jobcards Intermediate* (Creative Publications) | Observation |
| 22 | Small-group (accelerated students) investigation | Chips<br>*Probability Jobcards Intermediate* (Creative Publications) | Evaluation of a class presentation by Lisa in which she explains basic probability (using chips) to her classmates<br>Observation |
| 23 | Small-group (accelerated students) investigation | *Probability Jobcards Intermediate* (Creative Publications)<br>*TOPS Communication Deck I* (Dale Seymour Publications) | Assessment of Math Log entries<br>Observation |
| 24 | Individual investigation | Battleship (game)<br>*Developing Skills with Tables and Graphs: Book A* (Dale Seymour Publications)<br>Graph paper<br>*Grid and Bear It* (Cuisenaire) | Teacher-designed achievement test |
| 25 | Individual investigation | Battleship (game)<br>*Developing Skills with Tables and Graphs: Book A* (Dale Seymour Publications)<br>Graph paper<br>*Grid and Bear It* (Cuisenaire)<br>*Used Numbers—Sorting: Groups and Graphs* (Dale Seymour Publications) | Teacher-designed achievement test |

## LINKS TO READING, SCIENCE, AND SOCIAL STUDIES:

### *Reading*

1. *Creating and Solving Computation Story Problems*

   Anno, Mitsuma. *Anno's Mysterious Multiplying Jar* (New York: Putnam Publishing Group, 1983)

2. *Creating and Solving Logic Problems*

   Base, Graem. *The Eleventh Hour: A Curious Mystery* (New York: Abrams, 1990)

   Sobol, Ronald. *Encyclopedia Brown* (New York: Bantam, 1963)

3. *Constructing Maps and Graphing*

   Burnett, Frances Hodgson. *The Secret Garden* (New York: Lippincott/Harper Collins, 1912)

   Selden, George. *The Cricket in Times Square* (New York: Dell/Yearling, 1960)

   White, E. B. *The Trumpet of the Swan* (New York: Harper & Row, 1970)

4. *Probability and Experimentation*

   Van Allsburg, Chris. *Jumanji* (Boston: Houghton Mifflin Publishing Co., 1981)

### *Science*

The grade-2 science program focuses on classroom plants, air, flotation, and life cycles of the moth and butterfly. Lisa will be encouraged to use these science topics and related data as the bases for her creation of tables, pictographs, and bar graphs and for computation and logic story problems. During the grade-2 study of plants, Lisa will examine patterns occurring in the seed arrangements and the numbers of seeds, petal formation in flowers, and the arrangement and number of woody scales on pine cones. While students study air and construct and experiment with gliders, Lisa will read about and present reports on Orville and Wilbur Wright and Amelia Earhart.

### *Social Studies*

The grade-2 social studies program focuses on world geography; economics, history, and sociology of the community; and government and citizenship. When creating computation and logic story problems, Lisa will use data from maps of Chelsea, Boston, and the United States. She will plan trips from Chelsea to other Massachusetts locations, map the routes, compute the distances using the map scale, and calculate the costs of the trips by bus, taxi, and private automobile (including gasoline and tolls). Using Chelsea population information, Lisa will construct a profile of the city that will include pictographs, bar graphs, tables, and lists. To accompany these displays, Lisa will create computation and logic story problems. Lisa will also coordinate students' development of a scale model (three-dimensional map) of downtown Chelsea.

# Index